T0240120

Lecture Notes in Physics

Founding Editors

Wolf Beiglböck
Jürgen Ehlers
Klaus Hepp
Hans-Arwed Weidenmüller

Volume 1015

Series Editors

Roberta Citro, Salerno, Italy
Peter Hänggi, Augsburg, Germany
Morten Hjorth-Jensen, Oslo, Norway
Maciej Lewenstein, Barcelona, Spain
Luciano Rezzolla, Frankfurt am Main, Germany
Angel Rubio, Hamburg, Germany
Wolfgang Schleich, Ulm, Germany
Stefan Theisen, Potsdam, Germany
James D. Wells, Ann Arbor, MI, USA
Gary P. Zank, Huntsville, AL, USA

The series Lecture Notes in Physics (LNP), founded in 1969, reports new developments in physics research and teaching - quickly and informally, but with a high quality and the explicit aim to summarize and communicate current knowledge in an accessible way. Books published in this series are conceived as bridging material between advanced graduate textbooks and the forefront of research and to serve three purposes:

- to be a compact and modern up-to-date source of reference on a well-defined topic;
- to serve as an accessible introduction to the field to postgraduate students and non-specialist researchers from related areas;
- to be a source of advanced teaching material for specialized seminars, courses and schools.

Both monographs and multi-author volumes will be considered for publication. Edited volumes should however consist of a very limited number of contributions only. Proceedings will not be considered for LNP.

Volumes published in LNP are disseminated both in print and in electronic formats, the electronic archive being available at springerlink.com. The series content is indexed, abstracted and referenced by many abstracting and information services, bibliographic networks, subscription agencies, library networks, and consortia.

Proposals should be sent to a member of the Editorial Board, or directly to the responsible editor at Springer:

Dr Lisa Scalone
lisa.scalone@springernature.com

Alessio Zaccone

Theory of Disordered Solids

From Atomistic Dynamics
to Mechanical, Vibrational,
and Thermal Properties

 Springer

Alessio Zaccone
Dipartimento di Fisica
Universita' di Milano
Milano
Italy

ISSN 0075-8450   ISSN 1616-6361   (electronic)
Lecture Notes in Physics
ISBN 978-3-031-24705-7   ISBN 978-3-031-24706-4   (eBook)
https://doi.org/10.1007/978-3-031-24706-4

© The Editor(s) (if applicable) and The Author(s), under exclusive license to Springer Nature Switzerland AG 2023

This work is subject to copyright. All rights are solely and exclusively licensed by the Publisher, whether the whole or part of the material is concerned, specifically the rights of translation, reprinting, reuse of illustrations, recitation, broadcasting, reproduction on microfilms or in any other physical way, and transmission or information storage and retrieval, electronic adaptation, computer software, or by similar or dissimilar methodology now known or hereafter developed.

The use of general descriptive names, registered names, trademarks, service marks, etc. in this publication does not imply, even in the absence of a specific statement, that such names are exempt from the relevant protective laws and regulations and therefore free for general use.

The publisher, the authors, and the editors are safe to assume that the advice and information in this book are believed to be true and accurate at the date of publication. Neither the publisher nor the authors or the editors give a warranty, expressed or implied, with respect to the material contained herein or for any errors or omissions that may have been made. The publisher remains neutral with regard to jurisdictional claims in published maps and institutional affiliations.

This Springer imprint is published by the registered company Springer Nature Switzerland AG
The registered company address is: Gewerbestrasse 11, 6330 Cham, Switzerland

Dedicated to my mother.

Preface

The 1977 Nobel prize in Physics was famously awarded to Sir Neville Mott, Phil W. Anderson and John H. van Vleck for their pioneering work on the electronic theory of disordered solids. By "disordered" or "amorphous" solids, solid-state materials were typically defined as those which contain defects and impurities of various degrees, up to the limit of "strong disorder," typically identified with structural glasses. An important ramification of this definition was represented by solids with disorder in the spin orientations, which are of relevance for magnetism. The work of these pioneers concentrated almost exclusively on the electronic (and to some extent, magnetic) properties of amorphous solids, leading to such paradigms as the Anderson localization, Mott's variable range hopping, and the Coulomb glass, among others, which revolutionized our understanding of electrons and charge transport in structurally disordered solids. Significant input came, around the same time, from Soviet scientists such as B. I. Shklovskii and A. L. Efros. In parallel with these theoretical efforts, our understanding of "real solids" progressed also thanks to the fast development of ab initio computational methods for the electronic structure of solids. In spite of these tremendous advances in the physics of electrons in real solids, the progress on the side of atomic dynamics, phonons, elasticity, thermal properties, and lattice dynamics of disordered solids, in general, has been comparatively slower until more recent times. The central problem of clarifying the nature of the "glass transition" has catalyzed the most energy of researchers in this field until the 1990s, without, however, coming to a definitive theoretical framework or consensus, although the 2021 Nobel prize winning efforts of Giorgio Parisi for the replica theory of glass transition in higher dimensions was a significant step forward. Several different theories of the glass transition exist, each capturing a piece of truth, but a conclusive theoretical framework able to explain the phenomenon in all its facets will probably remain out of reach, also due to the impossibility of quantitatively estimating the configurational entropy of a supercooled liquid (an impossible task, in fact, not only for any mathematical theory, but even for computer simulations).

In spite of the lack of a conclusive theory of the glass transition, our understanding of glasses and disordered solids, and their physical properties, has spectacularly progressed over the past decades, making "amorphous solids" one of the most active and lively areas of condensed matter physics, with many implications and ramifications in soft matter physics, complex systems, materials science, and biophysics. This progress has often been facilitated by conceptual developments in the area of athermal amorphous solids (or granular solids), where the absence of thermal fluctuations makes it easier to develop mathematical descriptions of elasticity and mechanical behavior in disordered environments. Models and paradigms such as the "jamming" of random sphere packings, or gels formed of colloidal particles, have greatly contributed to clarifying outstanding questions about the microscopic rigidity, elasticity, viscosity, and vibrational excitations in disordered solids and soft materials. The model systems have also played an important role as benchmarks for quantitatively testing mathematical theories in great detail, including numerical prefactors, and are amenable to computer simulations. Often these concepts developed for athermal amorphous solids have served as useful starting point to develop a mathematical description of physical properties at finite temperature by suitably adding in the thermal fluctuations.

The goal of this book is to present a consistent mathematical theory of the non-electronic properties of amorphous solids, starting from the atomic-level dynamics and leading to descriptions of the macroscopic properties such as elastic and viscoelastic moduli, vibrational spectra, thermal properties, without worrying about the ultimate nature of the glass transition phenomenon. By focusing on atomic-level dynamics, issues related to the frozen-in disorder can be managed under the main assumption about the (undeniable) existence of an underlying "amorphous lattice." This allows us to relegate the theoretical uncertainties about the ultimate nature of the glass transition to a subsidiary role, and to take a more pragmatic approach for the modelling of physical properties. The predictions about physical properties can thus be successfully tested against experimental or computer simulation data, without having to solve the glass transition problem itself, which requires sophisticated theoretical approaches that would overshadow the physical understanding and would make the comparison with experiments much more challenging. A main justification for this approach is given by the experimental observation that macroscopic properties of glassy disordered materials change little with temperature - from few Kelvins above absolute zero up to few degrees below the glass transition temperature - and thus within a very broad range of several hundreds of Kelvins. The unifying perspective is that of nonaffine particle dynamics and nonaffine elasticity, which reflect the author's contributions to the field and expertise. I apologize for not including or touching upon all the various approaches and topics that have been developed and studied in recent years, which would be an impossible task.

Another important goal of this book is to introduce graduate students and researchers not only to the subtle physical concepts underlying the dynamics, mechanics, and statistical physics of amorphous solids, but also to the main mathematical and numerical methods used in the field. These cannot be learned from the specialized literature since they are spread out among many often technically demanding papers. These methods are sufficiently general, in that they allow for the mathematical or numerical description of novel physical phenomena observed across many different types of amorphous solids, regardless of the molecular details and of the particular chemistry of the material. The lack of textbooks and monographs where these theoretical methods for amorphous solids can be found, and the need to provide self-consistent mathematical derivations that are otherwise scattered in hardly accessible papers (which often omit the key steps), represents one of the main motivations of this monograph. A similar role, mutatis mutandis, historically has been played by the 1953 monograph by M. Born and K. Huang for the lattice dynamics of perfect crystals, which still serves as a key reference for researchers who need to set up calculations of elastic moduli of crystals or their thermal and vibrational properties. While crystals played a pivotal role for the development of solid state physics, amorphous solids are certainly no less important, given the fact that the overwhelming majority of solids in the universe are non-crystalline, including biological matter, rocks and soil, plastics, and many engineering materials, and represent the forefront of modern solid state and condensed matter physics.

This monograph thus aims to continue the systematic extension of lattice dynamics from crystalline to disordered solids, in the wake of the Born-Huang monograph, and under the working assumption (originally due to Shlomo Alexander) about the existence of an amorphous lattice around which a lattice dynamic description can be built by including key concepts of disorder, equilibrium and nonequilibrium statistical mechanics, dissipative systems, and internal stress. The unifying concept around which the lattice dynamics of amorphous solids is derived is that of nonaffine motions, from which nonaffine elasticity, and viscoelasticity, and nonaffine response theory are developed. These same concepts are used to provide a mathematical description of the main wave-damping phenomena in disordered solids, as well as to provide a mechanism and microscopic description of plasticity. The emphasis is on mathematical derivations in the style of applied mathematics/theoretical physics, with first-principle derivations (wherever possible) and self-contained mathematical steps leading to expressions that can be used to make sense of experimental or simulations data on either real materials or model coarse-grained simulated systems.

While aiming at a comprehensive treatment of all the most important topics in the area of disordered solids, it is inevitable that some, equally important, topics must be left out. The most prominent topic which is deliberately omitted in this monograph is that of glassy relaxation and aging phenomena, and the closely related dielectric response phenomena. The chief reason for this omission lies in the fact that relaxation phenomena in glasses have been elusive in terms of mathematical descriptions from first principles, whereas phenomenological models have proliferated, and therefore would not fit well with the mathematically inclined style of this book. Also, there exist excellent recent monographs on glassy relaxation, such as the one by Prof. Kia Ngai (K. L. Ngai, Relaxation and Diffusion in Complex Systems, Springer) to which the interested reader is referred to.

This monograph is aimed at physics, chemistry, materials science and engineering graduate students, PhD students, and post-docs and will complement standard monographs and textbooks in condensed matter, soft matter, statistical physics and disordered systems, and continuum mechanics. Although it is aimed at the graduate and postgraduate level, it will also be useful for post-docs and both beginning and experienced researchers working in solid state, condensed matter physics and materials science, soft matter, and applied mathematics. An undergraduate educational background in physics, theoretical chemistry, engineering, or applied mathematics is required with good basic understanding of classical physics, calculus, tensors, quantum mechanics, and continuum mechanics. A good understanding of basic solid state physics and of condensed matter physics at the level of the Kittel textbook (C. Kittel, Introduction to Solid State Physics, Wiley), and of statistical physics at the level of the Blundell and Blundell textbook (S. J. Blundell and K. M. Blundell, Concepts in Thermal Physics, Oxford University Press) are also required. The book will also be of interest to many researchers in the areas of disordered systems, soft matter, statistical physics and complexity, continuum mechanics, plasticity, and solid mechanics as well. Also researchers working on molecular dynamics simulations, molecular coarse-grained simulations, as well as on ab initio atomistic and density functional theory (DFT) methods for solid-state and materials science may find it useful.

Milano, Italy   Alessio Zaccone
September, 2022

Acknowledgments

I am grateful to my mentors, in particular to Eugene Terentjev who had a great influence on the development of my own approach to theoretical problems in condensed matter physics. I am indebted to my other advisors and collaborators on several projects, including both theorists and experimentalists, and in particular to: Massimo Morbidelli, Emanuela Del Gado, Matteo Baggioli, Konrad Samwer, Kostya Trachenko, Peter Schall, Daniel Bonn, Matthias Ballauff, Josep-Lluis Tamarit, Stefan Egelhaaf, Joris Sprakel, Marco Laurati, Daan Frenkel, Stefano Martiniani, Laurence Noirez, Zach Evenson, Gerhard Wilde, Harald Roesner, Amelia Liu, Valeriy Ginzburg, Hua Wu.

Contents

1 A Bird's-Eye View of Amorphous Solids 1
  1.1 Microscopic Bonding and Interactions in Disordered Solids 1
    1.1.1 Central-Force (Non-covalent) Interactions 1
    1.1.2 Covalent Bonding 3
    1.1.3 Metallic Bonding 5
    1.1.4 Anharmonicity 8
  1.2 The Structure of Disordered Solids 15
    1.2.1 Static Structure Factor and Radial Distribution Function 15
    1.2.2 Structural Paradigms for Disordered Solids 18
    1.2.3 The Fractal Model 30
  1.3 Dynamic Correlation Functions 33
  1.4 The Glass Transition 35
  1.5 Structural Order Parameters 38
    1.5.1 Bond-Orientational Order Parameter 38
    1.5.2 Inversion Symmetry 40
  1.6 Constraint Counting and Isostaticity 44
  1.7 Anharmonic Potential of Mean Force in Glasses 47
  References 49

2 Elasticity 53
  2.1 Introduction 53
  2.2 The Concept of Affine and Nonaffine Deformations 54
  2.3 Born-Huang Formulae for the Affine Elastic Moduli 57
  2.4 Elastic Moduli of Amorphous Solids 59
    2.4.1 General Theory 60
    2.4.2 The Shear Modulus of Random Jammed
          Sphere Packings 66
    2.4.3 Elasticity of Random Networks with Internal Stresses 73
  2.5 The Shear Modulus of Glasses 80
    2.5.1 The Shear Modulus of Polymer Glasses 80
    2.5.2 The Shear Modulus of Lennard-Jones Glasses 87
  2.6 The Shear Modulus of Colloidal Gels 88
    2.6.1 Elasticity of Fractal Gels 89
    2.6.2 Intermediate Dense Gels, and Cluster Glasses 91

2.7   The Bulk Modulus 96
2.8   Stress-Fluctuation Formalism for the Elastic Moduli
       of Thermal Systems 105
       2.8.1   General Formalism 106
       2.8.2   The Temperature-Dependent Shear Modulus of
               Perfect Crystals 109
       2.8.3   The Case of Liquids 109
       2.8.4   The Temperature-Dependent Shear Modulus of
               Glasses Revisited 111
       References 116

3   Viscoelasticity 119
     3.1   Fundamentals of Linear Viscoelasticity 119
     3.2   Microscopic Nonaffine Theory of Viscoelasticity 126
            3.2.1   Nonequilibrium Dissipative Equation of Motion 127
            3.2.2   Derivation of Microscopic Viscoelastic Moduli 128
     3.3   Case Study: Polymer Glasses 132
     3.4   Case Study: Metallic Glasses 140
            3.4.1   Application to $Cu_{50}Zr_{50}$ Metallic Glass 141
            3.4.2   Experiments 141
            3.4.3   MD Simulations with EAM Potentials 141
            3.4.4   Memory Kernel for the Friction 142
            3.4.5   Dynamic Viscoelastic Young's Moduli of $Cu_{50}Zr_{50}$ 144
     3.5   Microscopic Theory of Relaxation Modulus and Creep 147
            3.5.1   Theory of Power-Law Creep in Disordered Solids 149
     References 151

4   Wave Propagation and Damping 153
     4.1   Sound Attenuation 153
     4.2   Akhiezer Damping 156
     4.3   Microscopic Theory of Sound Attenuation in Amorphous Solids ... 160
            4.3.1   Linear Response Theory 161
            4.3.2   Dynamic Response Function: From the Stress to the
                    Displacement Correlator 162
            4.3.3   From Nonaffine Motions to the Susceptibility 163
            4.3.4   Acoustic Wave Propagation 166
            4.3.5   Acoustic Wave Attenuation 168
     4.4   Ioffe-Regel Crossover 174
     References 176

5   Phonons and Vibrational Spectrum 179
     5.1   The Debye Model of Solids 179
            5.1.1   The Debye Density of States 179
            5.1.2   Debye Frequency, Momentum, and Temperature 181
            5.1.3   Van Hove Singularities 182
     5.2   The Boson Peak in the VDOS 183

5.3   Case Study: Simple Lattices with Randomness 190
5.4   Theoretical Models 192
        5.4.1   The Random Matrix Theory (RMT) Model.................. 193
        5.4.2   Singular Behavior Near Marginal Stability 199
        5.4.3   The Damped Phonon Model of the VDOS................... 201
References 209

6   Thermal Properties 213
6.1   Specific Heat.. 213
6.2   Thermal Conductivity 215
6.3   The Tunneling Model 217
References 219

7   Viscosity of Supercooled Liquids 221
7.1   The Viscosity of Liquids 221
7.2   The Vogel-Fulcher-Tammann Empirical Expression 224
7.3   Fragility 224
7.4   The Adam-Gibbs Model and Growing Length Scales 225
7.5   Microscopic Theory of Viscosity 226
        7.5.1   Frenkel's Theory 226
        7.5.2   Dyre's Shoving Model 229
        7.5.3   Linking Viscosity to Bonding and Structure: The KSZ
                Model 231
7.6   First-Principles Frameworks 237
        7.6.1   Green-Kubo Formalism 238
        7.6.2   Nonaffine Response Theory 238
References 241

8   Plastic Deformation 243
8.1   Plasticity of Crystals.. 243
        8.1.1   Dislocations 244
        8.1.2   Schmid's Law 247
        8.1.3   Dislocations as Topological Defects 248
        8.1.4   Volterra Construction 249
8.2   Theory of Plastic Deformation in Amorphous Solids Mediated
        by Dislocation-Type Defects... 249
        8.2.1   Dislocation-Type Defects in Amorphous Solids from
                Nonaffine Displacements 251
8.3   Plasticity Mediated by Dislocation-Type Defects
        in Amorphous Solids: Polymer Glasses 259
8.4   Shear Banding and Eshelby-Like Quadrupoles 261
References 265

9   Confinement Effects 267
9.1   Elasticity and Waves Under Confinement 267
9.2   Comparison with Experimental and Simulations Data 272
        9.2.1   Confined Liquids 272

      9.2.2   Confined Amorphous Solids 274

   9.3   Vibrational Density of States of Confined Solids 274

   References 277

A A Brief Reminder of Elasticity Theory 279

B Lattice Dynamics of Metallic Glasses with the EAM Potential 285

C Generalized Langevin Equation Derived from Caldeira-Leggett
Hamiltonians 289

   C.1   Caldeira-Leggett Hamiltonian 289

   C.2   Derivation of the Generalized Langevin Equation 290

   C.3   The Fluctuation-Dissipation Theorem 291

   C.4   The Memory Kernel 292

   References 293

Glossary 295

Index 297

Acronyms

bcc	body centered cubic
BP	boson peak
CRN	continuous random network
CRR	cooperatively rearranging regions
CS	Carnahan-Starling
DHO	damped harmonic oscillator
DMA	dynamic mechanical analysis
DTDs	dislocation-type defects
EAM	embedded atom method
EOS	equation of state
fcc	face-centered cubic
HET	heterogeneous elasticity theory
INMs	instantaneous normal modes
INS	inelastic neutron scattering
IR	Ioffe-Regel crossover
IS	inherent structures
KG	Kremer-Grest, coarse-grained polymer model
KSZ	Krausser-Samwer-Zaccone viscosity relation
LAMMPS	Large-scale atomic/molecular massively parallel simulator
MD	molecular dynamics
MG	metallic glass
NA	nonaffine
NN	nearest-neighbor
PY	Percus-Yevick
r.h.s.	right-hand side
RCP	random close packing
RN	random network
TF	Thomas-Fermi
TLS	two-level states
VDOS	vibrational density of states
VFT	Vogel-Fulcher-Tammann viscosity relation

A Bird's-Eye View of Amorphous Solids

1

Abstract

In this chapter, we set the groundwork for the mathematical description of disordered or amorphous solids by introducing the fundamental concepts of microscopic bonding and structure.

1.1 Microscopic Bonding and Interactions in Disordered Solids

Atomic and molecular forces in disordered solids do not differ significantly from their counterparts in crystalline solids. However, they play a very important role in determining the structure and mechanical stability, possibly even more crucially than in crystals. Indeed, the structure in amorphous solids is often directly linked to the mechanical stability (in fact, it is a function of stability) in the sense that the building blocks self-organize into structures until stability is reached such that rigidity prevents further rearrangements. The single most important physical quantity, which determines the overall structure and its associated stability, is therefore the type of microscopic bonding and interactions between the building blocks. That is why we start from a brief discussion of the most common atomic and molecular forces, which give rise to bonding and cohesion in disordered solids.

1.1.1 Central-Force (Non-covalent) Interactions

In a condensed matter system, any given interaction implies a change of the system's energy upon changing the relative distance between two building blocks (atoms, molecules or grains/colloidal particles). When only changes in the absolute value of the distance $r = |\mathbf{r}|$ lead to corresponding changes of energy of the system,

© The Author(s), under exclusive license to Springer Nature Switzerland AG 2023
A. Zaccone, *Theory of Disordered Solids*, Lecture Notes in Physics 1015,
https://doi.org/10.1007/978-3-031-24706-4_1

the corresponding interactions are referred to as central-force interactions. Typical examples are van der Waals (vdW) interaction forces due to dipole-dipole London forces between two noble gas atoms or two colloidal particles. A full (anisotropic) dependence on the whole $\mathbf{r}$, including the orientation of the separation vector $\mathbf{r}$, is instead necessary to describe vdW forces between two molecules. vdW attractive forces are ultimately due to the shift of electronic clouds around atoms or molecules, which induces, by sheer electrostatics, a corresponding shift in a nearby atom or molecule, with a consequent attraction between the dipoles (or multipoles). They are typically weak with a characteristic energy in the range from 4 to 40 meV. Also, intermolecular vdW attractive forces typically decay with inverse sixth power of the intermolecular distance, $\sim r^{-6}$; hence, they are much more short-ranged compared to, e.g., the electrostatic Coulomb force, which decays only with $\sim r^{-1}$. For colloidal particles, many such dipole-induced dipole interactions need to be summed in order to compute the overall interaction between two particles that are comprised of very many atoms and molecules. Nonetheless, compact closed-form expressions for the interaction energy as a function of r can be obtained also in this case, the derivation of which can be found, with an excellent unsurpassed discussion of the underlying physics, in [1].

Another example of central-force bonding is provided by ionic forces, which are due to the bare Coulomb attraction between two ions with opposite charge (e.g., Na^+ and Cl^- in sodium chloride crystals). In this case, the bonding energy can be much more substantial, in the range from 150 kJ/mol up to 1500 kJ/mol. Ionic bonding as the main or exclusive type of bonding force is not extremely common in amorphous solids, since such glassy solid would be the result of ionic liquids (e.g., molten salts) undergoing a glass transition. More commonly encountered materials, of great interest for technological applications in energy conversion and storage, are so-called ionic glasses or conductive glasses, where a certain concentration of free ions is able to freely diffuse, thus providing electric conductive properties to the material.

The most widely used way of describing attractive vdW forces coexisting with a shorter-ranged hard-core repulsion, the latter due ultimately to Pauli exclusion principle as valence electrons start to overlap, is provided by the Lennard-Jones potential:

$$U(r) = 4\epsilon \left[\left(\frac{\sigma}{r} \right)^{12} - \left(\frac{\sigma}{r} \right)^6 \right] \tag{1.1}$$

where σ denotes the hard-core diameter of the particle and r is the center-to-center separation. The energy scale is set by ϵ, which is the depth of the attractive potential well, sometimes referred to as "dispersion energy." The hard-core repulsion is modelled via the repulsive (plus sign in the above equation) term which goes as $\sim r^{-12}$. This is an empirical way of modelling the steep short-range repulsion due to Pauli exclusion and certainly not the only way. Alternatively, one can use an exponential decay with r: $U(r) = A \exp(-br)$, which is more grounded in the

Fig. 1.1 The potential energy of interaction according to the Lennard-Jones potential profile as a function of the radial center-center distance between the two atoms or colloidal particles, given by Eq. (1.1). The energy minimum of the attractive well, obtained by setting $dU/dr = 0$ in Eq. (1.1), is located at $r = 2^{1/6}\sigma$, where $U = -\epsilon$

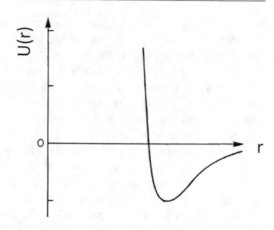

fundamentals of quantum mechanics and is sometimes referred to as the Born-Mayer repulsion term.

The LJ potential profile is sketched in Fig. 1.1.

1.1.2   Covalent Bonding

Covalent bonding is the most important type of bonding for all oxide network glasses (e.g., silica glass) as well as for other amorphous ceramic materials, amorphous carbon materials (including disordered graphene), etc. It is also the type of bonding active between molecular subunits in polymers.

Covalent bonds form between two atoms which have incomplete octets, that is, their outermost electron shells have fewer than eight electrons each. A covalent bond between two atoms is then realized through sharing a pair of electrons, which become rather delocalized over the two atoms, so that the octet rule is now satisfied for each atom. The sharing of the electrons pair can lead to formation of different orbitals, e.g., σ or π being the most common. The σ orbital is the strongest, with the probability of finding the electrons being concentrated in the gap between the two atoms. In the π orbital, instead, the highest probability is above and below a plane that joins the two atoms. Orbital hybridization is also common.

The bonding pair of electrons spends most of their time between the two atomic nuclei, thus screening the positive charges of the two nuclei from one another and enabling the nuclei to come closer together, compared to the case where the bonding electrons were absent. The negative charges of the electron pair attract the two nuclei, thus holding them together in a bond. For the electron sharing to be possible, the two atoms must have comparable electronegativity, with typical differences in electronegativity values in the range 0.0–0.4, hence much lower than for ionic bonds where the difference in electronegativity is >1.7.

In all cases, the covalent bond is highly directional in space, which means that the interaction energy is not just a function of the radial distance $|\mathbf{r}| = r$, and, for example, there is a strong resistance to sliding of the two atoms past each other (bending rigidity). This resistance, on the other hand, is completely absent, e.g., in ionic bonds given by the isotropic Coulomb interaction. This is intuitive if one thinks, e.g., of a π orbital, where the electron cloud is concentrated in specific regions of space above and beneath a well-defined plane. Upon moving one of the two atoms along an arc of circle, by keeping r fixed, the electronic configuration quickly deviates from being optimal, which is associated with a high energy cost. This is the reason why covalent bonds possess "bending rigidity," or tangential rigidity, unlike ionic bonds. This means that, by viewing the covalent bond as a rigid stick, there is an energy cost in order to "bend" the stick. Or, viewed from yet a different perspective, the energy cost to change the angle θ between three atoms connected by two covalent bonds is non-zero.

The latter consideration forms the basis for a convenient way to parameterize the non-central rigidity (i.e., the bending rigidity) of covalent bonds in numerical simulations and lattice dynamical calculations of many-atom systems. While a detailed description of covalent bonding must resort to the quantum mechanics of the sharing orbitals, this is of course no longer convenient when one has to deal with tens of thousands of atoms or more. To reproduce the macroscopic properties of materials, it often suffices to take a molecular mechanics approach and parameterize the covalent bond in terms of the characteristic (rest) angle of the triplet of atoms, θ_0, cfr. Fig. 1.2. A commonly used form of the three-body interaction energy among

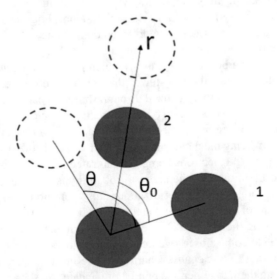

Fig. 1.2 Schematic representation of the difference between central-force and bond-bending interactions between atoms/particles. Particles 1 and 2 are connected to the particle at the center via covalent bonds which possess bending rigidity. The angular part of the covalent bond interaction energy is parameterized via the rest angle θ_0. Upon sliding particle 2 past the particle at the center, the angle changes from θ_0 to θ, which implies an increase of potential energy given by Eq. (1.2). This is very different from central-force interaction potentials, where only motions that imply a change in the radial coordinate r contribute a change of energy of the system

three atoms i, j, k with, e.g., j at the center being covalently bonded to both i and k is as follows:

$$U_{bend}(\theta) = \epsilon_{bend}[1 - \cos(\theta - \theta_0)] \tag{1.2}$$

where ϵ is the characteristic energy scale to change the angle from its rest value θ_0. To this bond-bending contribution one still has to add a central-force bond-stretching term, which can be conveniently written, e.g., using the Morse potential (a function of r only).

In conclusion, the covalent bond represents the strongest type of bonding in nature, and the hardest materials known are made exclusively of covalent bonds (which include diamond and the recently discovered ultrastrong 2D polymer materials [2]). Typical bonding energy in common glassy materials (e.g., C-C in polymers or Si-O in window glass) ranges between 300 and 500 kJ/mol, with bonding lengths in the range 1–2Å.

1.1.3 Metallic Bonding

Metals are traditionally found in crystalline states, given their high propensity to crystallize from the melt upon cooling. Since crystallization is always favored (over vitrification) by the size monodispersity of the constituent atoms or ions (or any building blocks including colloids and grains), this is always true for pure metals. For alloys made of different atomic species, however, modern vitrification techniques at high cooling rates have made it possible to circumvent crystallization and to achieve amorphous solid states called metallic glasses or amorphous metals. Metallic glasses, like crystalline metals, are good conductors of electric current because the external electron wavefunctions of their atoms largely overlap to form a nearly free electron gas. These delocalized electrons, i.e., the conduction electrons, interact with the positively charged ions via the Coulomb attraction at separation distances larger than a cutoff radius. The latter is approximately of the same size of the ion. Within this cutoff distance (i.e., within the ion), however, the true electronic wavefunction displays large oscillations since it is required (by the Pauli exclusion principle) to be orthogonal to the wavefunctions of the core electron shells inside the ion, which effectively "cancels" the strong electron-ion interaction (Phillips-Kleinman cancellation theorem). An extensive literature since the 1950s, with key contributions by J.C. Phillips, N. Ashcroft, M. L. Cohen, and V. Heine, among others, has been devoted to the mathematical modelling of the electron-ion pseudopotential. What is certain is that the large oscillations overall strongly reduce the electron-ion interaction within the cutoff radius. Ashcroft, with a bold assumption, went up to argue that one could take an "average" over the strong oscillations within the cutoff radius, which thus would give a zero net effect or an interaction energy equal to constant within the cutoff. Ashcroft was then able, with this simple ansatz, to accurately model an extensive set of experimental data of sound speed in various metals.

In general, with the pseudopotential method, one writes down the conduction electron wavefunction in terms of a pseudowavefunction and an additional contribution written as an expansion in the orthogonal basis set of core-electron states. Upon then writing the Schrödinger equation for the pseudowavefunction, the contribution written in terms of orthogonalized plane waves provides an additional energy contribution to the Hamiltonian, i.e., the actual *pseudopotential*.

With the advent of density functional theory (DFT) and its numerical implementations, it is now possible to construct pseudopoentials for a given material in a fully ab initio way, by solving the all-atom problem. This is a key topic in the area of electronic structure theory.

An even more complex problem than determining the electron-ion interaction is that of determining the ion-ion interaction energy. At sufficiently short-range inter-ion separations, the interaction is dominated by the screened-Coulomb repulsion and hence reduces to the Thomas-Fermi repulsion of the screened ion cores. However, at larger separation distance, a bonding minimum appears as a result of many-body interactions between the ions and the electron cloud. A very efficient numerical methodology to compute the ion-ion interaction energy is the so-called embedded atom method (EAM) [3], which can be used to perform molecular dynamics (MD) simulations of metallic liquids and glasses within the high-performance computing environment LAMMPS.

Within the EAM framework, the total potential energy acting on a tagged atom i is given by:

$$U_i = F_A \left(\sum_{j \neq i} \rho_{AB}(r_{ij}) \right) + \frac{1}{2} \sum_{j \neq i} \psi_{AB}(r_{ij}). \tag{1.3}$$

Here r_{ij} represents the radial distance between i and j atoms; ρ_{AB} is the contribution to the electronic charge density from particle j of type (i.e., metal element) B at the location of particle i of type A; ψ_{AB} is a pairwise potential between an atom of type A and an atom of type B; and F_A is the embedding function that gives the energy required to place the tagged particle i of type A into the electron cloud. Hence, the total potential is the sum over all particles, $U = \sum_i U_i$.

The many-body nature of the EAM potential is a result of the embedding energy term. Both summations in the formula are over all neighbors j of particle i within the cutoff distance [4].

A schematic picture of the many-body ion-ion interaction energy in amorphous metals is shown in Fig. 1.3.

The bonding minimum in the effective ion-ion pseudopotential in metals has a shape which resembles that of a Lennard-Jones potential (sketched above in Fig. 1.1), or at least this is the case for alkaline metals [5], even though, of course, the fundamental origin is completely different. The energy scale of the bonding minimum is in the range 500–1500 $k_B T$, where k_B is Boltzmann's constant, while the separation distance is in the range 4–6Å [5].

Fig. 1.3 Schematic representation of the ion-ion interaction energy between two ions in a liquid metal or in a metallic glass

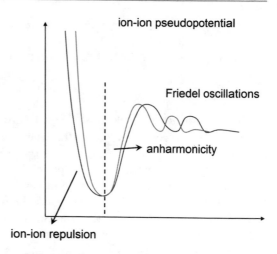

As mentioned earlier, at sufficiently short-ranged separations, the ion-ion interactions becomes analytically manageable because the many-body effects of the electronic cloud and of the other atoms become less important. This can be achieved at the level of empirical modelling (i.e., with parameters to be calibrated by fitting some measurable macroscopic property) by using an Ashcroft-type pseudopotential for the Thomas-Fermi screened-Coulomb repulsion [3] and, in addition, a Born-Mayer interaction term which accounts for the effect of Pauli exclusion repulsion between core-electron shells of the two interacting ions. With these two contributions, the ion-ion Ashcroft-Born-Mayer pseudopotential is given by:

$$U_{\mathrm{ii}}(r) = \frac{A\, e^{-q_{\mathrm{TF}}(r-2a_0)}}{r-2a_0} + B\, e^{-C(r-\bar{\sigma})}, \tag{1.4}$$

where a_0 is the Bohr radius, at which the repulsion energy diverges, and $\bar{\sigma}$ the average ionic core diameter of the alloy, which corresponds to the average size of the ionized atoms constituting the alloy. The average ionic core diameter is obtained through an averaging procedure of the respective ionic core diameters of the constituents with weights given by their volume ratios in the alloy. The values for the ionic core diameters of the atoms constituting the alloys are taken from Ref. [6]. The quantities A and B set the energy scales for the repulsive interaction from the Ashcroft and Born-Mayer terms, respectively. The parameter q_{TF} is the inverse of the Thomas-Fermi screening length given by Thomas-Fermi theory, and its value is known for different types of alloys [7]. We choose a representative value for q_{TF} as $1.25\ \mathring{A}^{-1}$ according to the values reported in Ref. [7]. The ionic core diameter $\bar{\sigma}$ is obtained by a weighted average of the core diameters of the atoms constituting the alloys, taken from [6], where the weights correspond to the ratios of the respective atoms.

The characteristic range $1/C$ of the core-electron shells overlap repulsion is not known a priori. However, its typical values are less sensitive to the atomic composition than the parameters $\bar{\sigma}$, A, and B. Different atoms have very similar values typically in the range 1–$2.5\,\text{Å}^{-1}$ [8].

Finally, A is the prefactor to the Ashcroft pseudopotential,

$$A = Z_{\text{ion}}^2 e^2 \cosh^2(q_{\text{TF}} R_{\text{core}}) \tag{1.5}$$

which is the product of the electrostatic nuclear repulsion, $Z_{\text{ion}}^2 e^2$, and the Ashcroft factor defined as $\cosh^2(q_{\text{TF}} R_{\text{core}})$ [9], where R_{core} is a typical value for the atom-specific core radius and Z_{ion} the effective ionic charge number. The latter cannot easily be estimated from first principles or from the literature. Similarly, the prefactor B of the Born-Mayer term can be rigorously evaluated only from the exchange integrals of the various overlapping electrons, which belong to the valence shells of the two interacting ions. This calculation, even in approximate form, is not feasible except for simple crystals. Hence, one can take both A and B as adjustable parameters in the mapping between the Ashcroft-Born-Mayer pseudopotential and some macroscopic property, as we shall see in Chap. 7. We shall remark that the Born-Mayer prefactor B typically has nontrivial large variations from element to element across the periodic table, as shown in many ab initio simulations studies [8, 10]. Consistent with this known fact, it turns out that B is the most sensitive parameter in the analysis, in the sense that small variations in B can lead to large deviations in the fitting of the experimental data. Conversely, the Ashcroft prefactor A is not as much a sensitive parameter, and its values are not crucial for the fitting of experimental data.

As shown in Fig. 1.4, the overall short-range repulsion (i.e., the sum of Ashcroft and Born-Mayer terms) is well approximated, empirically, by a logarithmically decaying function. This is a useful fact that can be used for the empirical modelling of the short-range structure of disordered solids and of its relation to macroscopic properties, such as the viscosity, as we shall see in Chap. 7.

1.1.4   Anharmonicity

Anharmonic lattices are most conveniently described in second quantization, which allows for significant simplifications in the mathematical treatment. Alternatively, as we shall see next, one has to resort to effective approximate descriptions.

We start considering the following standard Hamiltonian for the anharmonic crystal in second quantization [12]:

$$H = H_0 + H_A \tag{1.6}$$

where $H_0 = \sum_\lambda \hbar\omega_\lambda [b_\lambda^\dagger b_\lambda + \frac{1}{2}]$ is the harmonic part of the Hamiltonian and $b_\lambda^\dagger, b_\lambda$ the creation and annihilation operators, respectively. The compound index λ compactly represents the pair of indices $(\mathbf{k}j)$ where $\mathbf{k}$ is the wavevector and j is the branch index. Hence, $\omega_\lambda \equiv \omega_j(\mathbf{k})$. The anharmonic part can be described, in the

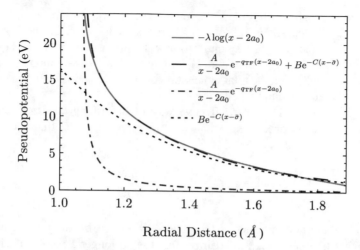

Fig. 1.4 The two main contributions to the short-range part of the ion-ion repulsion in liquid metals/metallic glasses, represented by the core-electron shell repulsion modelled through the exponentially decaying Born-Mayer term and the screened Coulomb-dominated Ashcroft-Thomas-Fermi term. The overall sum of the two contributions is well approximated by an empirical logarithmic decay with a single repulsion-steepness parameter λ that can be related to macroscopic properties such as the viscosity measured as a function of temperature (cfr. Chap. 7 for more details). Adapted from Ref.[11] with permission from the National Academy of Sciences of the USA

standard way, with terms of cubic and quartic order:

$$H_A = \sum_{n=3,4} \frac{1}{n!} \sum_{\lambda_1 \ldots \lambda_n} \zeta \prod_{i=1}^{n} [(b_{\lambda_i} + b^{\dagger}_{-\lambda_i})]. \tag{1.7}$$

Here, $\zeta \equiv \zeta(\lambda_1 \ldots \lambda_n)$ are coefficients related to the n-th order derivatives of the interatomic pair potential with respect to the lattice displacements from equilibrium position, while the factors $[(b_{\lambda_i} + b^{\dagger}_{-\lambda_i})]$ arise upon replacing the atomic displacements with the corresponding expressions in second quantization. Hence, the above equation is nothing but the usual potential energy expansion of the lattice about the rest positions of the atoms (cfr. Eq. (1.15) at the classical level). Importantly, the momenta involved in the three-phonon and four-phonon interactions are conserved. For example, following [13], for a three-phonon term, this allows us to write $\zeta_3 \sum_{p,q} (b^{\dagger}_{p-q} b^{\dagger}_q b_p)$. In a typical three-phonon process, in the initial state, there is one phonon with momentum p, which, as a result of anharmonic interaction, decays into two other phonons, with momenta q and $p-q$, respectively.

For quartic terms, typically we have:

$$\sum \zeta_4 \left[\sum_{q_1 q_2 q_3 q_4} (b^{\dagger}_{q_1} b^{\dagger}_{q_2} b_{q_3} b_{q_4}) + \sum_{k_1 k_2 k_3 k_4} (b^{\dagger}_{k_1} b^{\dagger}_{k_2} b_{k_3} b_{k_4}) \right] \tag{1.8}$$

where the momenta must obey conservation laws of the type $q_1 + q_2 = q_3 + q_4$ and $k_1 = k_2 + k_3 + k_4$ and so on. Here a typical process is the mutual scattering of two phonons with momenta k_1 and k_2 into two others, with momenta k_3 and k_4, or the decay of one phonon with momentum k_1 into three phonons with momenta k_2, k_3, k_4, and so on. In any case, the total momentum of created phonons must be equal to the total momentum of annihilated ones.

There is, however, a simpler, effective way of describing anharmonicity and to deduce important consequences for material properties. The free energy of a harmonic solid is given by:

$$F = k_B T \sum_p \ln \left[2 \sinh \frac{\hbar \omega_p}{k_B T} \right] \tag{1.9}$$

where ω_p are the normal modes of vibration. When the interatomic interactions are no longer harmonic (e.g., the pair potential is no longer a parabola), then the solid is *anharmonic*, and the above expression is no longer valid. However, it can still be used in the so-called quasi-harmonic approximation, provided that one puts care in re-defining the vibrational frequencies ω_p. In the anharmonic crystal, the vibration frequencies become dependent on volume V; this is intuitive, because the eigenfrequencies ω_p are given by $\sqrt{\kappa/m}$ where $\kappa = \frac{d^2 U_{ij}}{dr_{ij}^2}$ is the spring constant defined as the curvature of the interatomic interaction U_{ij}. If the potential U_{ij} between two atoms i and j is not quadratic in r_{ij}, it is clear that the curvature $\frac{d^2 U_{ij}}{dr_{ij}^2}$ will change with the interatomic distance r_{ij}; hence, it will change when the total volume V changes.

Grüneisen was the first to notice this fact and proposed to define a coefficient that measures the dependence of normal mode frequencies on the volume as:

$$\gamma_p = -\frac{\partial \ln \omega_p}{\partial \ln V} = -\frac{V}{\omega_p} \frac{\partial \omega_p}{\partial V} \tag{1.10}$$

which is now known as the Grüneisen parameter or Grüneisen coefficient. More generally, the Grüneisen parameter, γ_G, represents an average over all the normal modes of the material, so that γ_G normally differs from $1/3$ in materials having more realistic intermolecular interactions. Usually, in ordinary crystals, one has $\gamma_G \approx 1 - 2$, while larger values are indicative of strong anharmonicity.

The total free energy as a function of volume then becomes:

$$F(V) = \frac{1}{2} K \left(\frac{\delta V}{V} \right)^2 + k_B T \sum_p \ln \left[2 \sinh \frac{\hbar \omega_p(V)}{k_B T} \right] \tag{1.11}$$

where the first term is an elastic energy term for isotropic volume change δV (expansion or compression), with K the bulk modulus (see Appendix A). The second term represents the phonon contribution to the free energy, which is calculated by treating the lattice as if it were harmonic but at a given volume V. The main difference is that while in the harmonic case the phonon frequencies are constant, independent of the distance between atoms and hence independent of V, now they do depend on V because of the anharmonic potential. In this way, anharmonic effects can be taken into account via the Grüneisen parameter.

The equilibrium volume V in this scheme can be determined by minimizing the free energy. Since the second term in the above expression is temperature dependent, clearly, this also implies that the volume will change with temperature, which is the well-known phenomenon of thermal expansion (note that with the ideal harmonic solid, the thermal expansion is identically zero).

Upon differentiating Eq. (1.11) with respect to V and using the definition of the Grüneisen parameter, we obtain:

$$K \frac{\delta V}{V} = \sum p \gamma_G \frac{\hbar \omega_p}{2} \coth \frac{\hbar \omega_p}{2 k_B T} = \gamma_G E(T) \tag{1.12}$$

where $E(T)$ denotes the average energy of the lattice:

$$E(T) = \sum_p \frac{\hbar \omega_p}{2} \coth \frac{\hbar \omega_p}{2 k_B T}. \tag{1.13}$$

Upon further differentiating with respect to temperature, we can estimate the volume variation upon changing the temperature, i.e., the thermal expansion coefficient α_T as:

$$\alpha_T = \frac{1}{V} \frac{\partial V}{\partial T} = \frac{\gamma_G}{K} \frac{\partial E(T)}{\partial T} = \frac{\gamma_G C_V}{K} \tag{1.14}$$

which is a remarkable relationship between fundamental properties of a solid and sometimes referred to as the Grüneisen equation.[1] This equation clearly tells us that thermal expansion is a good measure of the anharmonicity of the lattice.

The thermal expansion coefficient and the Grüneisen parameter can be more quantitatively related to the shape of the interatomic or intermolecular potential. Using the Taylor expansion of the interaction potential U, following Kittel [14], we can write the pair potential as a function of the displacement x between two particles, up to 4th order as:

$$U(x) = \zeta_2 x^2 - \zeta_3 x^3 - \zeta_4 x^4 \tag{1.15}$$

---

[1] Note that the linear thermal expansion coefficient is defined as $\alpha_l = \alpha_T/3$, which introduces a factor $1/3$ in the above relation if one uses it instead of the volumetric one.

with ζ_2, ζ_3, and ζ_4 all real and positive. The linear thermal expansion coefficient α_l is nothing but the average displacement $\langle x \rangle$ from the equilibrium position, i.e., from the bonding minimum of the interaction. The average displacement can be evaluated in the Boltzmann ensemble as:

$$\langle x \rangle = \frac{\int_{-\infty}^{\infty} x \, \exp[-\beta U(x)] dx}{\int_{-\infty}^{\infty} \exp[-\beta U(x)] dx} \tag{1.16}$$

with the Boltzmann factor $\beta = \frac{1}{k_B T}$. Assuming that the cubic and quartic anharmonic correction terms are small compared to $k_B T$, we can Taylor expand $\exp[\beta \zeta_3 \, x^3 + \beta \zeta_4 \, x^4] \approx 1 + \beta \zeta_3 x^3 + \beta \zeta_4 x^4$ and then approximate the integrals as follows:

$$\int_{-\infty}^{\infty} x \, \exp[-\beta U(x)] dx \approx \int_{-\infty}^{\infty} \exp(-\beta \zeta_2 x^2)(x + \beta \zeta_3 x^4 + \beta \zeta_4 x^5) dx$$

$$= \frac{3\pi^{1/2}}{4} \frac{\zeta_3}{\zeta_2^{5/2}} \beta^{-3/2}$$

$$\int_{-\infty}^{\infty} \exp[-\beta U(x)] dx \approx \int_{-\infty}^{\infty} \exp(-\beta \zeta_2 x^2) dx = \left(\frac{\pi}{\beta \zeta_2}\right)^{1/2} \tag{1.17}$$

from which we get:

$$\alpha_l = \langle x \rangle \approx \frac{3\zeta_3}{4\zeta_2^2} k_B T. \tag{1.18}$$

This relation, valid for weak anharmonicity, clearly shows us that the thermal expansion coefficient is directly proportional to the cubic coefficient ζ_3 in the Taylor expansion of the interatomic potential Eq. (1.15). Furthermore, in this lowest order, the thermal expansion does not depend on the quartic coefficient ζ_4. A similar dependence is expected for the Grüneisen parameter, γ_G, in this approximation.

In turn, the Grüneisen parameter γ_G can be directly related to the anharmonicity of the interatomic or intermolecular potential. For perfect crystals with pairwise nearest neighbor interaction, the following relation holds [15]:

$$\gamma_G = -\frac{1}{6} \frac{U'''(a)a^2 + 2[U''(a)a - U'(a)]}{U''(a)a + 2U'(a)} \tag{1.19}$$

where a is the equilibrium lattice spacing between nearest neighbors and $U'''(a)$ denotes the third derivative of the interatomic potential $U(r_{ij})$ evaluated in $r_{ij} = a$.

The Grüneisen parameter can be computed numerically in MD simulations in an atom-resolved way [16] and even in an eigenmode-resolved way [17]. This is important as it allows one to assess the degree of anharmonicity of individual atoms or individual eigenmodes (phonons or vibrational excitations), as a way to evaluate the role of anharmonicity (or its absence) in a given phenomenon or in a given sector of the vibrational spectrum of a material under consideration.[2]

In general, computing phonon dispersion relations and spectra of anharmonic solids is a formidable task in modern solid-state theory and computational materials science. A well-established approach to anharmonicity is the self-consistent method historically introduced by Born and Hooton [18], leading to the concept of renormalization of phonon frequencies in the quasiharmonic or self-consistent phonon approximation. Within this scheme, the renormalized phonon frequencies arise from an effective vibrational dynamics within a region near equilibrium, which takes anharmonic terms of the potential into account via adjustable parameters obtained from a self-consistent solution to the many-body problem [19].

At a qualitative level and skipping the subtleties of the full many-body description, the phonon renormalization due to anharmonicity can be understood in the following way. As is well known from classical mechanics, the frequency ω of a driven damped harmonic oscillator gets lowered, compared to the undamped (fully harmonic) oscillator limit ω_0, by a softening correction controlled by the damping coefficient Γ: $\omega = \sqrt{\omega_0 - \Gamma^2}$. In a typical scattering experiment, one sends an incident wave of some radiation (typically X-rays, light, or neutrons) to the material sample. The incident wave excites the resonant amplitude of a vibrational mode, which, just as for the simple damped harmonic oscillator, is given by a Lorentzian function. The Lorentzian is peaked at the characteristic frequency of the phonon mode, which, due to the ubiquitous anharmonic damping in all real solids, is not the harmonic frequency ω_0, but it is the "corrected" or renormalized frequency $\omega = \sqrt{\omega_0 - \Gamma^2}$. The full width at half maximum (FWHM) of the Lorentzian is proportional to Γ and is an experimentally measurable quantity in all inelastic scattering measurements, also called the linewidth.

In the quasi-particle approximation, the harmonic phonon frequency ω_0 suffers a complex shift $\Delta - i\Gamma$ due to anharmonic phonon-phonon interactions, where the damping parameter Γ is the inverse of the phonon lifetime and 2Γ is the FWHM of the (Lorentzian) phonon peak. The corresponding propagator (retarded Green's function) is given by:

$$G = \frac{1}{\omega^2 - (\omega_0 + \Delta - i\Gamma)^2} \tag{1.20}$$

---

[2] For example, this analysis of the atom-resolved and eigenmode-resolved Grüneisen parameter has helped establish that the ubiquitous boson peak phenomenon in glasses and in certain crystals, i.e., the excess of vibrational modes above the Debye-model quadratic law, is associated with strongly anharmonic vibration modes and strongly anharmonic atomic trajectories in [17], characterized by large values of the Grüneisen parameter. Cfr. Chap. 5 for the phenomena and terminology discussed in this note.

and the Dyson equation for the propagator is given by:

$$G = \frac{1}{\omega^2 - \omega_0^2 - \Sigma(\omega)} \qquad (1.21)$$

where $\Sigma(\omega)$ is the (retarded) self-energy.

The quasi-particle approximation is valid if:

$$|\Delta - i\Gamma| \ll \omega \qquad (1.22)$$

which, from Eq. (1.20), implies:

$$\text{Re}\Sigma(\omega) \approx 2\omega_0\Delta$$
$$\text{Im}\Sigma(\omega) \approx -2\omega_0\Gamma = -\frac{2\omega_0}{\tau} \qquad (1.23)$$

where τ is the quasi-phonon lifetime. The phonon renormalization and damping/lifetime parameters Δ and Γ, respectively, can be calculated by solving the Feynman diagrams for the various phonon-phonon processes specified by the chosen Hamiltonian, according to self-consistent phonon theory. The details of the theory can be found, e.g., in [20]. Here we shall limit ourselves to some introductory qualitative considerations.

Under the conditions specified by Eq. (1.22), the propagator can be simplified as:

$$G = \frac{1}{\omega^2 - \omega_0^2 - 2\omega_0(\Delta - i\Gamma)}. \qquad (1.24)$$

The response function is peaked at a frequency $\omega^2 \approx \omega_0^2 + 2\omega_0\Delta$. The frequency renormalization parameter Δ is proportional to the occupation numbers of the final states into which the phonon is going to decay. For cubic anharmonicity and considering only the lowest-order contributions, one has $\Delta \sim (n_1 + n_2 + 1)$, if the phonon decays into two phonons with occupation numbers n_1 and n_2, or $\Delta \sim (n_1 - n_2)$ for scattering with absorption of a phonon [20]. If, instead, one considers only quartic interactions, as this is often the dominant contribution to the phonon renormalization [21], we have, to leading order [20], $\Delta \sim (2n_2 + 1)$. In all cases, with $n_{1,2} = [\exp \hbar\omega/k_B T - 1]^{-1}$, at high temperature $\hbar\omega \ll k_B T$, the final result for Δ will be $\Delta = 2^{-1}\omega_0^{-1}CT$, where C is a constant. This result is confirmed by more quantitative calculations of the self-consistent equations.[3]

---

[3] At soft mode (e.g., ferroelectric) phase transitions, $T = T_c$, the harmonic frequency ω_0 becomes unstable, $\omega_0^2 = -CT_c$. This is, for example, what happens with a double-well potential $V(\xi) = A\xi^2 + B\xi^4$, where ξ is the normal mode coordinate, with $A < 0$ and $B > 0$. As a result, upon approaching the phase transition, one has $\omega^2 = C(T - T_c)$, with the phonon frequency, which becomes "frozen" and purely imaginary in the low-temperature phase.

Since disorder and anharmonicity in solids are almost always inextricably related, these considerations will be useful throughout this book, in particular when discussing the wave propagation phenomena and the vibrational spectra, in Chaps. 4 and 5.

1.2 The Structure of Disordered Solids

In disordered solids, the microscopic building blocks (atoms, molecules, colloids, grains) do not occupy regular positions on an ordered lattice. As a consequence, standard results of solid-state theory, which are valid for crystalline solids, such as Bloch theorems relying on periodicity of the lattice or Bragg's law of diffraction from a well-defined lattice spacing, which repeats itself periodically in space, are not applicable. In the absence of periodicity, one has then to rely on statistical tools in order to quantify the spatial arrangements of the particles.

The main tool is offered by the pair correlation function, $g(\mathbf{r})$. In a spherically symmetric environment, this reduces to the radial distribution function (rdf), $g(r)$— i.e., a function of radial distance r only—which is defined as the probability of finding a particle at a radial separation $r + dr$ from a particle placed at the center of the (spherical) coordinate frame.

1.2.1 Static Structure Factor and Radial Distribution Function

While for crystalline solids the lattice structure, as quantified by group theory (crystallographic point groups) and experimentally measured by X-ray diffraction, represents the full structural characterization of the solid, for disordered solids, this role is played by the radial distribution function (rdf) $g(r)$ and, by its Fourier transform, the static structure factor $S(q)$, as measured by radiation scattering experiments, defined by:

$$S(q) - 1 = \frac{4\pi\rho}{q} \int_0^\infty r[g(r) - 1]\sin(qr)dr \tag{1.25}$$

where ρ is the particle density and q is the wavevector. For a typical (light, neutron, or X-ray) scattering experiment, $q = \frac{4\pi n}{\lambda}\sin(\theta/2)$, where λ is the wavelength of incident radiation, n the refractive index of the medium, and θ is the scattering angle determined by the detector's position relative to the sample. A derivation, based on elastic scattering, of the above relation is typically done within the Born approximation and the assumption of isotropicity and can be found, e.g., in [22].

An example of the experimentally measured static structure factor of metallic glass $Mg_{70}Zn_{30}$ is displayed in Fig. 1.5.

At low momentum transfer q the structure factor probes long length scales in the material. In the hydrodynamic limit $q \rightarrow 0$, as the system is probed over very large length scales, the structure factor contains thermodynamic information. In

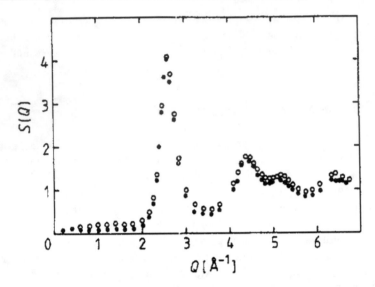

Fig. 1.5 Static structure factor of the metallic glass $Mg_{70}Zn_{30}$, as a function of momentum transfer (denoted as q in the text) measured experimentally by neutron scattering at room temperature using two different techniques. Open circles are obtained as the zeroth moment of the dynamical structure factor in a typical inelastic neutron scattering experiment, whereas black circles are obtained from static structure factor measurements using a time-of-flight technique. Adapted from Ref. [23]. ©IOP Publishing. Reproduced with permission. All rights reserved

particular, it is related to the isothermal compressibility χ_T of the liquid via the so-called compressibility equation:

$$\lim_{q \to 0} S(q) = \rho\, k_B T\, \chi_T = k_B T \left(\frac{\partial \rho}{\partial p} \right) \tag{1.26}$$

where ρ is density and p is pressure. Also, the compressibility is the inverse of the bulk elastic modulus, $\chi_T = K^{-1}$, and is therefore also related to the longitudinal speed of sound $v_L = \sqrt{K/\rho}$. The behavior of the structure factor in the limit $q \to 0$ also contains important information about the mesoscopic structure of the material. For example, an upturn or even divergent tail of $S(q)$ at $q \to 0$ is indicative of a highly non-uniform or inhomogeneous structure of the system.[4] Conversely, if, instead, $S(q) \to 0$ at $q \to 0$, then the system is *hyperuniform* [24], because this implies unusually low fluctuations in density on large length scales. For example, all perfect crystals are, trivially, hyperuniform, whereas amorphous solids that exhibit hyperuniformity are quite special exotic states of matter with interesting and novel physical properties (in particular, optical properties).

---

[4] In complex liquids and soft matter systems, this is typical, for example, of systems undergoing demixing transitions or spinodal decomposition.

Amorphous solids, just like liquids, present a first large or main peak in $S(q)$, which is a manifestation of atomic-scale density oscillations at the level of second and third coordination shells beyond nearest neighbors (the main peak is visible in Fig. 1.5 in the range 2–3 Å). In liquids, proteins, colloids, and polymers, the main peak in $S(q)$ is also telling the range of q where the dynamics is slow and most cooperative, a phenomenon known as "de Gennes narrowing." In practice [25], the characteristic linewidth of excitations in this q-range, $\Gamma(q)$, decreases as $[S(q)]^{-1}$, or, in other words, is found to "narrow" at the position of the peak. This implies that the characteristic lifetime or relaxation time grows in proportion to $S(q)$ and has its largest values in correspondence of the peak.

For multi-component systems with different types of atoms, the above quantities become, for example, for metal alloys, respectively, $g_{\alpha\beta}$ and $S_{\alpha\beta}$, also sometimes referred to as partial rdf and partial structure factors. Here, α and β may denote different atomic species present in the alloy.

The radial distribution function is related to the coordination number z, i.e., the number of nearest neighbors of a given particle. As is well known, indeed, a standard definition of the coordination number z is based on the rdf $g(r)$, the latter representing the probability of finding the center of a particle at a distance r, within dr, from a test particle set as the origin of the reference frame [26]. The average number dz of spheres lying in the range $r + dr$ is then given by $dz = 4\pi\rho g(r)r^2 dr$. By introducing the quantity $\sigma^+ \equiv \sigma + \epsilon$, where ϵ is an arbitrarily small number $\epsilon \to 0^+$, the average number of particles in contact (just touching) with the test particle is given by

$$z = 4\pi\rho \int_0^{\sigma^+} g(r)r^2 dr. \tag{1.27}$$

In liquids or glasses where the nearest neighbors are not exactly touching the particle at the center, it is more common to take the upper bound of the integral as the distance of the first minimum in the $g(r)$, denoted as r_{min} (see Fig. 1.6), and thus write:

$$z = 4\pi\rho \int_0^{r_{min}} g(r)r^2 dr. \tag{1.28}$$

The radial distribution function of a binary metal alloy is shown in Fig. 1.6. The main first peak represents the first coordination shell of nearest neighbor atoms around an atom at the center of the spherical frame. The subsequent peaks represent so-called Friedel oscillations due to the probability of finding further coordination shells of atoms farther away. The main first peak decreases with increasing the temperature T, which reflects the spreading out and loosening of the short-range structure upon increasing the temperature.

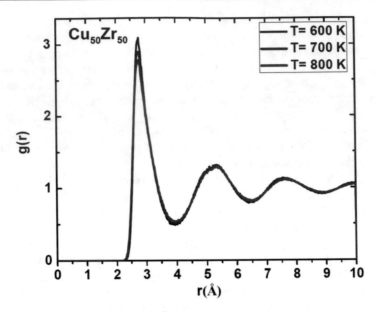

Fig. 1.6 Radial distribution function (rdf) of a binary alloy $Zr_{50}Cu_{50}$ from MD simulations using the EAM method, at three different temperatures. The first minimum, r_{min}, occurs at $r \approx 4$Å. Adapted, with modifications, from Ref. [27]

1.2.2 Structural Paradigms for Disordered Solids

Historically, two structural paradigms for disordered solids have dominated the stage: (i) the so-called random network model (sometimes also referred to as continuous random network, CRN, model), which goes back to Zachariasen's work, and (ii) the random close packing (RCP) model. The former applies to amorphous materials constituted, at the microscopic level, by atoms that are connected by covalent bonds into a system-spanning covalent network that lacks periodicity. The RCP model instead applies to systems where the microscopic building blocks (atoms, molecules, colloids, grains) are instead bonded via central-force non-covalent interactions, the typical example being van der Waals attractive forces coexisting with short-ranged hard-core repulsion (the standard Lennard-Jones potential). In the limit of granular materials, the particles are often interacting solely via the hard-core repulsion with no attraction or at most some frictional contacts.

A third paradigm, which may be considered a hybrid of the first two, is the packing of random coils. Here the random coils represent polymer chains where the subunits (monomers) are oriented randomly while still being bonded to the adjacent units via covalent bonds. A close packing of such random coils provides a good description of the structure of polymer glasses. In this paradigm, covalent bonds along the polymer chain coexist with non-covalent van der Waals or LJ-type interactions between monomers that are sufficiently close to be within the range of dispersion forces. Hence, non-covalent interactions can exist also between monomers that belong to different polymer chains. In this sense, a random packing of random coils owes its cohesion to both covalent and non-covalent bonding, with the latter playing a crucial role, as we shall see in Chap. 2 when dealing with the elasticity of polymer glasses.

Another frequently used structural model in disordered condensed matter systems is the fractal model. When spatial correlations between particles or building blocks follow a power-law dependence on distance, there typically is no preferred or characteristic length scale, and the system is thus self-similar, i.e., it looks the same, or approximately the same, on all length scales. This is the defining property of fractals, also known as scale invariance (when self-similarity is indeed exact). Fractal structures are encountered in disordered solids such as colloidal gels and aerogels from aggregated aerosols and in general in all disordered solids, which are formed as the result of a percolation process, by which the building blocks aggregate from a dispersed state into a system-spanning fractal network.

Below we shall look more closely at the three main paradigms, namely, the random network model, the random close packing, and the fractal model.

1.2.2.1 The Random Network Model

The random network model of glasses was introduced by Zachariasen (an expert of X-ray diffraction physics) in 1932 based on chemical bonding concepts in order to rationalize the structure and properties of oxide network glasses. These materials include, e.g., silica glass and other glasses where a metallic-type (or semiconducting) atom (a cation or electron donor) is bonded to oxygen atoms (more electronegative, acting as electron acceptors).

In this model, the building blocks are polygons whose geometry is dictated by the chemical bonding and valence of the cations (K) and oxygens (O). In turn, these polygons are then connected to other polyhedra in a random system-spanning network. For example, in silica glass, one has that each silicon (Si) atom is bonded to four oxygen (O) atoms and each oxygen atom, in turn, is bonded to two Si atoms, hence SiO_2.

The four Zachariasen's rules for glass formation (i.e., for network formation) of a generic oxide glass K_mO_n (where each oxygen is bonded to m cations and each cation is bonded to n oxygens) are as follows [28]: (i) an oxygen atom is linked to no more than two cations K; (ii) the number of oxygen atoms around each cation K is limited to three or four; (iii) among the oxygen-containing polyhedra, a polyhedron cation K shares corners, but does not share sides or faces; and (iv) for 3D networks of oxygen-containing polyhedra, at least three corners must be shared at a network junction.

The CRN model has served as a useful paradigm also for monoatomic amorphous materials, such as amorphous silicon, a-Si. In this case, each silicon atom is bonded to 4 other silicon atoms. Also in this case, in order to form a system-spanning network, "rings" (typically with 5, 6, and 7 members) are formed. While the CRN is perhaps an idealization of real covalently bonded disordered solids, its modifications to account for local (possibly paracrystalline) inhomogeneities still provide a useful

Fig. 1.7 A random jammed
packing of hard spheres with
packing fraction $\phi \approx 0.64$ as
generated by computer
simulations. Figure courtesy
of Dr. Carmine Anzivino

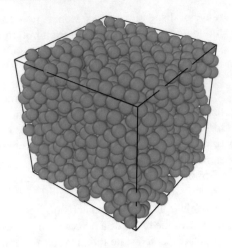

paradigm to quantitatively describe the radial distribution function extracted from experimental diffraction data [29].

All in all, CRN is widely used in the modelling of experimental data, most of the time in the context of computational/numerical models supported by Monte Carlo methods. Since analytical results are scarce or non-existent for this type of models, we shall focus on analytically tractable results that can instead be obtained, for example, with the random close packing model, to be discussed in the next section.

1.2.2.2 The Random Close Packing Paradigm

Since Bernal's work, the most important model for the random structure of non-crystalline condensed matter states (liquids and glasses) has been provided by the random close packing (RCP) paradigm of hard spheres.

A computer-generated random close packing of hard spheres with packing fraction $\phi \approx 0.64$ is shown in Fig. 1.7.

Hard spherical particles, which do not attract each other and only interact by excluded volume, represent one of the most important reference systems in statistical mechanics. While the hard-sphere system was initially devised to (just) model the short-range, infinitely steep, repulsive forces of an idealized mono-atomic liquid, subsequent pioneering experiments by Pusey, van Megen, Vrij, and others showed that colloidal systems, such as polymethyl methacrylate (PMMA) and silica particles coated with polymers, can be approximately modelled by the hard-sphere potential. Since then, the hard-sphere system has been studied in great detail, and its phase behavior is now well understood.

A well-established fact, proved first by simulations [30, 31], is that a system of identical hard spheres has a well-defined freezing transition driven by purely entropic effects. The bulk thermodynamically stable crystalline phase for hard spheres is the so-called face-centered-cubic (fcc) crystal, which coincides with the close-packed arrangement, yielding a packing fraction $\phi_{fcc} = \pi/3\sqrt{2} \approx 0.74$. Interestingly, the value of ϕ_{fcc} was correctly predicted by Kepler in his famous conjecture, later exactly computed by Gauss, and formally proved by Hales [32–34] in very recent times. If a system of hard spheres is compressed quickly, however, crystallization can be avoided. In this case, the maximum density ϕ_{fcc} is not reached and the spheres arrange in a disordered "jammed" configuration at a lower density, as further compression would lead to overlaps or deformation [35]. An open central problem in contemporary physics and mathematics is the determination of the so-called random close packing (RCP) density ϕ_{RCP}, which corresponds to the highest fraction of volume occupied by hard spheres jammed in a disordered assembly.

In 1960, a visionary experiment by Bernal and Mason [36] showed that random packings of spheres in $d = 3$ have a volume fraction around $\phi_{RCP} \approx 0.64$ and a coordination number $z = 6$. This crucial observation was understood by Bernal as a necessary consequence of *mechanical stability*. Indeed, since Maxwell, it is well known that a lattice with solely central-force interactions is rigid only if $z \geq 6$, regardless of the lattice structure [37].

On the theoretical side, an unambiguous determination of ϕ_{RCP} is plagued by at least two aspects. First of all, ϕ_{RCP} depends largely on the protocol used to form the packing [38, 39]. This protocol dependence is manifested in the relatively broad range of ϕ_{RCP} that have been reported in the literature, e.g., $\phi_{RCP} = 0.60$–0.69 in $d = 3$ [38], and $\phi_{RCP} = 0.81$–0.89 in $d = 2$ [40]. Second, it is difficult to provide a clear-cut definition of "randomness" of the packing. To overcome this problem, the authors of Refs. [41, 42] proposed ϕ_{RCP} to be a maximally random jammed state corresponding to some minimum value of a structural order parameter. Another possible solution to this problem was suggested in Ref. [43], where the idea of ϕ_{RCP} as a singularity in a set of metastable branches of pressure was proposed. The idea proposed in Ref. [43] echoes the findings of high-dimensional space theoretical descriptions provided by replica-symmetry breaking approaches, which are exact only in the limit $d \to \infty$ [44].

More recently, an analytical theory was introduced, which is able to predict sensible values for ϕ_{RCP} in both $d = 2$ and $d = 3$ dimensions [45]. The starting point of this theory is the main finding of Ref. [46], concerning the mechanical stability of random packings. As shown in Ref. [46], mechanical stability arises in a hard-sphere system in $d = 2$ and $d = 3$ when the condition $z = 2d$ for the coordination number z is satisfied, a result that we shall derive with all steps by means of nonaffine elasticity theory in Chap. 2, Sect. 2.4.2. As a consequence, any amorphous packing with average coordination number $z < 2d$ is liquid-like, while it is mechanically stable (i.e. rigid) for $z \geq 2d$, in full agreement with the abovementioned Maxwell criterion. By adopting Bernal's view of RCP, it was argued [45] that the threshold $z = 2d$ is the only rigorous criterion to define ϕ_{RCP} for monodisperse hard spheres. Combining this idea with a suitably modified liquid state theory for the radial distribution function, values for ϕ_{RCP} were predicted, which fall well within the range of values reported in the literature using different methods (experiments and simulations) and protocols [45]. Here we present a sketch of this analytical derivation.

1.2.2.3 Monodisperse Random Close Packing

Inspired by Bernal's work [36], the starting point of the theoretical scheme proposed
in Ref. [45] is the introduction of an operative definition for the random close
packing (RCP) density based on mechanical stability conditions. To this aim, the
exact solution of the elasticity problem of random packings of spheres in $d = 2$ and
$d = 3$ found in Ref. [46] is exploited. In Ref. [46], indeed, an accurate closed-form
expression for the shear modulus G of a d-dimensional (with $d = 2, 3$) system of
elastic spheres with diameter σ was shown to be given by:

$$G = \alpha(d)\rho\kappa\sigma^2(z - 2d), \tag{1.29}$$

where z is the coordination number, $\rho = N/V$ is the number density (obviously,
$\rho = N/S$ in $d = 2$), κ is the spring constant of the nearest neighbor interaction, and
α is a constant value given by $\alpha(d = 2) = 1/18$ in $d = 2$ and by $\alpha(d = 3) = 1/30$
in $d = 3$. See Chap. 2, for a detailed derivation of this equation from first principles.

As shown in Ref. [46], and later in Chap. 2, Sect. 2.4.2, Eq. (1.29) is in excellent
parameter-free agreement with simulations data of jammed packings from Ref. [47].

Based on Eq. (1.29), the onset of mechanical stability identified by the vanishing
of the shear modulus G arises in a d-dimensional (with $d = 2, 3$) random packing
of spheres when a critical coordination number $z_c = 2d$ is reached, in agreement
with the Maxwell isostaticity criterion presented in more detail in Sect. 1.6. While
the system is fluid for $z < z_c$, it is jammed for $z \geq z_c$. This is due to the fact
that for $z < z_c$, the potential energy of deformation is "spent" on sustaining the
so-called nonaffine motions (cfr. Chap. 2), and no energy is left to support the
elastic response to deformation [46, 48]. As a consequence, the system is still able
to undergo substantial rearrangements and to find denser configurations at larger
packing fraction. By contrast, for $z \geq z_c$, the particle contacts are able to balance
the energy cost of nonaffine relaxations. In other words, the particles come at fixed,
closest contacts with one another, blocking relative motion and depriving the system
of any further internal dynamics, i.e., the system is jammed. It follows that the
critical coordination number $z_c = 6$ can be used as an operative definition of the
RCP density of a system of hard spheres.

In order to take advantage of Eq. (1.29) for computing the RCP density ϕ_{RCP},
we employ suitably modified liquid state theory for the radial distribution function
(rdf). Recalling Eq. (1.27), it was noticed in Ref. [45] that the total rdf $g(r)$ can
be treated as a *partially continuous* probability distribution function (PDF). In
probability theory, besides fully continuous and fully discrete PDFs, one can also
define partially continuous distributions, also known as mixed distributions or mixed
random variables [49]. As an example of a fully discrete distribution, the PDF $f_d(x)$
of a distribution consisting of a set of points $x_i = \{x_1, \ldots, x_n\}$, with corresponding
probabilities $p_i = \{p_1, \ldots, p_n\}$, can be written as $f_d(x) = \sum_{i=1}^n p_i\delta(x - x_i)$.
A partially continuous (PC) distribution can be written as [49] $f_{PC}(x) = c(x) +
\sum_{i=1}^n p_i\delta(x - x_i)$ where $c(x)$ is the continuous part and the second term is the
discrete part. The latter implies that the distribution returns exactly the value x_i with

probability p_i. Upon normalizing to 1 over the relevant domain, $\int_0^\infty f_{PC}(x)dx = 1$, it follows that $f_{PC}(x)$ is indeed a valid PDF [50].

We use the considerations above to write:

$$g(r) = g_c(r) + g_{BC}(r) = g_0 g(\sigma)\delta(r - \sigma) + g_{BC}(r), \tag{1.30}$$

i.e., to split $g(r)$ into a continuous part $g_{BC}(r)$, describing the probability of finding particles in the region of space beyond contact (BC) $r > \sigma^+$, and a discrete part $g_c(r) \equiv g_0 g(\sigma)\delta(r - \sigma)$, describing the probability of having nearest neighbors in direct contact with the test particle. In the definition of the discrete part $g_c(r)$ of the rdf $g(r)$, $g(\sigma)$ is the contact value of the $g(r)$ [51], i.e., the probability of finding particles at exactly $r = \sigma$, and g_0 is a normalization factor to be determined later.

This model for the radial distribution function $g(r)$ of random sphere packings is schematically depicted in Fig. 1.8.

The total $g(r)$ is a generalized partially continuous PDF, which obeys the usual normalization condition $\int_0^\infty 4\pi\rho g(r)r^2 dr = N$. It follows that, by using Eqs. (1.27) and (1.30) and by recalling that the packing fraction ϕ is related to the number density ρ through $\phi \equiv \frac{4}{3}\pi(\sigma/2)^3\rho$, one finds:

$$z = 24\phi\frac{g_0}{\sigma}g(\sigma). \tag{1.31}$$

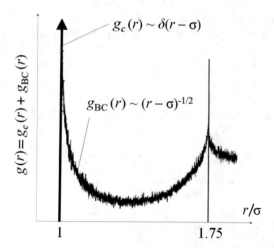

Fig. 1.8 Radial distribution function (rdf) $g(r)$ of a system of hard spheres with diameter σ in $d = 3$ dimension, at the random close packing density ϕ_{RCP}. The black curve represents the *continuous* part $g_{BC}(r)$ of the rdf, while the thick vertical arrow represents the Dirac delta in the *discrete* part $g_c(r)$ of the rdf (see Eq. (1.30)). The black curve represents numerical data adapted from [52], this is the *continuous* part. This picture is also consistent with all the previous literature on the $g(r)$ of random aggregates of spheres, e.g., [53] and [54]

The RCP density ϕ_{RCP} can then be found from Eq. (1.31) by imposing $z = z_c = 6$. However, we first need to find $g(\sigma)$ and g_0.

To compute the contact value of $g(r)$, we take advantage of the statistical theory of hard-sphere liquids, which provides a way to find $g(\sigma)$ analytically up to the (unphysical) packing fraction $\phi = 1$ while remaining agnostic about the possible onset of ordering. Closed-form expressions for $g(\sigma)$ of hard spheres are given by the Percus-Yevick (PY) and the Carnahan-Starling (CS) theories [26,55]. The former is given by:

$$g_{PY}(\sigma) = \frac{1 + \phi/2}{(1 - \phi)^2},$$ (1.32)

while the latter is given by:

$$g_{CS}(\sigma) = \frac{1 - \phi/2}{(1 - \phi)^3}.$$ (1.33)

As the last step to solve Eq. (1.31) and thus find ϕ_{RCP}, a condition to determine the normalization factor g_0 is needed. First of all, it is worth noticing that, based on dimensional analysis, $g_0 \propto \sigma$, to ensure dimensional consistency of Eq. (1.31). Then we observe that a possible condition that can be chosen to determine the numerical prefactor $g_0 \propto \sigma$ is based on a known reference state, for example, the closest packing (CP) value, which (monodisperse) spherical objects can never exceed. This limit can be used as an effective "boundary condition" in our problem to determine the unknown prefactor. In Ref. [45], the point at which the hard-sphere system has perfect (fcc) ordering was chosen. This point is identified by the coordination number $z_{fcc} = 12$ and closest (fcc) packing $\phi_{fcc} = \pi/3\sqrt{2}$ [38]. Imposing into Eq. (1.31) $z = z_{fcc} = 12$, $\phi = \phi_{fcc} = 0.74$ and choosing the PY approximation (1.32) for the contact value of the rdf $g(\sigma) = g_{PY}(\sigma)$, one finds $g_0/\sigma = 0.0331894$. Inserting the obtained g_0/σ into Eq. (1.31) and solving the latter with $g(\sigma) = g_{PY}(\sigma)$ and imposing $z = z_c = 6$, one finds:

$$\phi_{RCP}^{(3D)} = \frac{2\sqrt{648 + \pi[\pi(54 - 24\sqrt{2}\pi + 5\pi^2) - 108\sqrt{2}]}}{36\sqrt{2} + \pi(\sqrt{2}\pi - 36)} +$$
$$+ \frac{2(36\sqrt{2} - 48\pi)}{36\sqrt{2} + \pi(\sqrt{2}\pi - 36)} - 3$$ (1.34)
$$= 0.658963$$

If instead of $g_{PY}(\sigma)$, $g_{CS}(\sigma)$ is used, one finds $g_0/\sigma = 0.0187416$ and $\phi_{RCP} = 0.677376$. The assumption that $g(\sigma)$ can be well approximated with equilibrium-like crowding models has been later justified, also numerically, in Ref. [56].

A different effective boundary condition, or reference state, to determine g_0/σ has been suggested by Likos [57]. Instead of using the point at which the hard-sphere system displays fcc ordering, the one at which the system has bcc ordering can be used. This point is identified by the coordination number $z_{bcc} = 8$ and packing density $\phi_{fcc} = \pi\sqrt{3}/8$. This may be a better choice of a reference state than fcc since it is clearly closer to the RCP, being $z = 8$ of bcc much closer to $z = 6$ of RCP compared to $z = 12$ of fcc.

Imposing into Eq. (1.31) $z = z_{bcc}$, $\phi = \phi_{bcc}^{CP}$ and choosing the PY approximation (1.32) for the contact value of the rdf $g(\sigma) = g_{PY}(\sigma)$, one finds $g_0/\sigma \sim 0.0374068$. Inserting the obtained g_0/σ into Eq. (1.31) and solving the latter while using $g(\sigma) = g_{PY}(\sigma)$ and imposing $z = z_c = 6$, one finds:

$$\phi_{RCP} = 0.643320. \qquad (1.35)$$

If instead of $g_{PY}(\sigma)$, $g_{CS}(\sigma)$ is used, one finds $g_0/\sigma = 0.0242946$ and $\phi_{RCP} = 0.650594$.

We observe that the values of ϕ_{RCP} obtained by solving Eq. (1.31) with $z = z_c = 6$ while considering several approximations either for the contact value $g(\sigma)$ and for the boundary condition are all well within the range 0.61–0.69 for RCP observed with different experiments and simulations [38]. The closest value to the most quoted $\phi \approx 0.644$ is obtained in case $g(\sigma) = g_{PY}(\sigma)$, and the bcc boundary condition are used, which gives $\phi = 0.643$. For the sake of clarity, the values of g_0/σ and the corresponding ϕ_{RCP} are listed in Table 1.1.

A similar derivation can be done, mutatis mutandis, in $d = 2$. Using the hexagonal close packing condition, with $z = z_c = 4$, as the effective boundary condition to determine g_0, and minding the different dimension-dependent factors and metric factor in the integral over the rdf, one obtains $\phi_{RCP} \approx 0.889$, which is within the range of reported values based on numerical methods, although somewhat close to the upper limit (≈ 0.89). It should be mentioned that, in general, the situation in $d = 2$ is much more complex, due to the proximity to crystallization (which, in 2D, is further complicated by the Berezinskii-Kosterlitz-Thouless scenario, including hexatic phases, etc.), as discussed, e.g., in [58].

Recently also the problem of random close packing in higher dimensions has attracted considerable interest in view of the fact that the replica theory of glasses is exact in $d \to \infty$ and can make predictions for the RCP in high-dimensional spaces [44]. The above approach is based on "kissing contacts," since the RCP condition is derived from a mechanical stability criterion where the kissing contacts provide the local mechanical stability. It should be noted that in higher-dimension spaces, there

Table 1.1 Values of g_0 and ϕ_{RCP} obtained for a monodisperse system of hard spheres with diameter σ, computed by using different approximations (Percus-Yevick (PY) or Carnahan-Starling (CS)) for the contact value $g(\sigma)$ of the radial distribution function and different configurations (fcc or bcc) as boundary conditions

	PY + fcc	PY + bcc	CS + fcc	CS + bcc
$10^2 \cdot g_0/\sigma$	3.31894	3.74068	1.87416	2.42946
ϕ_{RCP}	0.658963	0.643320	0.677376	0.650594

is a significant gap of empty space between nearest neighbors, even in the case of ordered close packing. This can be evaluated exactly via group theory in the case of $d = 8$, where one gets $4 \binom{8}{2} + 2^7 = 112 + 128 = 240$ particles that pack $\sqrt{2}$ away from the origin and from each other (those happen to be the 240 root vectors of the 8-dimensional Euclidean E_8 Lie group) [59]. As a consequence of there being such large gaps, approaches such as the one above based on kissing contacts may not be easily transferable to higher-dimensional systems.

The mathematically inclined reader interested in knowing more about the group theoretical aspects of packings, in the broader context of discrete geometry, will find the excellent textbook by Conway and Sloane on these topics [60] very useful.

1.2.2.4 Polydisperse Random Packings

While certainly important from the point of view of discrete geometry, statistical mechanics, and disordered systems, as well as a structural paradigm for liquids and glasses, the monodisperse random close packing is still, certainly, a mathematical idealization. Any experimental realization of RCP necessarily involves a distribution of size of the spheres, since it is physically impossible to manufacture many spheres that have identically the same size, hence the motivation to understand and predict how ϕ_{RCP} changes upon changing the size distribution of the spheres.

The analytical theory presented in the previous section can be adapted to predict ϕ_{RCP} for random packings of spheres with distributed size, i.e., for polydisperse random packings.

The starting point is the relationship [61, 62]

$$Z(\phi) = 1 + 4\phi g(\sigma), \tag{1.36}$$

where $Z \equiv p/\rho k_B T$ is the so-called compressibility factor, p is pressure, ρ is the number density, and T and k_B are the (absolute) temperature and the Boltzmann constant, respectively.

Equation (1.36) provides a useful route to find an expression for the contact value $g(\sigma)$ of the radial distribution function, without having to derive the full $g(r)$. This equation clearly shows that $g(\sigma)$ can be obtained by the equation of state (EOS) of the system, expressed by the dependence $Z(\phi)$ of the compressibility factor on the packing fraction ϕ. This is precisely how the Carnahan-Starling expression for $g(\sigma)$ of monodisperse hard spheres was first obtained.[5]

---

[5] For a monodisperse hard-sphere system, two routes can be followed to analytically derive an EOS [26]. From the fluctuations in the grand canonical ensemble, the compressibility EOS $Z_{\mathrm{PY}}^c(\phi)$ can be derived, while from differentiation of the logarithm of the configuration integral, the virial EOS $Z_{\mathrm{PY}}^v(\phi)$ can be obtained. Thiele [63] and Wertheim [64] independently found the compressibility and the virial equations of state to be given by $Z_{\mathrm{PY}}^c(\phi) = (1+\phi+\phi^2)/(1-\phi)^3$ and $Z_{\mathrm{PY}}^v(\phi) = (1+2\phi+3\phi^2)/(1-\phi)^2$, respectively. Later, Carnahan and Starling [65] showed that a more accurate EOS for monodisperse hard spheres is given by a linear combination of $Z_{\mathrm{PY}}^v(\phi)$ and $Z_{\mathrm{PY}}^c(\phi)$ and introduced the so-called Carnahan-Starling (CS) EOS $Z_{\mathrm{CS}}(\phi) \equiv \frac{2}{3}Z_{\mathrm{PY}}^c(\phi) + \frac{1}{3}Z_{\mathrm{PY}}^v(\phi) = (1 + \phi + \phi^2 - \phi^3)/(1 - \phi)^3$.

By way of the EOS, it is particularly handy to extend the description to a multi-component system of m components (where each "component" is just a sphere with its own peculiar size), which in the limit $m \to \infty$ represents a polydisperse packing of spheres. Each sphere's size is drawn from a given size distribution.

Equation (1.36) can be extended to the case of an m-component mixture as [66–68]:

$$\frac{Z^{(m)}(\phi) - 1}{4\phi} = \sum_{i=1}^{m} \sum_{j=1}^{m} x_i x_j \frac{\sigma_{ij}^3}{\langle \sigma^3 \rangle} g_{ij}(\sigma_{ij}), \tag{1.37}$$

which was originally derived by Lebowitz [66]. Here, $g_{ij}(\sigma_{ij})$ gives the probability to find a sphere of species j at a distance σ_{ij} from a reference sphere of species i set at the origin of the reference frame. Again, here, a species i simply means a particle with its own size.

Equation (1.37) can be further (easily) generalized to the case of a polydisperse system of hard spheres whose diameter follow a continuous distribution $f(\sigma)$, by considering the limit $m \to \infty$. In this case, Eq. (1.37) becomes [69]:

$$\frac{Z^{(m\to\infty)}(\phi) - 1}{4\phi} = \frac{1}{\langle \sigma^3 \rangle} \int_0^\infty d\sigma \int_0^\infty d\sigma' f(\sigma) f(\sigma')$$
$$\times \left(\frac{\sigma + \sigma'}{2} \right)^3 g(\sigma, \sigma'), \tag{1.38}$$

where $\langle \sigma^n \rangle = \int_0^\infty d\sigma f(\sigma) \sigma^n$. We observe that in case $f(\sigma) = \sum_{i=1}^{m} x_i \delta(\sigma_i - \sigma)$, Eq. (1.38) reduces to Eq. (1.37), such that the EOS of an m-component mixture and that of a polydisperse system are strictly related.

Equation (1.38) represents the extension of Eq. (1.36) to polydisperse systems. We declare:

$$g^{(m\to\infty)}(\sigma) \equiv \frac{1}{8 \langle \sigma^3 \rangle} \int_0^\infty d\sigma \int_0^\infty d\sigma' f(\sigma) f(\sigma')(\sigma + \sigma')^3 g(\sigma, \sigma') \tag{1.39}$$

as the generalization to polydisperse systems of the $g(\sigma)$ used in Eq. (1.36).

The protocol to analytically compute the RCP volume fraction ϕ_{RCP} of a polydisperse hard-sphere fluid is, therefore, the following. We start from an approximate expression for the EOS $Z^{(m\to\infty)}(\phi)$ of the system under study and determine $g^{(m\to\infty)}(\sigma)$ through Eq. (1.38). By analogy with the monodisperse case, we find ϕ_{RCP} by substituting $g^{(m\to\infty)}(\sigma)$ into Eq. (1.31) and imposing the critical condition for jamming $z = z_c \equiv 6$. Importantly, the onset of mechanical stability occurs at $z = z_c \equiv 6$ independently of whether the system is monosdispere or polydisperse and independently of the chosen size distribution [70].

Since Eq. (1.38) correctly reduces to Eq. (1.36) in the limit of a one-component system, we use the values listed in the upper row of Table 1.1 for the normalization factor g_0 appearing in Eq. (1.31).

In the following, we choose three different approximations for the EOS of a polydisperse hard-sphere fluid, which reduce, in the monodisperse limit, either to the $Z_{PY}^v(\phi)$ or to the $Z_{CS}(\phi)$ equations of state considered in Ref. [45]. The first approximation considered here is the famous Boublík-Mansoori-Carnahan-Starling-Leland (BMCSL) EOS [67,71], which reads:

$$Z_{BMCSL}(\phi) = \frac{1}{1-\phi} + \frac{3\phi}{(1-\phi)^2}\frac{\langle\sigma\rangle\langle\sigma^2\rangle}{\langle\sigma^3\rangle} + \frac{\phi^2(3-\phi)}{(1-\phi)^3}\frac{\langle\sigma^2\rangle^3}{\langle\sigma^3\rangle^2}, \tag{1.40}$$

and reduces to the Carnahan-Starling (CS) EOS, $Z_{CS}(\phi)$, in the monodisperse limit. We then follow the recipe introduced by Santos et al. in Ref. [68] to derive the EOS $Z^{(m\to\infty)}(\phi)$ of a polydisperse mixture of additive hard spheres in terms of the EOS $Z(\phi)$ of a one-component system. These authors found $Z^{(m\to\infty)}(\phi)$ to be given by:

$$Z^{(m\to\infty)}(\phi) = 1 + [Z(\phi) - 1]\frac{\langle\sigma^2\rangle}{2\langle\sigma^3\rangle^2}(\langle\sigma^2\rangle^2 + \langle\sigma\rangle\langle\sigma^3\rangle)$$

$$+ \frac{\phi}{(1-\phi)}\left[1 - \frac{\langle\sigma^2\rangle}{\langle\sigma^3\rangle^2}(2\langle\sigma^2\rangle^2 - \langle\sigma\rangle\langle\sigma^3\rangle)\right], \tag{1.41}$$

where $Z(\phi)$ is the EOS of the pure hard-sphere fluid.

One can consider the cases $Z(\phi) = Z_{CS}(\phi)$ and $Z(\phi) = Z_{PY}^v(\phi)$, respectively. In the first case, we obtain the so-called extended Carnahan-Starling (eCS) EOS, which reads as:

$$Z_{eCS}(\phi) = Z_{BMCSL}(\phi) + \frac{\phi^3}{(1-\phi)^3}\frac{\langle\sigma^2\rangle}{\langle\sigma^3\rangle^2}(\langle\sigma\rangle\langle\sigma^3\rangle - \langle\sigma^2\rangle^2). \tag{1.42}$$

We will denote the EOS obtained in the second case as Z_{ePY} in the following. By construction, the $Z_{eCS}(\phi)$ and $Z_{ePY}(\phi)$ equations of state reduce to the $Z_{CS}(\phi)$ and $Z_{PY}(\phi)$ EOS, in the monodisperse limit, respectively. The closed-form expression of $Z_{BMCSL}(\phi)$, $Z_{eCS}(\phi)$, or $Z_{ePY}(\phi)$ is known when the moments of a certain distribution are computed.

We now suppose the particle diameter σ to follow a continuous probability distribution $f(\sigma)$. We focus on the case of a log-normal distribution [72], for which results from numerical simulations are available in literature [39,73], although other distributions can also be handled by the present theory.

The log-normal distribution, $f_{\log}(\sigma)$, is defined as: [72]

$$f_{\log}(\sigma) = \frac{1}{\sigma\sqrt{2\pi\alpha^2}} e^{-(\ln\sigma-\mu)^2/2\alpha^2}, \tag{1.43}$$

where α and μ are arbitrary parameters.

The n-th moment $\langle\sigma^n\rangle \equiv \int_{-\infty}^{\infty} d\sigma f_{\log}(\sigma)\sigma^n$ of $f_{\log}(\sigma)$ is given by:

$$\langle\sigma^n\rangle = e^{n\mu+n^2\alpha^2/2}, \tag{1.44}$$

such that the average value $\langle\sigma\rangle$ and the variance $\mathrm{VAR}[\sigma] \equiv \langle\sigma^2\rangle - \langle\sigma\rangle^2$ of the distribution are $\langle\sigma\rangle = e^{\mu+\sigma^2/2}$ and $\mathrm{VAR}[\sigma] = e^{\sigma^2-1}e^{2\mu+\sigma^2}$, respectively, and the relative standard deviation can be written as:

$$s_\sigma^{\log} \equiv \frac{\left(\langle\sigma^2\rangle - \langle\sigma\rangle^2\right)^{1/2}}{\langle\sigma\rangle} = (e^{\alpha^2} - 1)^{1/2}. \tag{1.45}$$

Replacing Eq. (1.44) in the $Z_{\mathrm{BMCSL}}(\phi)$, $Z_{\mathrm{eCS}}(\phi)$, and $Z_{\mathrm{ePY}}(\phi)$ approximations introduced in the previous section enables us to find three distinct analytical (though approximate) expressions for the EOS $Z^{(m\to\infty)}(\phi)$ of our polydisperse system. As it is easy to check, in all cases, the obtained $Z^{(m\to\infty)}(\phi)$ does not depend on the parameter μ but only on the parameter α.

As clear from Eq. (1.45), also the relative standard deviation $s_\sigma^{\log}$ of the $f_{\log}(\sigma)$ distribution depends on α only. It follows that the effect of the polydispersity on the system can be fully described by varying α, for any arbitrary value of μ.

In Fig. 1.9, we plot the obtained RCP packing fraction as a function of the polydispersity parameter α. To characterize our findings, rather than ϕ_{RCP}, we plot $\Delta\phi_{\mathrm{RCP}} \equiv \phi_{\mathrm{RCP}} - \phi_{\mathrm{RCP}}^{\mathrm{mono}}$, where $\phi_{\mathrm{RCP}}^{\mathrm{mono}}$ indicates the RCP packing fraction obtained in the previous section for monodisperse hard spheres and reported in Table 1.1.

The analytical theory (with no adjustable parameters) appears to reproduce the simulations data reasonably well. Interestingly, the analytical theory predicts a plateau in the limit $\alpha \to \infty$, with limiting values $\phi_{\mathrm{RCP}} \approx 0.95 - 0.97$, which is physically meaningful since, of course, the increase in ϕ_{RCP} brought about by polydispersity must saturate at a value of ϕ necessarily lower than 1.

Finally, the above approach can also be easily applied to binary mixtures of hard spheres, i.e., packings of spheres that can only have two different sizes, σ_1 and σ_2. So the first component consists of N_1 spheres of diameter σ_1, while the second component consists of N_2 spheres of diameter σ_1, and the total number of spheres is $N = N_1 + N_2$. The application of the method of [45] was done in Ref. [74] and in Ref. [56]. It was found that the analytical theory can reproduce fairly well, with no fitting parameters, curves of the RCP volume fraction as a function of $\frac{N_1}{N_1+N_2}$, i.e., the fraction of the small component ($\sigma_1 < \sigma_2$). In particular, the theory is able to correctly reproduce how these curves change upon changing the size ratio σ_1/σ_2.

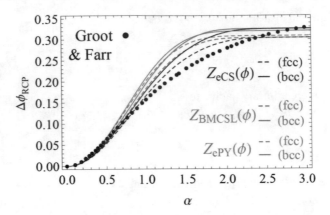

Fig. 1.9 Random close packing (RCP) volume fraction of a polydisperse hard-sphere system, with particle diameters distributed according to the log-normal distribution, plotted as a function of the size polydispersity parameter α (see Eq. (1.45)). $\Delta\phi_{\text{RCP}} \equiv \phi_{\text{RCP}} - \phi_{\text{RCP}}^{\text{mono}}$ is the RCP volume fraction ϕ_{RCP} of the polydisperse system minus the RCP volume fraction $\phi_{\text{RCP}}^{\text{mono}}$ of the monodisperse system. Green, violet, and orange lines indicate results obtained when using the $Z_{\text{BMCSL}}(\phi)$, $Z_{\text{eCS}}(\phi)$, and $Z_{\text{ePY}}(\phi)$ approximations for the equation of state of the system, respectively. Solid and dashed lines represent results obtained by using the fcc and bcc ordered packings, respectively, as a boundary condition to determine g_0/σ. While for dashed violet and dashed green curves $\phi_{\text{RCP}}^{\text{mono}} = 0.677376$ and for continuous violet and full green curves $\phi_{\text{RCP}}^{\text{mono}} = 0.650594$. For dashed orange and full orange curves $\phi_{\text{RCP}}^{\text{mono}} = 0.658964$ and $\phi_{\text{RCP}}^{\text{mono}} = 0.643320$, respectively. Black symbols are simulation results from Ref. [73]. In all cases we find $\Delta\phi_{\text{RCP}} = 0$, i.e., $\phi_{\text{RCP}} = \phi_{\text{RCP}}^{\text{mono}}$, at $\alpha = 0$. The polydisperse theory correctly recovers the corresponding monodisperse limits from Refs. [45, 57] upon setting the standard deviation of the size distribution to zero. By increasing α, instead, $\Delta\phi_{\text{RCP}}$ increases monotonically as a function of α until a plateau limiting value of α is reached where $\Delta\phi_{\text{RCP}}$ remains constant with further increasing α. Figure courtesy of Dr. Carmine Anzivino

1.2.3   The Fractal Model

The fractal model is particularly useful to describe systems whose growth occurs at a rate lower than the growth rate of the embedding Euclidean space and is often the result of some percolation process, e.g., in the vicinity of a critical point of a continuous second-order phase transition. For example, let us consider spherical particles (e.g., Brownian particles or colloids) that aggregate from a fully dispersed state and stick upon contact, thanks to some (typically short-ranged) attractive interaction force (e.g., dispersion forces). If the particles were to aggregate into a spherically shaped "drop" in $d = 3$, the number of particles in the spherical drop would be $N \sim \left(\frac{R_g}{a}\right)^3$, where R_g is the gyration radius of the drop (which obviously coincides with the drop radius) and a the radius of the spherical particle. Clearly, as more particles get into the drop, the latter grows or fills the space at the same rate as the embedding Euclidean space, $\sim r^d$.

Let us now assume that, instead, the aggregation process is such that particles stick upon contact, but this time, the attraction is so strong that particles have effectively no chance to rearrange after sticking upon contact. This is the typical situation encountered in diffusion-limited aggregation of Brownian particles. In this case we have:

$$N \sim \left(\frac{R_g}{a}\right)^{d_f}$$

(1.46)

where $d_f < d = 3$ is the fractal dimension of the aggregate. For diffusion-limited cluster aggregation (DLCA) of colloidal particles in a solvent driven by van der Waals/dispersion forces, one has $d_f \approx 1.7 - 1.8$, indeed significantly lower than the dimension of the embedding Euclidean space $d = 3$. If an electrostatic, e.g., electric double-layer, barrier is present, which slows down the aggregation, then $d_f \approx 2.0 - 2.1$, which is known as the reaction-limited cluster aggregation (RLCA) limit. Similarly, if there is no energy barrier for aggregation but the bonding is weaker (the attraction well is finite such that bonds can break up on the time scale of observation for at least particles that are weakly coordinated at the outer edge of the cluster), then also $d_f \approx 2$ is found. This value of fractal dimension, along with the cluster size distribution of the clusters and the growth of the largest cluster to form a system-spanning network, is predicted by kinetic models of nonequilibrium coagulation-fragmentation master kinetic equations [75, 76].

If the cluster aggregation and growth process are allowed to continue, the formation of a system-spanning network, with the same fractal dimension of the clusters, leads to a disordered solid with fractal structure. Typically this happens in a percolative way, i.e., with a largest cluster whose size eventually diverges (or, more practically, becomes equal to the system's size). Such a process is shown in Fig. 1.10 for the case of a colloidal gelation process.

The radial distribution function (rdf) of a fractal system can be related to the fractal dimension as follows. The gyration radius is defined as:

$$R_g^2 = a^2 + \frac{1}{2N^2} \sum_{m,j=1}^{N} r_{mj}^2$$

(1.47)

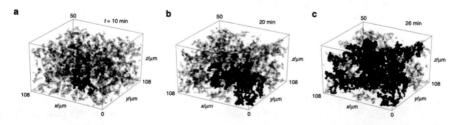

Fig. 1.10 Experimental observation of aggregating colloidal particles, interacting via a finite attraction energy provided by critical-Casimir solvent. The fractal structure with $d_f = 2$ is evident, and so is the largest cluster (highlighted) that grows to eventually span the whole system size. Adapted from Ref. [75]. From (**a**) to (**b**) to (**c**) the growth of the largest connected cluster (in red) is noticeable

where r_{mj} is the distance between two particles, m and j, in the system. Combining this with the definition of rdf, e.g., Eq. (1.28), which gives the number of particles within a certain radial distance from the particle at the center of the frame, one obtains:

$$R_g^2 = a^2 + \frac{4\pi}{2N} \int_0^\infty r^4 g(r) dr \tag{1.48}$$

Clearly, since R_g exhibits power-law scaling with distance, $\sim r^{d_f}$, this implies that also the $g(r)$ must exhibit a fractal power-law scaling governed by the fractal exponent d_f, known as the fractal dimension. This is indeed the case, as verified in countless experiments using static radiation scattering, or confocal microscopy, and in many simulation studies using Monte Carlo techniques.

For systems where the particles are immobilized due to a deep attraction well to their nearest neighbors, it is found that the $g(r)$ can be modelled at short range with a Dirac delta contribution, in pretty much the same spirit as we have seen for the RCP structure, cfr. the first term of Eq. (1.30) and Fig. 1.8. In particular, one has:

$$g(r) = \frac{z}{4\pi\sigma^2}\delta(r - \sigma), \quad r \approx \sigma^+. \tag{1.49}$$

At farther separations, starting from approximately $r = 6a = 3\sigma$, the $g(r)$ exhibits the hallmark power-law regime:

$$g(r) \sim r^{d_f - 3} \tag{1.50}$$

which is exactly what we would expect from the definition of the $g(r)$ and from the fractal scaling:

$$N \sim \int_0^{R_g} g(r) r^2 dr \sim \int_0^{R_g} r^{d_f - 3} r^2 dr, \tag{1.51}$$

which correctly yields $N \sim R_g^{d_f}$ upon performing the integral in the last step.

A schematic of the $g(r)$ of a fractal aggregate is shown in Fig. 1.11.

A corresponding fractal scaling occurs also in the static structure factor, according to Eq. (1.5),

$$S(q) \sim q^{-d_f} \tag{1.52}$$

in a certain window of momentum transfer, q, values.

Fractal scalings arise in numerous macroscopic properties of disordered solids, including elasticity of gels (as will be discussed in depth in Chap. 2), resistivity, and electric conductivity in amorphous semiconductors and random resistor networks [77, 78] and in other inhomogeneous materials with randomly percolating defects

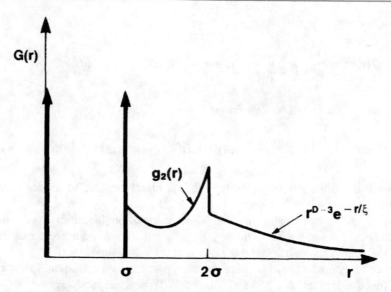

Fig. 1.11 Schematic of the radial distribution function $g(r)$ of a fractal aggregate. The Dirac delta at $r = \sigma$ corresponds to Eq. (1.49), while the fractal power-law decaying regime (1.50) sets in after $r = 2\sigma$ (more precisely after $r \approx 3\sigma$ as shown in [54]). The fractal regime at large r is then suppressed by an exponential (or stretched-exponential [54]) cutoff function, $e^{-r/\xi}$, where ξ is on the order of the aggregate size. Adapted from Ref.[53] with permission of the American Physical Society

or magnetic impurities or in materials with nano-scale textures arising from self-assembly of nanoparticles [79].

1.3 Dynamic Correlation Functions

While static correlations functions, such as the radial distribution function, provide information about the probability of finding particles at some distance from a tagged particle in a *static* snapshot, dynamic correlations functions contain additional information about the time evolution of these probabilities. These information are also contained in the measured intensity of inelastic scattering experiments through the key quantity known as the dynamic structure factor, which we shall introduce in this section.

Following [26], the van Hove correlation function for a spatially uniform system containing N point particles is defined as:

$$G(\mathbf{r}, t) = \frac{1}{N} \left\langle \sum_{i=1}^{N} \sum_{j=1}^{N} \int \delta[\mathbf{r} - \mathbf{r}_j(t) + \mathbf{r}_i(0)] d\mathbf{r} \right\rangle \tag{1.53}$$

which can be rewritten as:

$$G(\mathbf{r}, t) = \frac{1}{N} \left\langle \int \sum_{i=1}^{N} \sum_{j=1}^{N} \delta[\mathbf{r}' + \mathbf{r} - \mathbf{r}_j(t)]\delta[\mathbf{r}' - \mathbf{r}_i(0)]d\mathbf{r}' \right\rangle \tag{1.54}$$

and, in terms of density correlations, as:

$$G(\mathbf{r}, t) = \frac{1}{N} \left\langle \int \rho(\mathbf{r}' + \mathbf{r}, t)\rho(\mathbf{r}', 0)d\mathbf{r}' \right\rangle = \frac{1}{\rho}\langle \rho(\mathbf{r}, t)\rho(\mathbf{0}, t)\rangle. \tag{1.55}$$

At $t = 0$ and for an isotropic system, this is clearly related to the static radial distribution function (rdf) presented in Sect. 1.2.1.

The physical meaning of the van Hove correlation function is as follows: $G(\mathbf{r}, t)d\mathbf{r}$ is the number of particles j in the shell $d\mathbf{r}$ around a point $\mathbf{r}$ at time t given that there was a particle i at the origin at the initial time $t = 0$.

Depending on whether i and j are the same particle, or not, one can split the van Hove function into its "self" (s) and "distinct" (d) parts, respectively:

$$G(\mathbf{r}, t) = G_s(\mathbf{r}, t) + G_d(\mathbf{r}, t) \tag{1.56}$$

where:

$$G_s(\mathbf{r}, t) = \frac{1}{N} \left\langle \sum_{i=1}^{N} \delta[\mathbf{r} - \mathbf{r}_i(t) + \mathbf{r}_i(0)] \right\rangle \tag{1.57}$$

and:

$$G_d(\mathbf{r}, t) = \frac{1}{N} \left\langle \sum_{i=1}^{N} \sum_{j=1}^{N} \delta[\mathbf{r} - \mathbf{r}_j(t) + \mathbf{r}_i(0)] \right\rangle. \tag{1.58}$$

The Fourier transform in space of the van Hove function yields the intermediate scattering function (ISF) $F(\mathbf{k}, t)$:

$$F(\mathbf{k}, t) \equiv \int G(\mathbf{r}, t) \exp(-i\mathbf{k} \cdot \mathbf{r}) \, d\mathbf{r} \tag{1.59}$$

in wavevector space $\mathbf{k}$, which can be accessed experimentally by means of inelastic scattering experiments. Also the ISF can be split into a self and a distinct part. The ISF contains important information about the relaxation processes in supercooled liquids and glasses, such as the α relaxation and the β relaxation. When both are present, the ISF plotted as a function of time for a representative k value exhibits two subsequent decays, one at a shorter time-scale indicative of β relaxation and a later one at larger value of time corresponding to the α relaxation.

By performing also a Fourier transform in the time domain, one finally arrives at the *dynamic structure factor*:

$$S(\mathbf{k}, \omega) \equiv \frac{1}{2\pi} \int_{-\infty}^{\infty} F(\mathbf{k}, t) \exp(i\omega t)\, dt, \qquad (1.60)$$

which is one of the most important quantities in physics. The dynamic structure factor can be directly related to the scattering cross section in inelastic scattering experiments.

Another definition of the dynamic structure factor can be obtained from the quantum theory of neutron scattering and reads as:

$$S(\mathbf{k}, \omega) = \frac{1}{2\pi N} \int_{-\infty}^{\infty} \sum_{i,j} \langle \exp\left(i\mathbf{k} \cdot \mathbf{r}_j(t) - i\mathbf{k} \cdot \mathbf{r}_i(0)\right)\rangle \exp\left(-i\omega t\right) dt. \qquad (1.61)$$

For many collective excitations in condensed matter, for example, phonons, near the resonance frequency $\omega \sim \Omega(k)$, the dynamic structure factor can be approximated in terms of a damped harmonic oscillator (DHO):

$$S(k, \omega) \propto \frac{\omega \Gamma(k)}{(\Omega^2(k) - \omega^2)^2 + \omega^2 \Gamma(k)^2}. \qquad (1.62)$$

with the characteristic Lorentzian shape and where $\Gamma(k) \sim \tau^{-1}$ represents the linewidth (inverse of the lifetime) and $\Omega(k)$ provides the dispersion relation of the excitation.

1.4 The Glass Transition

Condensed bodies can be distinguished, characterized, and classified based on their macroscopic properties, as well as their structure and internal dynamics. Crystallization and melting phase transformations bring the liquid state into the crystalline solid state and vice versa, respectively. The transformation from liquid to solid is accompanied by spontaneous symmetry breaking of translations, which, according to Goldstone's theorem, sees the emergence of gapless modes for each broken generator of continuous Lie group symmetries. In this case, for a 3D system, translations in three spatial directions are spontaneously broken upon crystallization leading to three acoustic (phonon) modes, one longitudinal (LA) and two transverse (TA).

The transformation from liquid state into the amorphous solid state, i.e., into glass, is more subtle and is the object of endless debate. This transformation is accompanied by the emergence of rigidity, i.e., of a non-zero low-frequency shear modulus, and by a kink (change of slope) in the thermal expansion coefficient vs temperature. The latter method is operationally a good methodology to identify the glass transition temperature T_g with the change of slope, from a higher (linear) slope as a function of T in the liquid state to a significantly lower (but still linear) slope in the glass state. Calorimetric methods have been used extensively also, although the temperature dependence of the specific heat on temperature is more complex.

In general, a well-defined amorphous glassy state is one where the building blocks are localized to a small well-defined portion of space (the "glassy cage") within which they perform thermal motion and collide with their neighbors. Their nearest neighbors are long-lived and represent the physical borders of the cage. This is different from the situation in the liquid where the particles remain only transiently in the cage of their nearest neighbors from which they escape after a finite time interval to explore farther away regions of space.

The caging in glasses also provides a spontaneous symmetry breaking of translations at the level of short- and medium-range order, which is enough to generate acoustic phonons and the corresponding finite elastic moduli (remember that speed of sound is $v = \sqrt{G/\rho}$ with G the elastic modulus and ρ the density) also in this case. Another symmetry which is broken upon going from liquid to crystal or to glass is the permutation symmetry of the atoms as labelled by their spatial coordinates [37]. In the liquid, the permutation symmetry is active since each atom, on sufficiently long times, can explore all available positions in the space occupied by the system (In other words, the system is ergodic). This is no longer possible when atoms are localized, be it on a regular crystalline lattice or on a disordered "lattice" as in the glass. This is however not a continuous symmetry, so it is not directly linked to the emergence of rigidity, which is instead, as pointed out above, a direct consequence of the spontaneous symmetry breaking of continuous translations, in both crystals and glasses.

The breaking of ergodicity, however, plays an important role in our understanding of the glass transition, since it is the phenomenon that leads to localization and caging. The caging as a result of ergodicity breaking is successfully predicted by the mode-coupling theory (MCT) of glass transition as developed by Wolfgang Goetze and coworkers [80]. When applied to real systems, the MCT predicts transition temperatures that are systematically higher than the glass transition temperature, as measured, e.g., by calorimetry, etc. Other approaches are instead based on the entropy of cooperatively rearranging regions with a first-order like scenario [81, 82] and related developments using replica theory applied to cloned liquid configurations [83].

The physics of the glass transition is such a vast topic that several books would be needed to cover it in detail. Here we shall adopt the viewpoint on amorphous solids introduced by Alexander [37] and relegate the existence and details of the glass transition to a subsidiary role.

The main assumption will be that such glass transition, or crossover, exists and occurs at a certain temperature denoted as T_g or T_c in the following. Below this temperature, spatial localization of the building blocks within the glassy cage is such that the particles occupy positions on an "amorphous lattice". around which they vibrate due to thermal motion and collisions with their neighbors. At zero temperature, as for granular packings and alike, their position is completely fixed by

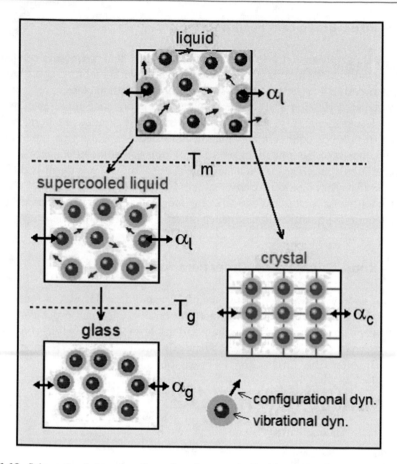

Fig. 1.12 Schematic of phase transformations between the liquid and the solid state. The extents of thermal expansion and vibrational dynamics in the different states are schematically shown. Courtesy of Dr. Peter Lunkenheimer

the mechanical contacts with their nearest neighbors, and no vibrations take place (unless the system is externally submitted to forced vibrations).

The distinction between crystalline solids, glassy solids, and liquids can be summarized as in Fig. 1.12. The scheme also shows how a single parameter, the thermal expansion coefficient α, can be used to discriminate between the different states. As shown in Sect. 1.1.4, α is proportional to the third-order coefficient in a Taylor expansion of the potential energy (or potential of mean force) around the local energy minimum; hence, it is also a measure of the anharmonicity of the system [14]. The latter is very large in the liquid state, and lower, but still significant, in the amorphous glassy state.

1.5 Structural Order Parameters

It is evident, also based on the above structural models, that disorder in condensed matter is multifaceted and can take different nuances within a broad spectrum that goes from perfect crystalline order (such as for a perfect face-centered cubic or a body-centered cubic crystal) to the maximally random state represented by the random close packing. It is clear, for example, that the continuous-random network model will produce configurations intermediate between those two limits of the spectrum, thanks to the limitations on the randomness that are effectively imposed by Zachariasen's rule. The question therefore arises: how can one quantify different degrees of disorder in a given material/system?

In the following, we discuss two different paradigms leading to normalized order parameters for the quantification of lattice order in disordered materials.

1.5.1 Bond-Orientational Order Parameter

The most widely used parameter to quantify crystalline order was introduced by Nelson et al. [84] and is referred to as the bond-orientational order parameter, denoted as F_6 or Q_6 or q_6 in the literature, and has been used many times on glasses and defective crystals [85, 86]. For each pair of nearest neighbor (NN) atoms i and j, one first defines the correlator of NN orientations:

$$S_6(i, j) = \frac{\sum_{m=-6}^{6} q_{6m}(i) q_{6m}^*(j)}{|\sum_{m=-6}^{6} q_{6m}(i)| \, |\sum_{m=-6}^{6} q_{6m}(j)|}, \tag{1.63}$$

where $q_{lm}(i)$ is defined as:

$$q_{lm}(i) = \frac{1}{z(i)} \sum_{j}^{z(i)} Y_{lm}(\mathbf{r}_{ij}) \tag{1.64}$$

with $z(i)$ the number of NNs of atom i and $Y_{lm}(\theta, \phi)$ are the usual spherical harmonics. Depending on the choice of l, these parameters are sensitive to different crystal structures. Through the spherical harmonics, they depend on the angles between the vectors to the neighboring particles only, and therefore these parameters do not depend on the choice of reference frame. Different approaches based on these local bond order parameters were developed to analyze the structure of the crystalline nuclei during nucleation and crystallization in metastable liquids [85]. Especially $l = 4$ and $l = 6$ are of widespread use as they are a good choice to distinguish between cubic and hexagonal structures [87]. In particular, there is general consensus on the use of $l = 6$ to characterize phase transformations in liquids of spherical particles interacting via spherically symmetric, central-force potentials.

Focusing on $l = 6$, one then defines the local bond-orientational order parameter for atom i as:

$$f_6(i) = \frac{1}{z(i)} \sum_j \Theta[S_6(i, j) - S_6^0],$$ (1.65)

where S_6^0 is a threshold equal to 0.7, as discussed in [85], while Θ is the Heaviside function. One then finally averages $f_6(i)$ over all atoms in the system to obtain F_6, which can be written as:

$$F_6 = \frac{1}{N} \sum_{i=1}^{N} \frac{1}{z(i)} \sum_{j=1}^{z(i)} S_6(i, j).$$ (1.66)

F_6 measures the degree of sixfold correlation among bond orientations, or in simple words, how many bonds are aligned along the same directions. Hence, F_6 has its largest value and is equal to 1 for fcc crystal lattices where all bonds are aligned along the same (few and selected) crystallographic orientations of the Bravais lattice. Introducing disorder, either in the form of defects or in the form of amorphous structure such as in glass, causes F_6 to take lower values.

While the fcc perfect crystal lattice limit with $F_6 = 1$ is rather intuitive, it is not so obvious what should the opposite limit of a fully amorphous system be, nor whether systems with $F_6 \approx 0$ can be realized in practice. Take, for example, a random network of harmonic springs all of the same length approximately, as shown in Fig. 1.13.

For such random network, F_6 was found to be non-zero and about 0.3 [88]. The fact that F_6 is non-zero in this random network is due to the fact that bonds (springs) have all the same length, which imposes somewhat of a constraint on the network structure that is reflected in a certain, statistical degree of preferential bond

Fig. 1.13 2D slice of a 3D random network of harmonic springs all of the same length. The requirement that springs are all of the same length introduces constraints on the bond orientations such that F_6 acquires a finite value

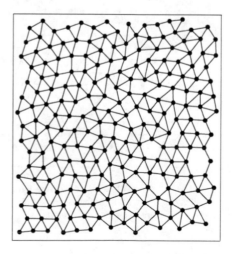

orientations. This is what actually happens, however, in real network glasses, where there is a typical length scale, i.e., that of covalent bonds, which provides such a constraint. This is also the case in random packings, where the excluded volume of the spheres imposes additional constraints on the bond orientations. Lower values of F_6 could, in theory, be obtained in networks where the bond length is randomly distributed in a uniform way, with bonds of any length being all equally probable.

1.5.2   Inversion Symmetry

In a sense, the bond-orientational order parameter informs, essentially, about the "spread" in bond orientations in a lattice or in an amorphous system. The bond-orientational order parameter introduced in the previous section may not always be useful as a quantifier of disorder. This will become clear with the following example.

Take a defective fcc crystal, where defects could be represented either by randomly missing bonds, i.e., the case of a randomly bond-depleted lattice, or by the standard case of a lattice with vacancies. These two situations are shown in Fig. 1.14.

In both cases, the allowed bond orientations are still the same as in the case of a perfect fcc lattice with no defects. Since F_6 is just a measure of the "spread" of the bond orientations in a given system, its values will still be close to 1 even though the systems of Fig. 1.14 have a considerable degree of structural disorder, as shown in [88].

What is evident from Fig. 1.14 is that the defects (either randomly removed bonds or randomly removed lattice points) break the inversion symmetry around lattice

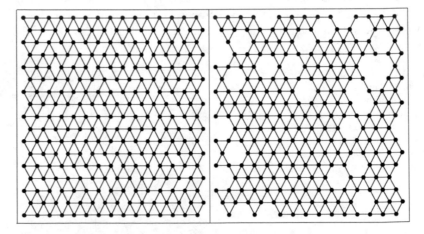

Fig. 1.14 2D slices of a 3D fcc lattice with randomly removed bonds (left panel) and with vacancies (right panel). In both cases, $F_6 \approx 1$, as computed numerically in Ref. [88], even though there is significant disorder as reflected, e.g., in the broken inversion symmetry around the lattice nodes

points. This, as we shall see also in Chap. 2, has profound consequences on the mechanical properties of the lattice as well as on its vibrational spectrum [88].

In these examples, therefore, the bond-orientational order parameter is not a good choice of a metric to quantify disorder, and one has to resort to an altogether different metric. A good choice, in this case, is provided by order parameters sensitive to the centrosymmetry or inversion symmetry of the lattice.

The simplest such metric was proposed by Plimpton and coworkers [89] and is the so-called CS or centrosymmetry parameter:

$$CS = \sum_{i=1}^{z/2} |\mathbf{r}_i + \mathbf{r}_{i+z/2}|^2 \tag{1.67}$$

where $\mathbf{r}_i$ and $\mathbf{r}_{i+z/2}$ are the vectors of bonds representing the $z/2$ pairs of diametrically opposed nearest neighbors of the atom at the center of the frame. By vectorially adding each pair of vectors together, the sum of the squares of the $z/2$ resulting vectors, one gets a scalar quantity, which is defined as the centrosymmetry parameter. Actually, the magnitude of the centrosymmetry parameter provides a metric of the departure from centrosymmetry in the immediate vicinity of any atom. The higher the centrosymmetry parameter, the stronger the structural disorder and the more non-centrosymmetric the local atomic environment.

While this metric was originally devised to quantify the breaking of inversion symmetry due to defects (in particular, dislocations) in crystalline lattices, it is also applicable to disordered solids such as glasses. An example is shown in Fig. 1.15, where this metric has been computed across the glass transition of a binary metallic glass alloy, $Cu_{50}Zr_{50}$.

A different metric to quantify the degree of local inversion symmetry can be constructed in terms of the local balance of forces on a given atom/particle in the system, which are transmitted by its nearest neighbors. Clearly, if the atom is a perfect center of inversion symmetry, forces transmitted to the atom by its nearest neighbors in response to an external strain, e.g., shear, cancel out perfectly to zero. If inversion symmetry is broken, clearly this cancellation does not occur, and a residual force will be acting on the atom. The greater the deviation from centrosymmetry, the greater this force will be.

As will be derived with full details in Chap. 2, this residual force per unit strain is given, for systems with central-force interactions, by:

$$\Xi_{xy,i} = -\sum_{j} \left(\kappa_{ij} r_{ij} - t_{ij} \right) n_{ij}^x n_{ij}^y \mathbf{n}_{ij}. \tag{1.68}$$

where κ_{ij} is the spring constant of the nearest neighbor interaction ij, t_{ij} the bond tension (force) which may be present away from the harmonic minimum, r_{ij} is the radial distance between neighbors i and j, and $\mathbf{n}_{ij}$ the unit vector along the orientation from atom i to j.

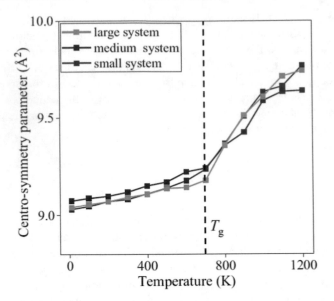

Fig. 1.15 The centrosymmetry order parameter CS defined as in Eq. (1.67), computed for a vitrifying $Cu_{50}Zr_{50}$ metal alloy across the glass transition T_g. The kink in CS and its subsequent lower values, and slope, in the solid amorphous state signal the higher statistical degree of inversion symmetry in the amorphous solid state, since CS is actually a measure of deviation from centrosymmetry. Molecular dynamics simulations data are shown for three different sample sizes. From Ref. [90], with permission from American Physical Society

A good starting point is the absolute value of the sum of all nearest neighbor force vectors (squared) $|\Xi|^2$, which is identically zero for perfect centrosymmetric crystal lattices and has its largest values for lattices where the local inversion symmetry is completely absent. To measure the degree of symmetry breaking independent of the direction of deformation, one can additionally sum over all possible Cartesian coordinate pairs $\alpha\beta$ (and not just $\alpha = x$ and $\beta = y$ as in Eq. (1.68)), $|\Xi|^2 \equiv \sum_{\alpha,\beta\in\{x,y,z\}} |\Xi_{\alpha\beta}|^2$. The order parameter for local inversion-symmetry is thus defined as [88]:

$$F_{IS} = 1 - \frac{\sum_{\alpha,\beta\in\{x,y,z\}} |\Xi_{\alpha\beta}|^2}{\sum_{\alpha,\beta\in\{x,y,z\}} |\Xi_{\alpha\beta}|^2_{ISB}}, \tag{1.69}$$

where $|\Xi_{\alpha\beta}|^2_{ISB}$ indicates the limit in which inversion symmetry is completely broken and there cannot be any correlations among bond orientations whatsoever. For the latter case, one finds $|\Xi_{\alpha\beta}|^2_{ISB} = \kappa^2 R_0^2 \sum_{ij} \left(n_{ij}^\alpha n_{ij}^\beta\right)^2$. Assuming that each lattice site has the same coordination number z, which is true at least on average, one can simplify the denominator to $\sum_{\alpha,\beta\in\{x,y,z\}} |\Xi_{\alpha\beta}|^2_{ISB} = \kappa^2 R_0^2 N z$. Hence,

$F_{IS} = 1$ for any *perfect* centrosymmetric lattice, while $F_{IS} = 0$ for the limiting configuration at which the local breaking of inversion-symmetry is maximum.

In [88] it has been shown quantitatively that F_{IS} is able to capture the difference between the situation in Fig. 1.13, and that of Fig. 1.14, whereas F_6 is completely insensitive to this difference. Upon varying the coordination number z of the lattice for, e.g., the random network and the bond-depleted fcc and computing the F_6 and the F_{IS}, one finds the trends shown in Fig. 1.16.

Remarkably, if z is the same, the random network and the bond-depleted fcc lattice turn out to have exactly the same shear modulus and the same vibrational spectrum [88], in spite of having a quite different structure. This is, precisely, because it is the inversion symmetry (as we shall see in much more detail in Chap. 2), which controls the shear modulus of the material and hence also the vibration spectrum. This fact is captured by the F_{IS} parameter being the same for the two systems, whereas the F_6 parameter yields very different values and is unable to correlate with the macroscopic properties.

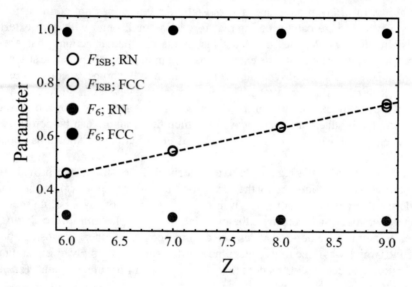

Fig. 1.16 Comparison between the bond-orientational order parameter F_6 and the inversion symmetry order parameter F_{IS} upon varying the coordination number z for the random network of Fig. 1.13 and the bond-depleted fcc lattice of Fig. 1.14 (left panel). The F_6 order parameter yields very different values for the two systems, in spite of the fact that their properties are very different, especially in terms of elasticity (shear modulus) and vibrational spectrum [88]. Courtesy of Dr. Rico Milkus

1.6   Constraint Counting and Isostaticity

Bonding and structure are intimately related in condensed matter, and this is true
also for disordered solids. The most basic link, perhaps, between bonding and
structure is provided by constraint counting, which is an extended version of
isostaticity concepts. In statics and structural mechanics, a structure is statically
determinate when its mechanical constraints are just enough to ensure stability.
In other words, the number of mechanical constraints (i.e., force and moment
equilibrium conditions) just matches the number of degrees of freedom. This is the
definition of *isostatic* structure or isostatic condition. If the number of mechanical
constrains were lower than the number of degrees of freedom, the structure would
be statically under-determinate, and mechanically, it would be floppy or unstable
(hypostatic). If, instead, the number of mechanical constrains were larger than the
number of degrees of freedom, the structure would be statically over-determinate,
and mechanically, it would be fully rigid or hyperstatic.

The isostatic point is therefore equivalent to the onset of rigidity, or the verge
of rigidity, for a given structure. The isostatic structure does not admit states of
self-stress, i.e., there cannot be internal forces in equilibrium with zero external
loads. In other words, the set of displacements or internal actions that are in
mechanical equilibrium with the external load is unique. For a hyperstatic structure,
instead, there exists a nontrivial (non-zero) solution to the homogeneous system of
mechanical equilibrium equations. This indicates that states of self-stress may exist,
where stress may persist even in the absence of an external load. For a hypostatic
structure, finally, there are many ways by which the structure can be deformed at
zero energy cost. These zero-energy modes of deformation are known as "floppy
modes."

In a disordered solid, e.g., a glass, the mechanical constraints are provided by
the bonds between atoms and by the corresponding spring-like Newton's equations,
while the degrees of freedom are given by the whole set of $3N$ possible motions
of the N atoms, which constitute the material. For a simple central-force lattice of
spherical atoms, each bond provides a mechanical constraint, then the isostatic point
is found (which was done for the first time by Maxwell [37]), simply by equating the
total number of degrees of freedom ($3N$) to the total number of constraints ($zN/2$):

$$3N = \frac{zN}{2} \tag{1.70}$$

which gives the isostatic condition in terms of critical coordination number for the
onset (or vanishing) of rigidity as:

$$z_c = 6 = 2d \tag{1.71}$$

and therefore, $z_c = 4$ in $d = 2$ etc., since in generic d-dimensions the total number
of degrees of freedom is dN. This value remarkably coincides with the coordination
number at the random close packing of spheres, which is not entirely unexpected

since hard spheres interact via central forces only. The above condition also implies that simple cubic lattices, with $z = 6$, are not fully rigid. The only element which crystallizes in simple cubic structure at ambient conditions is polonium and is indeed to be found in extremely mechanically unstable simple-cubic crystals.

This simple counting no longer suffices to describe covalent bonds due to the presence of the angular interaction terms given, e.g., by Eq. (1.2). In the lattice, deformations are possible in which the bond angles and bond lengths are unchanged. The number of constraints is now augmented by the additional constraints on the dihedral angles θ_0, cfr. Fig. 1.2 and Eq. (1.2). By simple counting, there are thus additional $2z - 3$ constraints associated with each z-coordinated atom. The isostatic point is then found to be, in this case:

$$3N = \frac{zN}{2} + 2Nz - 3N, \tag{1.72}$$

from which the critical coordination at the isostatic point follows as:

$$z_c = \frac{12}{5} = 2.4. \tag{1.73}$$

This is significantly lower than the $z_c = 6$ found for central-force networks, which is intuitively clear as bond-bending terms such as Eq. (1.2) in the potential energy of the system contribute to greatly enhancing the rigidity of the material. This simple constraint-counting argument for covalent networks explains a surprisingly good deal of observations for certain covalent glasses [91]. In particular, amorphous semiconductors, given the generally lower valency compared to, e.g., oxide glasses, were found to vitrify into random covalent networks with z marginally larger than $z_c = 2.4$ [92].

Similar constraint-counting concepts can be formulated for non-spherical particles, e.g., ellipsoids. For ellipsoids, the counting of degrees of freedom clearly differs from that of spheres, due to the additional rotational degrees of freedom that each ellipsoid possesses. For spheres, the $3N$ Newton's equations for translational motions in the three Cartesian directions suffice to completely describe the degrees of freedom of the system and thus require an equal number of constraints in terms of bonding spring-like equations. For ellipsoids, on top of that, we have additional $3N$ torque balance equations that need to be satisfied. For frictionless ellipsoids, this gives:

$$6N = \frac{zN}{2}, \quad z_c = 12 \tag{1.74}$$

which is indeed much larger than the isostatic point of frictionless spheres.

Finally, while the above arguments hold for both atoms and granular or colloidal particles, the situation in the latter cases becomes very different due to friction between the grains. While frictionless spheres and ellipsoids are an idealization (though very useful to obtain important mathematical results), in reality, friction is unavoidable for particles of size larger than ~ 10 nanometers. Friction between solid surfaces is unavoidable due to microscopic asperities due to the surfaces being non-smooth at the atomic scale [1]. Theories like Johnson-Kendall-Roberts (JKR) or Derjaguin-Muller-Toporov (DMT) provide a route to estimate the adhesion surface resulting from frictional contacts [1]. In turn, this adhesive surface provides an additional bonding to the particles with bending rigidity, which can be measured experimentally, e.g., in colloidal particles [93].

Importantly, however, the friction between two grains in contact can lie anywhere in a range between zero and a limiting value; this is because the Coulomb's law for solid frictional force is actually an inequality: $f_t \leq \mu f_n$, where f_t is the tangential force between two particles in contact, and f_n is the normal force, while μ is the friction coefficient. In the limit of strong friction μ, or mathematically infinite friction $\mu \to \infty$, the sliding of two particles past each other is severely hindered or, basically, impossible to occur. This practically means that there are extra mechanical constraints to be taken into account. These extra constraints are three (in $d = 3$) torque balance equations for the bond between two spheres, which can be visualized as a rod in d-dimensional space. Hence, there are $zN/2$ bonds/contacts, as before, but for each of them, there are extra d dynamical torque equations provided by the tangential friction between the two particles; hence, the total number of constraints is $dzN/2$. The total number of degrees of freedom is also different, since we need to take into account the extra rotational degrees of freedom for each "rod" representing a contact/bond. A rod in d-dimensional space has $d(d - 1)$ rotational degrees of freedom; hence, since a bond/contact is shared between two particles, we have a total of $dN + d(d - 1)N/2$ degrees of freedom. Equating constraints to degrees of freedom, we now obtain:

$$dN + \frac{d(d - 1)N}{2} = \frac{dNz}{2} \tag{1.75}$$

from which we get $z_c = d + 1$, in the limit of infinite friction. This is also known as the "loose random packing" (LRP) limit, as it provides the lowest density at which spheres can pack randomly. In general, the contact number z of a frictional random packing will fall within the two limits of LRP and RCP, and one has:

$$d + 1 \leq z \leq 2d \tag{1.76}$$

Establishing a more precise dependence of z upon friction μ is a topic of ongoing active research [94, 95]. Following an analogous reasoning, for random packings of frictional ellipsoids, one also finds $z_c \geq d + 1$, and typically therefore $d + 1 \leq z \leq d(d + 1)$, which means that in 3D packings of ellipsoids can have z falling in a broad range from $z = 4$ to $z = 12$, which provides a very large parameter space of packing conditions [96].

1.7 Anharmonic Potential of Mean Force in Glasses

In a disordered system, the potential of mean force is defined as:

$$w(r) = -k_B T \ln g(r) \tag{1.77}$$

and, physically, represents the reversible work to change the radial distance r between a particle 1 and a particle 2.

This can be demonstrated as follows:

$$
\begin{aligned}
-\left\langle \frac{d}{d\mathbf{r}_1} U(\mathbf{r}^N) \right\rangle_{\mathbf{r}_1,\mathbf{r}_2} &= \frac{-\int d\mathbf{r}_3 \ldots d\mathbf{r}_N \frac{dU}{d\mathbf{r}_1} \exp(-\beta U)}{\int d\mathbf{r}_3 \ldots d\mathbf{r}_N \exp(-\beta U)} \\
&= k_B T \frac{\left[\frac{d}{d\mathbf{r}_1} \int d\mathbf{r}_3 \ldots d\mathbf{r}_N \exp(-\beta U)\right]}{\int d\mathbf{r}_3 \ldots d\mathbf{r}_N \exp(-\beta U)} \\
&= k_B T \frac{d}{d\mathbf{r}_1} \ln \int d\mathbf{r}_3 \ldots d\mathbf{r}_N \exp(-\beta U) \\
&= k_B T \frac{d}{d\mathbf{r}_1} \ln \left[N(N-1) \frac{\int d\mathbf{r}_3 \ldots d\mathbf{r}_N \exp(-\beta U)}{\int d\mathbf{r}^N \exp(-\beta U)} \right] \\
&= k_B T \frac{d}{d\mathbf{r}_1} \ln g(|\mathbf{r}_1 - \mathbf{r}_2|) = k_B T \frac{d}{d\mathbf{r}_1} \ln g(r)
\end{aligned}
\tag{1.78}
$$

where in the last line we used one of the definitions of the pair correlation function or rdf $g(r)$ [97]. Furthermore, we used the short-hand notation $d\mathbf{r}^N = d\mathbf{r}_1 d\mathbf{r}_2 \ldots d\mathbf{r}_N$.

The averaging over the ensemble of all other particles but the two tagged ones is done by holding the positions of the two tagged particles, $\mathbf{r}_1$, $\mathbf{r}_2$, fixed. The factor $N(N-1)$ arises because there are N possible ways of picking the first particle, and, hence, $N-1$ ways of picking the second particle.

Equation (1.78) thus shows that the gradient of $k_B T \frac{d}{d\mathbf{r}_1} \ln g(r)$ gives the interaction force $-\langle \frac{d}{d\mathbf{r}_1} U(\mathbf{r}^N) \rangle_{\mathbf{r}_1,\mathbf{r}_2}$ between particles 1 and 2 in the averaged field of all the other particles. The corresponding potential (of which the force is equal to minus the gradient), $w(r)$, contains much richer information that the bare two-body pair potential. This is because the net interaction force considered here takes into account all the interactions that the two tagged particles entertain with all the other $N-2$ particles in the system, including all many-body effects, anharmonicities, etc. This proves very useful to extract effective interaction potentials in complex soft matter systems as discussed in depth in [98].

Besides providing a direct link between structure (encoded in $g(r)$) and effective interactions between two particles, the potential of mean force $w(r)$ defined in this way is a much more useful form of the pairwise interaction than the "bare" pair

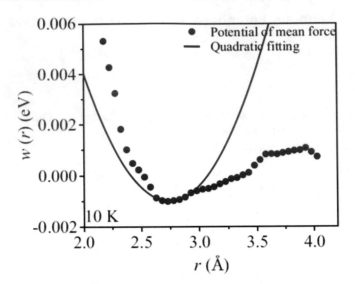

Fig. 1.17 Potential of mean force as a function of interatomic distance for the $Cu_{50}Zr_{50}$ metallic glass. Scattered points in red refer to the potential of mean force computed based on MD snapshots by means of Eq. (1.77). The blue line is the best quadratic fitting of points around the valley. Reproduced from Ref. [17] with permission from the Americal Physical Society

potential. For example, in a given system, the pair potential can be fairly harmonic-like, so one would expect a not too large value of the Grüneisen parameter. This is however in contradiction with simulation results for, e.g., metallic glasses [17] and amorphous silicon [99], which show that low-energy vibrational modes, in particular, have large values of γ_G, sometimes even larger than 4.

This apparent contradiction can be resolved if one considers the potential of mean force instead of the bare pair potential. While the bare pair potential can indeed be very weakly anharmonic or mostly harmonic-like, the potential of mean force may instead be very anharmonic.

This is exemplified in Fig. 1.17, for the case of the $Cu_{50}Zr_{50}$ metallic glass. The potential of mean force, computed from the radial distribution function according to Eq. (1.77), is plotted along with a parabolic fit.

It is clear that the potential of mean force is strongly anharmonic and strongly deviates from the best harmonic (parabolic) fitting. In particular, it is observed that the potential of mean force is much shallower, compared to the harmonic fit, upon moving farther apart from the bonding minimum between two atoms. This reflects the already mentioned existence of "soft channels," in disordered environments with a low local degree of inversion symmetry, along which particles can escape from a bonded position. This is a highly many-body effect that clearly cannot be captured by the bare pair potential. To the left of the minimum, instead, the potential of mean force is somewhat steeper than the parabolic fit, which reflects the strong Pauli exclusion repulsion between internal electron shells of the two atoms.

While it is questionable that the derivation in Eq. (1.78) may still hold in strongly out-of-equilibrium situations due to the use of canonical ensemble averaging, the potential of mean force Eq. (1.77) is still widely used in glasses and other complex nonequilibrium systems to obtain effective many-body interactions or as a way of coarse-graining smaller subsystems for accelerated MD simulations.

References

1. J. Israelachvili, *Intermolecular and Surface Forces* (Academic, Cambridge, 2011)
2. Y. Zeng, P. Gordiichuk, T. Ichihara, G. Zhang, E. Sandoz-Rosado, E.D. Wetzel, J. Tresback, J. Yang, D. Kozawa, Z. Yang, M. Kuehne, M. Quien, Z. Yuan, X. Gong, G. He, D.J. Lundberg, P. Liu, A.T. Liu, J.F. Yang, H.J. Kulik, M.S. Strano, Nature **602**(7895), 91 (2022)
3. M. Finnis, *Interatomic Forces in Condensed Matter* (Oxford University Press, Oxford, 2003)
4. A.P. Sutton, J. Chen, Philos. Mag. Lett. **61**(3), 139 (1990)
5. N. Kovalenko, Y. Krasny, U. Krey, *Physics of Amorphous Metals* (Wiley-VCH, Berlin, 2001)
6. R.D. Shannon, Acta Crystallogr. Sect. A **32**(5), 751 (1976)
7. W. Wang, C. Dong, C. Shek, Mater. Sci. Eng. R. Rep. **44**(2), 45 (2004)
8. V. Gaydaenko, V. Nikulin, Chem. Phys. Lett. **7**(3), 360 (1970)
9. T.E. Faber, *Introduction to the Theory of Liquid Metals* (Cambridge University Press, Cambridge, 1972)
10. J. Hafner, *From Hamiltonians to Phase Diagrams* (Springer, Berlin, 1987)
11. J. Krausser, K.H. Samwer, A. Zaccone, Proc. Natl. Acad. Sci. **112**(45), 13762 (2015)
12. H. Böttger, *Principles of the Theory of Lattice Dynamics* (Physik-Verlag, Weinheim, 1983)
13. D. Khomskii, *Basic Aspects of the Quantum Theory of Solids: Order and Elementary Excitations* (Cambridge University Press, Cambridge, 2010)
14. C. Kittel, *Introduction to Solid State Physics. Eighth Edition* (Wiley, Hoboken, 2005)
15. A.M. Krivtsov, V.A. Kuz'kin, Mech. Solids **46**(3), 387 (2011)
16. D. Cuffari, A. Bongiorno, Phys. Rev. Lett. **124**, 215501 (2020)
17. Z.Y. Yang, Y.J. Wang, A. Zaccone, Phys. Rev. B **105**, 014204 (2022)
18. M. Born, D.J. Hooton, Zeitschrift für Physik **142**(2), 201 (1955)
19. M.L. Klein, G.K. Horton, J. Low Temperat. Phys. **9**(3), 151 (1972)
20. T. Tadano, S. Tsuneyuki, Phys. Rev. Lett. **120**, 105901 (2018)
21. M.T. Dove, *Structure and Dynamics.* Oxford Master Series in Physics (Oxford University Press, Oxford, 2003)
22. J. Als-Nielsen, D. McMorrow, *Elements of Modern X-ray Physics* (Wiley, New York, 2011)
23. J.B. Suck, H. Rudin, H.J. Guntherodt, H. Beck, J. Phys. C Solid State Phys. **14**(17), 2305 (1981)
24. S. Torquato, Phys. Repor. **745**, 1 (2018). Hyperuniform States of Matter
25. P. De Gennes, Physica **25**(7), 825 (1959)
26. J. Hansen, I. McDonald, *Theory of Simple Liquids* (Elsevier Science, Amsterdam, 2006)
27. A.E. Lagogianni, J. Krausser, Z. Evenson, K. Samwer, A. Zaccone, J. Statist. Mech. Theory Exper. **2016**(8), 084001 (2016)
28. W.H. Zachariasen, J. Am. Chem. Soc. **54**(10), 3841 (1932)
29. M.M.J. Treacy, K.B. Borisenko, Science **335**(6071), 950 (2012)
30. W.W. Wood, J.D. Jacobson, J. Chem. Phys. **27**(5), 1207 (1957)
31. B.J. Alder, T.E. Wainwright, J. Chem. Phys. **27**(5), 1208 (1957)
32. T.C. Hales, Ann. Math. **162**, 1065 (2005)
33. T. Hales, J. Harrison, S. McLaughlin, T. Nipkow, S. Obua, R. Zumkeller, Discrete Comput. Geom. **44**(1), 1 (2010)
34. T. Hales, M. Adams, G. Bauer, T.D. Dang, J. Harrison, L.T. Hoang, C. Kaliszyk, V. Magron, S. Mclaughlin, T.T. Nguyen, et al., Forum of Math. Pi **5**, e2 (2017)

35. A. van Blaaderen, P. Wiltzius, Science **270**(5239), 1177 (1995)
36. J.D. Bernal, J. Mason, Nature **188**(4754), 910 (1960)
37. S. Alexander, Phys. Rep. **296**(2), 65 (1998)
38. S. Torquato, F.H. Stillinger, Rev. Mod. Phys. **82**, 2633 (2010)
39. M. Hermes, M. Dijkstra, EPL (Europhys. Lett.) **89**(3), 38005 (2010)
40. R. Blumenfeld, Phys. Rev. Lett. **127**, 118002 (2021)
41. S. Torquato, T.M. Truskett, P.G. Debenedetti, Phys. Rev. Lett. **84**, 2064 (2000)
42. T.M. Truskett, S. Torquato, P.G. Debenedetti, Phys. Rev. E **62**, 993 (2000)
43. R.D. Kamien, A.J. Liu, Phys. Rev. Lett. **99**, 155501 (2007)
44. P. Charbonneau, J. Kurchan, G. Parisi, P. Urbani, F. Zamponi, Annu. Rev. Condens. Matter Phys. **8**(1), 265 (2017)
45. A. Zaccone, Phys. Rev. Lett. **128**, 028002 (2022)
46. A. Zaccone, E. Scossa-Romano, Phys. Rev. B **83**, 184205 (2011)
47. C.S. O'Hern, L.E. Silbert, A.J. Liu, S.R. Nagel, Phys. Rev. E **68**, 011306 (2003)
48. A. Zaccone, E.M. Terentjev, Phys. Rev. Lett. **110**, 178002 (2013)
49. J.J. Shynk, *Probability, Random Variables, and Random Processes: Theory and Signal Processing Applications* (Wiley, New York, 2012)
50. H. Pishro-Nik, *Introduction to Probability, Statistics and Random Processes* (Kappa Research, LCC, Amherst, 2014)
51. S. Torquato, J. Chem. Phys. **149**(2), 020901 (2018)
52. A. Donev, S. Torquato, F.H. Stillinger, Phys. Rev. E **71**, 011105 (2005)
53. P. Dimon, S.K. Sinha, D.A. Weitz, C.R. Safinya, G.S. Smith, W.A. Varady, H.M. Lindsay, Phys. Rev. Lett. **57**, 595 (1986)
54. M. Lattuada, H. Wu, M. Morbidelli, J. Colloid Interf. Sci. **268**(1), 106 (2003)
55. Y. Song, E.A. Mason, R.M. Stratt, J. Phys. Chem. **93**(19), 6916 (1989)
56. C. Anzivino et al., J. Chem. Phys. **158**, 044901 (2023)
57. C. Likos, J. Club Condensed Matt. Phys. (2022). https://doi.org/10.36471/JCCM-March-2022-02
58. S. Meyer, C. Song, Y. Jin, K. Wang, H.A. Makse, Phys. A Statist. Mech. Appl. **389**(22), 5137 (2010)
59. E.S. Barnes, N.J.A. Sloane, Can. J. Math. **35**(1), 117–130 (1983)
60. J.H. Conway, N.J. Sloane, *Sphere Packings, Lattices, and Groups* (Cambridge University Press, Cambridge, 1993)
61. M.P. Allen, D.J. Tildesley, *Computer Simulation of Liquids: Second Edition*, 2nd edn. (Oxford University Press, Oxford, 2017)
62. S. Torquato, *Random Heterogeneous Materials: Microstructure and Macroscopic Properties* (Springer, New York, 2002)
63. E. Thiele, J. Chem. Phys. **39**(2), 474 (1963)
64. M.S. Wertheim, Phys. Rev. Lett. **10**, 321 (1963)
65. N.F. Carnahan, K.E. Starling, J. Chem. Phys. **51**(2), 635 (1969)
66. J.L. Lebowitz, Phys. Rev. **133**, A895 (1964)
67. G.A. Mansoori, N.F. Carnahan, K.E. Starling, T.W. Leland, J. Chem. Phys. **54**(4), 1523 (1971)
68. A. Santos, S.B. Yuste, M.L. de Haro, Molecular Phys. **96**(1), 1 (1999)
69. F. Lado, Phys. Rev. E **54**, 4411 (1996)
70. H. Mizuno, L.E. Silbert, M. Sperl, S. Mossa, J.L. Barrat, Phys. Rev. E **93**, 043314 (2016)
71. T. Boublík, J. Chem. Phys. **53**(1), 471 (1970)
72. H. Cramer, *Mathematical Methods of Statistics* (Princeton University Press, Princeton, 1954)
73. R.S. Farr, Powder Technol. **245**, 28 (2013)
74. S. Suo, C. Zhai, M. Xu, M. Kamlah, Y. Gan, arXiv e-prints arXiv:2205.01934 (2022)
75. J. Rouwhorst, C. Ness, S. Stoyanov, A. Zaccone, P. Schall, Nature Commun. **11**(1), 3558 (2020)
76. J. Rouwhorst, P. Schall, C. Ness, T. Blijdenstein, A. Zaccone, Phys. Rev. E **102**, 022602 (2020)
77. B.I. Shklovskiĭ, A.L. Éfros, Soviet Phys. Uspekhi **18**(11), 845
78. S. Kirkpatrick, Rev. Mod. Phys. **45**, 574 (1973)

79. K. Wegner, P. Piseri, H.V. Tafreshi, P. Milani, J. Phys. D Appl. Phys. **39**(22), R439 (2006)
80. W. Goetze, *Complex Dynamics of Glass-Forming Liquids: A Mode-Coupling Theory* (Oxford University Press, Oxford, 2009)
81. V. Lubchenko, P.G. Wolynes, Phys. Rev. Lett. **87**, 195901 (2001)
82. V. Lubchenko, P.G. Wolynes, Proc. Natl. Acad. Sci. **100**(4), 1515 (2003)
83. S. Franz, G. Parisi, P. Urbani, F. Zamponi, Proc. Natl. Acad. Sci. **112**(47), 14539 (2015)
84. P.J. Steinhardt, D.R. Nelson, M. Ronchetti, Phys. Rev. B **28**, 784 (1983)
85. S. Auer, D. Frenkel, J. Chem. Phys. **120**(6), 3015 (2004)
86. C.P. Goodrich, A.J. Liu, S.R. Nagel, Nat. Phys. **10**(8), 578 (2014)
87. W. Lechner, C. Dellago, J. Chem. Phys. **129**(11), 114707 (2008)
88. R. Milkus, A. Zaccone, Phys. Rev. B **93**, 094204 (2016)
89. C.L. Kelchner, S.J. Plimpton, J.C. Hamilton, Phys. Rev. B **58**, 11085 (1998)
90. D. Han, D. Wei, J. Yang, H.L. Li, M.Q. Jiang, Y.J. Wang, L.H. Dai, A. Zaccone, Phys. Rev. B **101**, 014113 (2020)
91. J. Phillips, M. Thorpe, Solid State Commun. **53**(8), 699 (1985)
92. K. Tanaka, Solid State Commun. **54**(10), 867 (1985)
93. J.P. Pantina, E.M. Furst, Phys. Rev. Lett. **94**, 138301 (2005)
94. K. Shundyak, M. van Hecke, W. van Saarloos, Phys. Rev. E **75**, 010301 (2007)
95. C. Song, P. Wang, H.A. Makse, Nature **453**(7195), 629 (2008)
96. W. Man, A. Donev, F.H. Stillinger, M.T. Sullivan, W.B. Russel, D. Heeger, S. Inati, S. Torquato, P.M. Chaikin, Phys. Rev. Lett. **94**, 198001 (2005)
97. David Chandler, *Introduction to Modern Statistical Mechanics* (Oxford University Press, Oxford, 1986)
98. C.N. Likos, Phys. Rep. **348**(4), 267 (2001)
99. J. Fabian, P.B. Allen, Phys. Rev. Lett. **79**, 1885 (1997)

Elasticity

2

Abstract

In Chap. 1, we have considered how atoms/molecules/grains interact and self-organize into forming solid systems with non-crystalline internal structure. In this chapter, we shall derive a mathematical theory of the microscopic elasticity of generic randomly organized solids, i.e., a theory of the elastic constants expressed in terms of the microscopic (atomic, molecular, particle) organization and interactions. The theory will then be applied to random jammed solids, Lennard-Jones glasses, polymer glasses, colloidal gels, and pre-stressed networks.

2.1 Introduction

Amorphous solids deform under small strains in a microscopically different way compared to crystalline solids. In perfect crystals, the elastic moduli (or elastic constants) at zero temperature can be analytically expressed in terms of lattice sums over suitable combinations of bond unit vectors dictated by the crystallographic structure. Under the harmonic approximation of solid state theory, the prefactor that multiplies the lattice sum is proportional to the lattice spring constant. For simple crystal structures, such as body-centered cubic (bcc) or face-centered cubic (fcc), compact formulae can be easily found [1]. All these approximations are called into question and need to be either relaxed or reformulated, when developing a microscopic theory of elasticity for amorphous solids, which is the main goal of this chapter.

The key approximation under which the atomistic formulae for elastic constants of crystals are derived is that of *affine* deformation of all material points, which in the continuum mechanics language is called *compatibility* of deformation. We shall discuss what this means, what deviations from this approximation imply, and why they

© The Author(s), under exclusive license to Springer Nature Switzerland AG 2023
A. Zaccone, *Theory of Disordered Solids*, Lecture Notes in Physics 1015,
https://doi.org/10.1007/978-3-031-24706-4_2

are so important in amorphous solids. We will start from elucidating this concept in light of empirical observations obtained in the numerical simulation (molecular dynamics) of model amorphous solids, such as Lennard-Jones glasses. We will then present a systematic discussion of the affine deformation method, leading to the derivation of the Born-Huang formulae for the microscopic elastic constants. This will set the stage for the systematic extension of microscopic elasticity theory to account for the role of disorder in amorphous lattices and, in general, deviations from centrosymmetry. We will thus present the general mathematical framework of nonaffine deformation theory, leading to analytical expressions for the elastic moduli of disordered systems, in particular the shear modulus of amorphous solids, with explicit derivations for jammed random packings and random spring networks, polymer glasses, Lennard-Jones glasses, and colloidal gels. The compression (bulk) modulus will also be considered, as well as different formulations of the nonaffine theory based on statistical mechanics, in particular the so-called stress fluctuation formalism.

The first sections of the chapter introduce the theoretical formalism for the elastic response of generic amorphous solids, with all the mathematical steps. The mathematical formalism is subsequently applied to produce analytical calculations on a few examples of disordered solids, as mentioned above. The most important application is to glasses. As discussed in Chap. 1, the glass transition is a phenomenon which remains poorly understood in spite of a tremendous number of theories and models that have been proposed over several decades [2]. Here, for the mathematical description of the elastic properties of glasses, we shall take an agnostic approach and model the glass as an amorphous lattice made of particles (atoms, molecules, colloids) frozen-in in random positions. The glass transition temperature is defined in the following as the temperature at which the shear rigidity vanishes, although of course this is a rather arbitrary definition, and other definitions (based on calorimetry and inflection points of various thermodynamic quantities) are available [2]. The issue of the dependence of the glass transition temperature on the cooling rate from the supercooled liquid is also largely ignored, as it is an open topic of research. Similarly, also for colloidal gels, we only briefly recap the main mechanisms of formation of these disordered solid states, inasmuch this information is needed as input for the building of elastic models. For the detailed description of structural aspects, percolation theories, etc., the interested reader is referred to [2].

2.2   The Concept of Affine and Nonaffine Deformations

Let us consider a shear deformation field, characterized by a shear angle γ, as schematically depicted in Fig. 2.1a. The deformation is represented by a deformation gradient tensor $\mathbf{F}$, see below for its analytical definition. If the solid is a perfect crystal, as schematically depicted in Fig. 2.1b, top panel, the forces acting on the central particle due to its nearest-neighbors cancel by inversion symmetry both in the rest configuration and, importantly, in the sheared configuration. Even if there are no forces in the rest configuration (which is the common case or

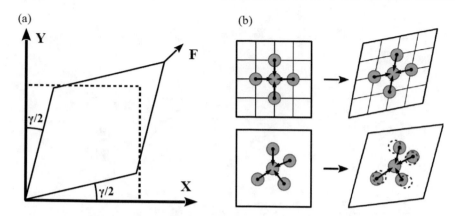

Fig. 2.1 Shear deformation field resolved in the 2D $x - y$ shearing plane (**a**). Panel (**b**) schematically shows how particles are displaced in a centrosymmetric crystal (top) where inversion symmetry guarantees mechanical equilibrium in the affine positions and in amorphous lattices (bottom) where extra displacements are needed to relax forces in the affine positions. Courtesy of Dr. Rico Milkus

assumption of interparticle bonds being in the energy minimum, thus resulting in zero bond tensions), in the sheared configuration there can be non-zero forces from the neighbor particles since these are also moving toward the positions (affine positions) prescribed by the external strain. Thanks to the inversion symmetry of the ordered crystalline lattice, however, the sum of all these forces is still zero also in the sheared configuration. In an amorphous lattice, instead, such as the one depicted in the lower panel of Fig. 2.1b, the particles in the shear configuration do not occupy the positions prescribed by the deformation tensor **F**, which are depicted as dashed circles, and are instead in some other positions. The motions from the dashed circles to the final particle positions are called nonaffine motions and arise from the total force that each particle receives from its nearest-neighbors and which remains unbalanced in the absence of inversion symmetry.

These additional nonaffine motions, as we shall see more in mathematical detail in the next subsections, are internal (atomic, molecular – in general, microscopic) rearrangements that are needed to maintain mechanical equilibrium. Hence, also from a thermodynamic point of view, they contribute a negative (internal) contribution to the work done by the solid in resisting the applied strain.

Upon summing up all these negative contributions, the free energy of deformation ends up being significantly reduced, which leads to lower values of the elastic constants. This is indeed what one observes in molecular dynamic (MD) simulations of the deformation of glasses. In Fig. 2.2 simulation data of deformation of athermal Lennard-Jones glasses are shown, from Ref. [3]. All the particles are first deformed in an affine way and subsequently allowed to relax into the nearest energy minimum. This procedure allows for the determination of the nonaffine displacements of all particles and for quantifying the elastic constants for affine and nonaffine deformations. In terms of the Lame' parameters, λ and μ, where $\mu \equiv G$

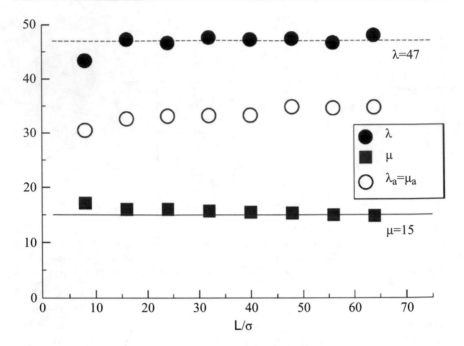

Fig. 2.2 Lame' coefficients λ and μ, measured in MD simulations of athermal Lennard-Jones glass, plotted as a function of the linear system size L divided by the particle diameter σ. The affine estimates λ_a and μ_a are computed by deforming the system based on the deformation tensor **F** and without letting the particles relax once they reach the affine positions. The actual (nonaffine) values are instead computed by allowing for additional relaxation of the system into the nearest energy minimum by means of conjugate gradient minimization. Reproduced from Ref. [3] with permission of the American Physical Society

coincides with the shear modulus, while the first Lame' constant can be related to the bulk modulus K (i.e., $K = \lambda + \frac{2}{3}\mu$), it is seen that the difference between the affine values and the actual values is significant. The difference is larger for the shear modulus μ than for the first Lame' parameter λ, for reasons that we shall discuss later in Sect. 2.7 devoted to the compression modulus. For the shear modulus, in particular, the difference can be larger than a factor of two and can be entirely ascribed to the nonaffine mechanism discussed above. In the next subsection we will develop a mathematical theory of the affine elastic constants, which will then be extended to include the nonaffine relaxation mechanism.

2.3   Born-Huang Formulae for the Affine Elastic Moduli

The central quantity in the harmonic theory of solids is the so-called Hessian matrix (the dynamical matrix familiar from solid state physics is just the Fourier transform of the Hessian) defined as[1]

$$H_{ij}^{\alpha\beta} = \frac{\partial U(r_{ij})}{\partial r_i^\alpha \partial r_j^\beta}, \tag{2.1}$$

where U is the internal energy of the solid expressed as a function of the set of pairwise interparticle distances $\{r_{ij}\}$ between any two bonded particles i and j. Given a pairwise interparticle potential $V(r_{ij})$, the internal energy U is just the sum over all pair contributions of the interaction energy $V(r_{ij})$. In the above equation, Latin indices are used to label particles, whereas Greek indices are used to denote the Cartesian components ($\alpha = x, y, z$).

We define the bond stiffness or spring constant

$$\kappa_{ij} \equiv \frac{d^2 U}{dr_{ij}^2}. \tag{2.2}$$

By means of this definition, the second derivative of the total energy U with respect to an external strain parameter can be evaluated, as we shall now demonstrate. For the case of a shear strain, the strain parameter is the angle γ as schematically shown in Fig. 2.1a. The corresponding elastic modulus is the shear modulus G, which, based on linear elasticity theory [4], is defined in terms of the second derivative of the free energy with respect to the applied strain:

$$G = \frac{1}{V}\frac{\partial^2 U}{\partial\gamma^2}, \tag{2.3}$$

where in our case of an athermal solid the free energy coincides with the total internal energy and the limit $\gamma \to 0$ of linear elasticity is implied. Upon further defining also the bond tension $t_{ij} \equiv \frac{dU}{dr_{ij}}$ and upon applying the chain rule of partial derivatives, we obtain the following expression for G:

$$G = \frac{1}{V}\frac{\partial^2 U}{\partial\gamma^2} = \frac{1}{V}\sum_{\langle i,j \rangle}\left(\kappa_{ij}\frac{\partial r_{ij}}{\partial\gamma}\frac{\partial r_{ij}}{\partial\gamma} + t_{ij}\frac{\partial^2 r_{ij}}{\partial\gamma^2}\right). \tag{2.4}$$

The sum in the above formula runs over all bonded pairs i, j. To complete our program, we now need to evaluate the derivatives that appear in Eq. (2.4).

---

[1] Note that each [ij] element of $H_{ij}^{\alpha\beta} \equiv \mathbf{H}_{ij}$ is a 3×3 tensor and, in general, a $d \times d$ tensor in a d-dimensional space.

We first note that the interparticle distance is a scalar quantity defined as $r_{ij} \equiv |\mathbf{r}_j - \mathbf{r}_i|$. Next, the following identities are valid:

$$\frac{\partial r_{ij}}{\partial r_n^\alpha} = \delta_{ji}^n n_{ij}^\alpha, \tag{2.5}$$

where $\delta_{ji}^n = \delta_{nj} - \delta_{ni}$, and

$$\frac{\partial r_{ij}}{\partial \gamma} = r_{ij} n_{ij}^x n_{ij}^y. \tag{2.6}$$

Here, $\mathbf{n}_{ij}$ indicates the unit vector along the straight line joining particle i to particle j. In the last identity, we have used, again, the usual linear elasticity assumption of small or vanishing strain γ and the fact that the interparticle distance transforms, under a deformation gradient tensor $\mathbf{F}$ according to

$$\mathbf{r}'_{ij} = \mathbf{F} \cdot \mathbf{r}_{ij} = \begin{pmatrix} r_{ij}^x \cos(\frac{\gamma}{2}) + r_{ij}^y \sin(\frac{\gamma}{2}) \\ r_{ij}^y \cos(\frac{\gamma}{2}) + r_{ij}^x \sin(\frac{\gamma}{2}) \\ r_{ij}^z \end{pmatrix}. \tag{2.7}$$

The above transformation Eq. (2.7) defines an affine deformation since it belongs to the generic class of vector transformations $\mathbf{v}' = \mathbf{F} \cdot \mathbf{v} + \mathbf{a}$, of the homogeneous type with $\mathbf{a} = 0$, where $\mathbf{F}$ is a transformation matrix. The deformation gradient tensor $\mathbf{F}$ is related to the Green-Saint Venant strain tensor $\boldsymbol{\eta}$ (cfr. Appendix A) via the following relation:

$$\boldsymbol{\eta} = \frac{1}{2}(\mathbf{F}^T \mathbf{F} - \mathbf{1}) = \frac{1}{2} \begin{pmatrix} 0 & \sin\gamma & 0 \\ \sin\gamma & 0 & 0 \\ 0 & 0 & 0 \end{pmatrix}, \tag{2.8}$$

where $\mathbf{1}$ denotes the identity matrix.

With the aid of Eq. (2.5), we can now get the following identity also needed for Eq. (2.4):

$$\frac{\partial^2 r_{ij}}{\partial \gamma^2} = -r_{ij} (n_{ij}^x n_{ij}^y)^2. \tag{2.9}$$

Using Eqs. (2.6) and (2.9) in Eq. (2.4), we finally obtain

$$G = \frac{1}{V} \sum_{\langle i,j \rangle} \left(r_{ij} \kappa_{ij} - t_{ij} \right) r_{ij} n_{ij}^x n_{ij}^y n_{ij}^x n_{ij}^y. \tag{2.10}$$

If the particles are in the minimum of the pair interaction potential, then $t_{ij} = 0$ identically, and one is left with the Born-Huang formula for the affine shear modulus:

$$G = \frac{1}{V} \sum_{\langle i,j \rangle} r_{ij}^2 \kappa_{ij} n_{ij}^x n_{ij}^y n_{ij}^x n_{ij}^y. \tag{2.11}$$

As discussed by Born and Huang in their original monograph [1], the above formula can be generalized to an arbitrary component of the elastic tensor:

$$C_{\iota\xi\kappa\chi} = \frac{1}{V} \sum_{\langle i,j \rangle} r_{ij}^2 \kappa_{ij} n_{ij}^\iota n_{ij}^\xi n_{ij}^\kappa n_{ij}^\chi. \tag{2.12}$$

The above Born-Huang formula can be much simplified for simple crystals, upon evaluating the lattice sum for a given crystallographic structure, as discussed in [1] (for the evaluation of the lattice sums, the Appendix of [5] is useful). Furthermore, it can also be extended to include next to nearest-neighbors and even a third shell of nearest-neighbors.

In general, the two main assumptions that have been used to derive the Born-Huang formulae for the elastic constants are:

(i) The assumption of *affine* deformation encoded in Eq. (2.7)
(ii) The assumption of pairwise central-force bonding forces.

While the latter assumption of pairwise central forces remains tenable to a good approximation also for certain amorphous solids, we shall see in the next sections that the assumption (i) of affine deformation is not satisfied in the presence of either structural disorder, as in glasses and in fact not even for non-centrosymmetric crystals such as quartz. In the continuum mechanics literature, the affine assumption is often referred to as "compatibility condition," "compatible deformation," or Cauchy-Born assumption [6, 7]. The concept of compatibility of deformation is linked to the single-valuedness of the elastic displacement field. Mathematically, the compatibility condition is expressed by the vanishing of the curl of the deformation gradient tensor **F** as discussed in [7].

If bond-bending interparticle interaction forces are present (as for, e.g., covalent bonds), similar derivations apply, leading to an additional contribution. The manipulations become however much lengthier and can be found in Ref. [8].

2.4 Elastic Moduli of Amorphous Solids

We have seen in the previous section that if a perfectly (centrosymmetric) crystalline system at zero temperature is subjected to a small external strain, such as shear or compression, the particles in the solid will undergo a homogeneous displacement following the external macroscopic strain. These displacements are called affine displacements and the final position of the particle in the deformed frame is called the affine position.

Things are very different in structurally disordered (amorphous) solids. The lack of inversion symmetry around a given particle (atom, molecule, colloidal or granular particle, or monomers in polymer chains) is responsible for a net force that would act on the particle in the affine position. This force has to be relaxed at every step during the deformation process in order to maintain the mechanical equilibrium. We shall see how these physical considerations can be written in mathematical form so as to produce lattice dynamical atomistic expressions for the elastic moduli of amorphous solids. We shall not dwell on detailed microscopic models of particles and their interactions and also not on detailed polymer models.

2.4.1 General Theory

Starting from some reference configuration, in which the coordinate of a material point is denoted as $\mathbf{r}_{i,0}$, and applying a pure shear of angle γ, the coordinate transformation between the initial and final configuration can be written as $\mathbf{r}_i = \mathbf{F}(\gamma) \cdot \mathbf{r}_{i,0}$, where $\mathbf{F}$ is the deformation gradient tensor introduced in the previous section.

If the same transformation $\mathbf{F}$ is applied to a disordered solid, where the particles do not align on an inversion-symmetric lattice, we encounter a different situation. As depicted in the bottom panel of Fig. 2.1b, initially the particles occupy random position and are in mechanical equilibrium. This means that the forces acting on the central particle sum up to zero in the disordered equilibrium configuration. Applying the pure shear transformation of the particle system displaces each particle to its affine position. However, the system is not in mechanical equilibrium anymore. In this new configuration, the forces on the central particle, for instance, do not cancel due to the lack of local particle symmetry. In order to reach the mechanical equilibrium of the deformed configuration, these residual forces have to be relaxed. Consequently, the particles move away from the affine position and the system relaxes to its final equilibrium configuration. These final equilibrium positions are the nonaffine position. In Fig. 2.3, the affine positions in the deformed configuration would still lie on the dashed lines, whereas the particles in the nonaffine positions appear as displaced from the dashed lines.

Clearly, due to the randomness of the disordered configuration, the relaxation of the particles to the equilibrium positions cannot be described by the action of a linear map like the deformation gradient tensor $\mathbf{F}$, which is why they are called *nonaffine* displacements.

We can set up the nonaffine quasi-static deformation formalism by considering the action of $\mathbf{F}$. Here quasi-static means that the deformation occurs through a series of re-equilibration steps represented by the nonaffine relaxations. In computer simulations of athermal amorphous solids, this protocol can be implemented by applying small strain steps of the simulation box followed by conjugate-gradient minimization of the forces acting on the particles [9].

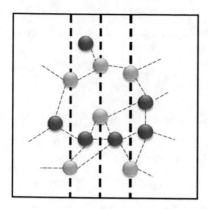 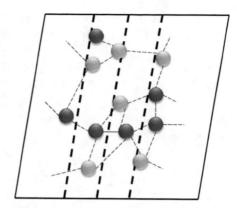

Fig. 2.3 Schematic illustration of nonaffine displacements in amorphous media. The scheme shows the rearrangements or displacements of atoms upon application of an external shear strain. If the deformation were affine, atoms that sit exactly on the dashed lines in the underformed frame (left) would still sit exactly on dashed lines also in the deformed frame (right). However, in a disordered environment, this does not happen: the atoms that were sitting on the dashed lines in the undeformed frame are no longer sitting on the dashed lines in the deformed frame but are displaced from them. The distance from the actual/final positions of the atoms to the dashed line corresponds to the nonaffine displacements

After the shear deformation of the reference configuration $\{\mathbf{r}_{i,0}\}$, the affine configuration is given by $\mathbf{r}_i = \mathbf{F}(\gamma) \cdot \mathbf{r}_{i,0}$. Since we want to incorporate the resulting nonaffine correction to the particle's trajectory, we generalize this relation by making the reference configuration γ-dependent, i.e.

$$\mathbf{r}_i = \mathbf{F}(\gamma) \cdot \mathring{\mathbf{r}}_i(\gamma), \tag{2.13}$$

where the new variable $\mathring{\mathbf{r}}_i$ does the book-keeping of the nonaffine displacements in the undeformed configuration. The ring notation thus indicates that the particle or material point coordinates are measured in the undeformed frame. This means that keeping $\mathring{\mathbf{r}}_i$ fixed with γ and changing $\mathbf{F}(\gamma)$ correspond to the affine part of the deformation, whereas keeping $\mathbf{F}(\gamma)$ constant and letting $\mathring{\mathbf{r}}_i(\gamma)$ vary result in the nonaffine correction. The potential energy of the system is then a function of the shear angle and the nonaffine coordinates $U(\mathbf{r}(\gamma)) = U(\mathring{\mathbf{r}}_i(\gamma))$ and of course is a scalar invariant upon going from one frame to the other.

In order to set up an equation of motion that takes the nonaffine deformation mechanism into account, we can implicitly define the trajectories of the nonaffine displacements by imposing the condition of mechanical equilibrium on the particle trajectories. This is done by requiring that the force $\mathbf{f}_i$ acting on the particle i is zero. In practice, we require for all particles i that

$$\mathbf{f}_i = \left.\frac{\partial U}{\partial \mathring{\mathbf{r}}_i}\right|_\gamma (\mathring{\mathbf{r}}_i, \gamma) = 0 \tag{2.14}$$

be fulfilled along the trajectory. The derivative is taken at fixed value of γ. Upon varying this equation with respect to the shear angle γ, we arrive at the expression:

$$\delta \mathbf{f}_i = \sum_j \frac{\partial U}{\partial \mathring{\mathbf{r}}_i \partial \mathring{\mathbf{r}}_j} \delta \mathring{\mathbf{r}}_j + \frac{\partial U}{\partial \mathring{\mathbf{r}}_i \partial \gamma} \delta \gamma = 0 \qquad (2.15)$$

where the summation convention over Latin (particle) indices is implied. This tantamount to write

$$\frac{\partial U}{\partial \mathring{\mathbf{r}}_i \partial \mathring{\mathbf{r}}_j} \frac{\mathcal{D} \mathring{\mathbf{r}}_j}{\mathcal{D}\gamma} + \frac{\partial U}{\partial \mathring{\mathbf{r}}_i \partial \gamma} = 0. \qquad (2.16)$$

The symbol $\mathcal{D}$ denotes a material derivative which is taken along a pathway of mechanical equilibrium.

This equation is the formal representation of the nonaffine relaxation mechanism explained above.

Since we want to obtain the linear response of the system to an infinitesimal strain $\partial \gamma$, we evaluate the above relation in the limit $\gamma \to 0$. We recognize that, in this limit, the second term in Eq. (2.16) represents the Hessian matrix,

$$\mathbf{H}_{ij} = \frac{\partial U}{\partial \mathring{\mathbf{r}}_i \partial \mathring{\mathbf{r}}_j}\bigg|_{\gamma \to 0} = \frac{\partial U}{\partial \mathbf{r}_i \partial \mathbf{r}_j}\bigg|_{\mathbf{r} \to \mathbf{r}_0} \qquad (2.17)$$

since $\mathring{r}(\gamma)|_{\gamma \to 0} = \mathbf{r}_0$.

The second term in Eq. (2.16) can be identified with the force acting upon the ith particle as the consequence of an infinitesimal affine displacement. We thus introduce the affine force Ξ_i (called "affine" because it is the force that would act on the particle in the affine position and which therefore triggers the ensuing nonaffine displacement) defined as

$$\Xi_i = -\frac{\partial U}{\partial \mathring{\mathbf{r}}_i \partial \gamma}\bigg|_{\gamma \to 0}. \qquad (2.18)$$

With these definitions, we can rewrite Eq. (2.16) more compactly as

$$\mathbf{H}_{ij} \frac{\mathcal{D} \mathring{\mathbf{r}}_j}{\mathcal{D}\gamma} = \Xi_i. \qquad (2.19)$$

Upon putting care in removing the three (Goldstone) zero modes (for rigid-body translation of the whole solid) so that the Hessian is invertible, the above equation becomes

$$\frac{\mathcal{D} \mathring{\mathbf{r}}_j}{\mathcal{D}\gamma} = \mathbf{H}_{ij}^{-1} \Xi_i. \qquad (2.20)$$

The solutions to this equation provide the tangents along which the nonaffine displacements are directed. This represents the linear response of the system to the extra forces which appear by the infinitesimal affine transformation characterized by γ. We can subsequently linearize the relation $\mathbf{r}_i = \mathbf{F}(\gamma) \cdot \mathring{\mathbf{r}}_i$ in the linear response regime, which yields

$$\mathbf{r}_i(\gamma) = \mathring{\mathbf{r}}_i(0) + \frac{\mathcal{D}}{\mathcal{D}\gamma}\left(\mathbf{F}(\gamma)\mathring{\mathbf{r}}_i(\gamma)\right)\Bigg|_{\gamma \to 0} \gamma + O(\gamma^2) \tag{2.21}$$

$$= \mathring{\mathbf{r}}_i(0) + \left(\frac{\mathcal{D}}{\mathcal{D}\gamma}\mathbf{F}(\gamma)\right)\mathring{\mathbf{r}}_i(\gamma)\Bigg|_{\gamma \to 0} \gamma + \mathbf{F}(\gamma)\left(\frac{\mathcal{D}}{\mathcal{D}\gamma}\mathring{\mathbf{r}}_i(\gamma)\right)\Bigg|_{\gamma \to 0} \gamma + O(\gamma^2). \tag{2.22}$$

Using that $\mathbf{F}(0) = \mathbf{1}$ and $\mathring{\mathbf{r}}_i(0) = \mathbf{r}_{i,0}$, we have [10],

$$\mathbf{r}_i(\gamma) = \left(1 + \frac{\mathcal{D}}{\mathcal{D}\gamma}\mathbf{F}(\gamma)\Bigg|_{\gamma \to 0}\gamma\right)\mathbf{r}_{i,0} + \left(\frac{\mathcal{D}}{\mathcal{D}\gamma}\mathring{\mathbf{r}}_i(\gamma)\right)\Bigg|_{\gamma \to 0}\gamma + O(\gamma^2) \tag{2.23}$$

$$= \mathbf{r}_{i,A} + \delta\mathbf{r}_{i,NA} + O(\gamma^2) \tag{2.24}$$

where we identified the affine position after the application of the deformation gradient tensor $\mathbf{r}_{i,A} = \mathbf{F}(\gamma) \cdot \mathbf{r}_{i,0}$ and the nonaffine correction given by $\delta\mathbf{r}_{i,NA}$ in the linear response regime.

The nonaffine displacements can be easily measured in numerical simulations of amorphous solids, an example of which is shown in Fig. 2.4.

Importantly, the extent of nonaffine displacements can also be quantitatively deduced from the shear distortion of the pair correlation function, $g(\mathbf{r})$. This allows for the experimental measurement of nonaffine displacements from the shear-distorted structure factor, as demonstrated in Ref. [12].

Fig. 2.4 Nonaffine displacements resulting from a shear in the x, y, plane, for a random tessellated lattice. Notice the characteristic vortex-like pattern. Figure reproduced with permission of the American Physical Society from [11]

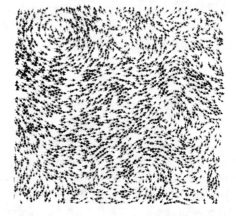

In order to evaluate the linear nonaffine elastic response of the solid, we have to consider the derivative of the total potential energy with respect to the shear angle γ. The material stress σ is defined as the first derivative with respect to the strain as usual in linear elasticity [4]. However, since we are interested in the stress which includes the nonaffine relaxation, the total derivative $\frac{\mathcal{D}}{\mathcal{D}\gamma}$ must apply, although it eventually reduces to the standard derivative:

$$\sigma = \frac{1}{V} \frac{\mathcal{D}U}{\mathcal{D}\gamma}\bigg|_{\gamma \to 0} \tag{2.25}$$

$$= \frac{1}{V} \left(\frac{\partial U}{\partial \gamma} + \frac{\partial U}{\partial \mathring{\mathbf{r}}_i} \frac{\mathcal{D}\mathring{\mathbf{r}}_i}{\mathcal{D}\gamma} \right)\bigg|_{\gamma \to 0} \tag{2.26}$$

$$= \frac{1}{V} \left(\frac{\partial U}{\partial \gamma} \right)\bigg|_{\gamma \to 0}, \tag{2.27}$$

where the summation convention over repeated Latin indices always holds and the last equality holds because we are differentiating under the constraint of mechanical equilibrium $\frac{\partial U}{\partial \mathring{\mathbf{r}}} = 0$.

The elastic moduli are defined as the second (again, total or material) derivative of the energy U with respect to the strain, where in particular the shear modulus G is given by

$$G = \frac{1}{V} \frac{\mathcal{D}^2 U}{\mathcal{D}\gamma^2}\bigg|_{\gamma \to 0}. \tag{2.28}$$

Upon noticing that $\frac{\mathcal{D}^2 U}{\mathcal{D}\gamma^2} = \frac{\mathcal{D}}{\mathcal{D}\gamma} \frac{\partial U}{\partial \gamma}$ again because mechanical equilibrium $\frac{\partial U}{\partial \mathring{\mathbf{r}}} = 0$ is always active, we then obtain

$$\frac{\mathcal{D}^2 U}{\mathcal{D}\gamma^2}\bigg|_{\gamma \to 0} = \frac{\mathcal{D}}{\mathcal{D}\gamma} \frac{\partial U}{\partial \gamma}\bigg|_{\gamma \to 0} = \left(\frac{\partial^2 U}{\partial \gamma^2} + \frac{\partial U}{\partial \mathring{\mathbf{r}}_i \partial \gamma} \frac{\mathcal{D}\mathring{\mathbf{r}}_i}{\mathcal{D}\gamma} \right)\bigg|_{\gamma \to 0}. \tag{2.29}$$

Using the definitions of Eqs. (2.18)–(2.19), we conclude that the shear modulus can be written as [9, 10, 13]

$$G = \frac{1}{V} \left(\frac{\partial^2 U}{\partial \gamma^2}\bigg|_{\gamma \to 0} - \Xi_i \mathbf{H}_{ij}^{-1} \Xi_j \right). \tag{2.30}$$

The first term in the above expression represents the variation of the stress with the strain γ when the particles in the system are constrained not to relax into the nonaffine equilibrium positions (i.e., the nonaffine relaxations are forbidden by construction). This is, therefore, the affine contribution to the shear elasticity, abbreviated as $G_A \equiv \frac{1}{V} \frac{\partial^2 U}{\partial \gamma^2}$, which corresponds to the Born-Huang formulae derived in the previous section. The second contribution on the r.h.s. of Eq. (2.30) gives the nonaffine correction to the shear modulus $G_{NA} \equiv -\frac{1}{V} \left(\Xi_i \mathbf{H}_{ij}^{-1} \Xi_j \right)$, which arises due to the additional relaxation processes arising due to the lack of local particle symmetry in the system. This is a contraction in a dN-dimensional space, where Ξ_i are dN-dimensional vectors and $\mathbf{H}_{ij}$ is a $dN \times dN$-dimensional matrix, and the summation over repeated indices is implied. Note that this nonaffine contribution is always negative and therefore makes the actual values of the elastic constants smaller than the affine Born-Huang estimate.

The reason why the nonaffine correction is always negative (meaning "softening" induced by nonaffine rearrangements) can be appreciated by considering the thermodynamic nature of internal displacements. Since a certain atom is not at mechanical equilibrium in the affine position, a net force acting on it has to be relaxed or released in order to maintain the mechanical equilibrium. Under the action of this force, the atom thus performs the additional *nonaffine* displacement, as explained above. Since a force times a displacement amounts to a work and since this work is done internally by the system to keep the mechanical equilibrium, it contributes negatively to the free energy of deformation. Therefore, it ultimately shows up as a negative correction to the elastic constants.

We now specialize to the case of central force interactions and look for an explicit microscopic form in terms of lattice sums. While the affine modulus has already been given explicitly in terms of lattice sums in the previous section (the Born-Huang formulae), we still need to express the affine force Ξ in terms of lattice sums.

We recall the definition of Eq. (2.18), $\Xi_i = \frac{\partial U}{\partial \mathbf{r}_i \partial \gamma}$, where for convenience we drop the ring notation and the limit of small strain is implicit. Using the chain rule, we rewrite it (in components) as

$$\Xi_n^\alpha = \frac{\partial^2 U(r_{ij})}{\partial r_n^\alpha \partial \gamma} = \kappa_{ij} \frac{\partial r_{ij}}{\partial r_n^\alpha} \frac{\partial r_{ij}}{\partial \gamma} + t_{ij} \frac{\partial^2 r_{ij}}{\partial r_n^\alpha \partial \gamma}. \tag{2.31}$$

Now we need to make use of the identities, Eqs. (2.5)–(2.6), found in Sect. 2.2, as well as of the following identity:

$$\frac{\partial^2 r_{ij}}{\partial r_n^\alpha \partial \gamma} = -\delta_{ji}^n \left(n_{ij}^x n_{ij}^y n_{ij}^\alpha - \left(\delta_{\alpha y} n_{ij}^x + \delta_{\alpha x} n_{ij}^y \right) \right), \tag{2.32}$$

where we recall that $\mathbf{n}_{ij} = \mathbf{r}_{ij}/r_{ij}$, and $\delta_{ji}^n = \delta_{nj} - \delta_{ni}$.

Hence, using Eqs. (2.5)–(2.6) together with Eq. (2.32) in Eq. (2.31), we finally obtain

$$\Xi_i = -\sum_j \left(\kappa_{ij} r_{ij} - t_{ij} \right) n_{ij}^x n_{ij}^y \mathbf{n}_{ij}. \tag{2.33}$$

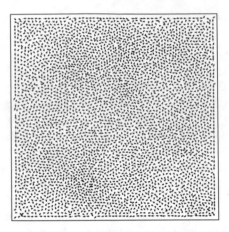

Fig. 2.5 Rendering of numerically computed affine force field Ξ_i for a random network of harmonic springs of narrow distributed length, computed based on numerical simulations using Eq. (2.33). The arrows represent the affine force vectors Ξ_i acting on each atom i. One can clearly see the typical random character of the field as there are very little correlations between neighboring components. Details about the simulated system can be found in [14]. Courtesy of Dr. Rico Milkus

Hence we managed to find also for Ξ an expression in terms of lattice sums that can be computed knowing the positions and interactions of, e.g., a numerically simulated system of particles.

The affine force field Ξ for a sheared amorphous solid presents a typical random structure, with no discernible patterns, due mainly to the random nature of local forces in a fully disordered environment. A typical numerical calculation rendering is shown in Fig. 2.5.

2.4.2   The Shear Modulus of Random Jammed Sphere Packings

As a first application of the above general theory to a specific system, we consider random packings of soft spheres, also known in the literature on disordered systems as "jammed packings." This system is important as it plays a central role as a paradigm for disordered solids, since many properties of glasses and other, especially granular, amorphous solids can be traced back to fundamental properties of jammed packings and jamming. This is true for elastic, plastic, and vibrational properties.

Without loss of generality, we can consider harmonically repulsive ("soft") spheres. The random close packing of spheres, as shown in seminal work by Bernal, occurs at a packing fraction $\phi_c \equiv \phi_{RCP} \approx 0.64$ (cfr. Chap. 1, Sect. 1.2.2.2) and

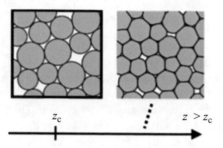

Fig. 2.6 Schematic of the jamming transition of soft spheres. Spheres are just touching at $z = z_c = 2d$, where d is the space dimension, which happens at the random close packing fraction $\phi_c = 0.64$. At the jamming point ϕ_c, the external pressure is identically zero. Larger values of ϕ and of z can be achieved by applying an external pressure which creates additional contacts between neighboring spheres through the deformation of the particle surface. Re-elaborated with modifications from Ref. [15]

coincides with a jamming transition, meaning that the system is overall floppy[2] below ϕ_c, and is rigid above ϕ_c. This rigidity transition is connected with the isostatic condition which is exactly fulfilled at ϕ_c.

As shown schematically in Fig. 2.6, the spheres are just touching their nearest-neighbors at ϕ_c where the external pressure is zero, whereas they get squeezed upon each other (with their surface being deformed) upon increasing the external pressure. At ϕ_c, known as the jamming transition, the coordination number z is found in numerical simulations to be $z = z_c = 2d$, i.e., $z_c = 6$ in 3D [16]. This coincides with the isostatic point where the number of degrees of freedom is just equal to the number of constraints. For spheres in 3D, each sphere has three degrees of freedom, and hence the total number of degrees of freedom is $3N$. The number of constraints to their motion is provided by the mechanical contacts with their nearest-neighbors: if there are on average z nearest-neighbors, this means a total of $\frac{zN}{2}$ constraints. Upon equating the number of degrees of freedom to the number of constrains, we thus obtain $z_c = 2d = 6$. This estimate was done by Maxwell in the nineteenth century and is referred to as the Maxwell rigidity criterion or isostaticity criterion (for more details, see Chap. 1 Sect. 1.6). This is always true in the absence of non-central interactions (hence the spheres must be frictionless as adhesional friction implies tangential forces and torque balance, which have to be included in the above counting) and for spherical particles.

The jammed packing of soft spheres with harmonic repulsion as outlined above can be mapped onto a random spring network where springs have all approximately the same length [14]. One obtains indeed the same results for the shear elasticity [17], although, as we shall see later in this chapter, there are

---

[2] This is true if the particles are athermal. For thermal hard spheres, there is a residual elasticity of entropic origin in the hard sphere glass state in the range of $0.58 < \phi < 0.64$ [2].

significant differences between the two systems insofar as the compression elasticity is concerned.

The Maxwell result must also be recovered from the analysis of microscopic elasticity of these systems, which we can now carry out by using the general theory presented in the previous section. In practice, all we need to do is to find an analytical expression for the shear modulus G and study where the shear modulus vanishes, which coincides with the rigidity (un)jamming transition. This was done in Ref. [10] and we shall present below the salient steps in the derivation.

Our goal is to bring Eq. (2.30) into an analytically manageable form for the system of Fig. 2.6. Hence, we work with harmonically (central-force) interacting soft spheres.

In order to simplify the nonaffine term in Eq. (2.30), we apply a normal mode decomposition, which is made possible by the fact that the Hessian matrix can be diagonalized to extract its eigenvectors and eigenvalues.

This will allow us to provide a characterization of the nonaffine displacements in terms of the eigenvalue spectrum of the system and to track the contributions to the nonaffine elastic moduli coming from different eigenmodes. We will drop for the economy of notation the particle indices.

Using the Dirac notation, we denote the eigenvectors of the Hessian as $|\mathbf{v}_p\rangle \equiv |\mathbf{p}\rangle$ which satisfy the eigenvalue problem $\mathbf{H}|\mathbf{p}\rangle = \lambda_p |\mathbf{p}\rangle$. Assuming that all particles have the same mass m, the eigenvalues are given by $\lambda_p = m\omega_p^2$, where ω_p is the eigenfrequency associated with the pth mode. We therefore expand the affine force field Ξ (note that $\Xi \equiv \Xi_i$ is a dN-dimensional vector) with respect to the eigenbasis as

$$\Xi = \sum_p |\mathbf{p}\rangle\langle\mathbf{p}|\Xi\rangle = \sum_p \hat{\Xi}_p |\mathbf{p}\rangle \tag{2.34}$$

with the expansion coefficients given by $\hat{\Xi}_p = \langle\mathbf{p}|\Xi\rangle$. Employing the same expansion, the nonaffine displacements can be written as

$$\delta\mathbf{r}_{NA} = \left(\frac{\mathcal{D}}{\mathcal{D}\gamma}\mathring{\mathbf{r}}(\gamma)\right)\bigg|_{\gamma\to 0} = \mathbf{H}^{-1}|\Xi\rangle = \sum_p |\mathbf{p}\rangle\langle\mathbf{p}|\mathbf{H}^{-1}|\Xi\rangle \tag{2.35}$$

$$= \sum_p \frac{\hat{\Xi}_p}{\lambda_p}|\mathbf{p}\rangle. \tag{2.36}$$

In the last step we made use of $\mathbf{H}^{-1}|\mathbf{p}\rangle = \lambda_p^{-1}|\mathbf{p}\rangle$. Note that in the sum over p the zero eigenvalues corresponding to the three Goldstone modes for rigid-body translation need to be excluded. Subsequently, the nonaffine correction to the shear modulus G_{NA} takes the form

$$G_{NA} = \frac{1}{V}\langle\Xi|\mathbf{H}^{-1}|\Xi\rangle = \frac{1}{V}\sum_p \frac{\langle\Xi|\mathbf{p}\rangle\langle\mathbf{p}|\Xi\rangle}{\lambda_p}. \tag{2.37}$$

With this equation at hand, we can also obtain insights into the contributions to G_{NA} coming from the different eigenvalues λ_p and the corresponding vibrational frequencies ω_p and we will make large use of this result also in Chap. 3 on viscoelasticity.

The formula that we obtained for G_{NA} in Eq. (2.37) is a good step forward, but it still requires the evaluation of a large sum over many essentially random quantities that cannot be easily evaluated without a computer. The next step is therefore to perform meaningful averages over the disorder and in particular averages over the various lattice sums that represent the key quantities at play.

Before doing that, we still need to express also the Hessian matrix in terms of lattice sums, so that all our quantities of interest will be in the form of lattice sums over which we can average. Recalling the definition of the Hessian in Eq. (2.1) and using the chain rule of differentiation, we have

$$H_{ij}^{\alpha\beta} = \frac{\partial U(r_{ij})}{\partial r_i^\alpha \partial r_j^\beta} = \kappa_{ij} \frac{\partial r_{ij}}{\partial r_i^\alpha} \frac{\partial r_{ij}}{\partial r_j^\beta} + t_{ij} \frac{\partial^2 r_{ij}}{\partial r_i^\alpha \partial r_j^\beta}. \tag{2.38}$$

Using the identity Eq. (2.5) as well as the following identity:

$$\frac{\partial^2 r_{ij}}{\partial r_n^\alpha \partial r_m^\beta} = \frac{\delta_{ji}^n \delta_{ji}^m}{r_{ij}} \left(\delta_{\alpha\beta} - n_{ij}^\alpha n_{ij}^\beta \right) \tag{2.39}$$

and replacing in Eq. (2.38), we finally obtain

$$H_{ij}^{\alpha\beta} = \delta_{ij} \sum_s \kappa c_{is} n_{is}^\alpha n_{is}^\beta - (1 - \delta_{ij}) \kappa c_{ij} n_{ij}^\alpha n_{ij}^\beta, \tag{2.40}$$

where c_{ij} is a random coefficient matrix with $c_{ij} = 1$ if i and j are nearest-neighbors and $c_{ij} = 0$ otherwise. c_{ij} is a matrix where each row and each column have on average z elements equal to 1 distributed randomly with the constraint that the matrix be symmetric. Furthermore, we assumed $\kappa_{ij} \equiv \kappa$ is the same for all bonds ij between nearest-neighbors. The above notation also makes explicit the diagonal terms of the Hessian (the first term in Eq. (2.40)) and the off-diagonal terms (the second contribution).

We are now ready to proceed with the disorder-averaging over the lattice sums. We start noting that the bond unit vectors n_{ij} are functions of the azimuthal and polar angles (ϕ_{ij} and θ_{ij}, respectively) which identify the orientation of the bond ij in the solid angle.

The Hessian is a random matrix and both $|p\rangle$ and λ_p depend on the realization of disorder. Also, being the Hessian a sparse matrix, there are no exact forms for its statistical spectral distributions. Nevertheless, the deterministic limit of

Eq. (2.37) can be evaluated analytically within the following approximations. In $d = 3$, it is $\mathbf{n}_{ij} = (\cos\phi_{ij}\sin\theta_{ij}, \sin\phi_{ij}\sin\theta_{ij}, \cos\theta_{ij})$ and the pair of angles ϕ_{ij} and θ_{ij} univocally specifies the orientation of the bond $\langle ij\rangle$. In order to keep the problem analytically tractable, we expand the bond-orientational term $n_{ij}^{\alpha}n_{ij}^{\beta}$ around its average value and neglect higher-order correlation terms. This is an effective medium approximation where we do an average over the (bond-orientational) disorder (but keeping the positional disorder unchanged) and only after averaging we proceed with the analytical calculation [18]. Therefore, we take the isotropic bond orientation-average over the entire solid angle Ω, $\frac{1}{4\pi}\int_{\Omega} n_{ij}^{\alpha}n_{ij}^{\beta}\sin\theta_{ij}d\theta_{ij}d\phi_{ij} = \delta_{\alpha\beta}/d$, equivalent to assuming that, locally, all bond orientations are *equally probable* (with probability density $\sin\theta_{ij}/4\pi$ in $d = 3$) and *statistically independent*. Within this approximation, the orientationally averaged Hessian reads as

$$H_{ij}^{\alpha\beta} = \frac{\kappa}{d}\left(\delta_{ij}\sum_j c_{ij} - (1 - \delta_{ij})c_{ij}\right)\delta_{\alpha\beta}. \qquad (2.41)$$

As one can easily verify by inspection, this matrix commutes with the dynamical matrix, which implies that its eigenvectors are also eigenvectors of the dynamical matrix.

According to Eq. (2.41), we can define $\mathbf{H} = \tilde{\mathbf{H}}\otimes\mathbf{1}$, where $\mathbf{1}$ is the $d \times d$ identity matrix (which represents the Kronecker $\delta_{\alpha\beta}$ in Eq. (2.41) and $\tilde{\mathbf{H}}$ is the $N \times N$ matrix which multiplies $\delta_{\alpha\beta}$ in Eq. (2.12).

Furthermore, the eigenvectors of this orientation-averaged matrix $\tilde{H}_{ij}^{\alpha\beta}$ are of the form: $\mathbf{v}_p = \mathbf{a}_q\otimes\mathbf{v}_l$. Denoting with $\{\mathbf{a}_{q=1..N}\}$ the set of eigenvectors of $\tilde{\mathbf{H}}$, which is an orthonormal basis (ONB) of $\mathbb{R}^N$, and with $\{\mathbf{e}_{l=1..d}\}$ the standard Cartesian basis of $\mathbb{R}^d$, it follows that $(\tilde{\mathbf{H}}\otimes\mathbf{1})(\mathbf{a}\otimes\mathbf{e}) = \lambda(\mathbf{a}\otimes\mathbf{e})$ and thus the dN dimensional set $\{\mathbf{a}_q\otimes\mathbf{e}_l\}$ is an ONB of eigenvectors of $\mathbf{H}$ as given by Eq. (2.41).

Upon recalling the lattice sum expression that we found for Ξ in Eq. (2.33), this allows us to write [10]

$$\langle\Xi|\mathbf{p}\rangle\langle\mathbf{p}|\Xi\rangle = \kappa^2 R_0^2\left(\sum_{i<j} a_i c_{ij}n_{ij}^{\alpha}n_{ij}^{x}n_{ij}^{y}\right)\left(\sum_{r<s} a_r c_{rs}n_{rs}^{\alpha}n_{rs}^{x}n_{rs}^{y}\right) \qquad (2.42)$$

$$= \kappa^2 R_0^2\sum_{i<j,r<s} a_i a_r c_{ij}c_{rs}n_{ij}^{\alpha}n_{ij}^{x}n_{ij}^{y}n_{rs}^{\alpha}n_{rs}^{x}n_{rs}^{y} \qquad (2.43)$$

where the sum runs over two distinct pairs of nearest-neighbor particles, ij and rs, and the summation over repeated Greek indices is implied. Furthermore, we assumed that the bond length r_{ij} is the same for all bonds and equal to R_0.

We still have an orientation-dependent factor $n_{ij}^\alpha n_{ij}^x n_{ij}^y n_{rs}^\alpha n_{rs}^x n_{rs}^y$ which needs to be averaged. For a generic element of the elastic tensor, this factor reads as $n_{ij}^\alpha n_{ij}^\iota n_{ij}^\xi n_{rs}^\alpha n_{rs}^\kappa n_{rs}^\chi$.

Upon taking the orientational average of these residual orientation-dependent factors in Eq. (2.43) with our randomly distributed bond-orientation approximation, we replace the orientation-dependent terms with their isotropic angular-averaged values which gives $n_{ij}^\alpha n_{ij}^\iota n_{ij}^\xi \, n_{rs}^\alpha n_{rs}^\kappa n_{rs}^\chi = \left(\delta_{ir}\delta_{js} - \delta_{is}\delta_{jr}\right) B_{\alpha,\iota\xi\kappa\chi}$, where the $B_{\alpha,\iota\xi\kappa\chi}$ are geometric coefficients resulting from the angular averaging. For $d = 3$ and $d = 2$ they are as follows [10]:

α	$d = 3$				$d = 2$		
	x	y	z	$\sum_\alpha$	x	y	$\sum_\alpha$
$B_{\alpha,xxxx}$	$\frac{1}{7}$	$\frac{1}{35}$	$\frac{1}{35}$	$\frac{1}{5}$	$\frac{5}{16}$	$\frac{1}{16}$	$\frac{3}{8}$
$B_{\alpha,xyxy}$	$\frac{1}{35}$	$\frac{1}{35}$	$\frac{1}{105}$	$\frac{1}{15}$	$\frac{1}{16}$	$\frac{1}{16}$	$\frac{1}{8}$
$B_{\alpha,xxyy}$	$\frac{1}{35}$	$\frac{1}{35}$	$\frac{1}{105}$	$\frac{1}{15}$	$\frac{1}{16}$	$\frac{1}{16}$	$\frac{1}{8}$

(2.44)

Substituting in Eq. (2.43), we obtain

$$\langle \Xi | \mathbf{p} \rangle \langle \mathbf{p} | \Xi \rangle = \kappa^2 R_0^2 \, B_{\alpha,\iota\xi\kappa\chi} \left(\sum_{rs} a_r^2 \, c_{rs} c_{rs} - \sum_{rs} a_r a_s \, c_{rs} c_{sr} \right)$$

$$= \kappa^2 R_0^2 \, B_{\alpha,\iota\xi\kappa\chi} \frac{d}{\kappa} \sum_{rs}^N a_r a_s \tilde{H}_{rs}$$

(2.45)

where we used that $c_{rs}^2 = c_{rs} c_{sr} = c_{rs}$ and the identities $\sum_r^N a_r^2 \sum_s c_{rs} - \sum_{rs} a_r a_s c_{rs} = \sum_{rs}^N a_r a_s [(\sum_j^N c_{rj})\delta_{rs} - c_{rs}(1 - \delta_{rs})] = \frac{d}{\kappa} \sum_{rs}^N a_r a_s \tilde{H}_{rs}$. Recalling that $\sum_{s=1}^N \tilde{H}_{rs} a_s = \lambda_r a_r$, we finally obtain

$$\langle \Xi | \mathbf{p} \rangle \langle \mathbf{p} | \Xi \rangle = d\kappa R_0^2 \lambda_p \sum_\alpha B_{\alpha,\iota\xi\kappa\chi}.$$

(2.46)

This equation is important because it shows that the square of the affine force field Ξ projected onto the p-th eigenvector of the Hessian is proportional to the corresponding p-th eigenvalue of the Hessian. This result will also play a role in building a microscopic theory of frequency-dependent viscoelasticity of amorphous solids in Chap. 3.

With this result, we can now go back to Eq. (2.37) and find a closed-form expression for the nonaffine correction to the elastic tensor:

$$C_{\iota\xi\kappa\chi}^{NA} = \frac{1}{V} \sum_{q=1}^N \sum_{\alpha=1}^d \frac{d\kappa R_0^2 \lambda_q B_{\alpha,\iota\xi\kappa\chi}}{\lambda_q} = d\frac{N}{V} \kappa R_0^2 \sum_{\alpha=1}^d B_{\alpha,\iota\xi\kappa\chi}.$$

(2.47)

For the case of shear strain, this becomes

$$C_{xyxy}^{NA} \equiv G_{NA} = \frac{1}{V} \sum_{q=1}^{N} \sum_{\alpha=1}^{d} \frac{d\kappa R_0^2 \lambda_q B_{\alpha,xyxy}}{\lambda_q} = d\frac{N}{V}\kappa R_0^2 \sum_{\alpha=1}^{d} B_{\alpha,xyxy}. \quad (2.48)$$

Upon reading off the values of the coefficients $B_{\alpha,xyxy}$ from Table 2.44, we thus arrive at quantitative expression for G_{NA}. All we still need to do is to evaluate also the affine part of the modulus, G_A, by means of the same orientation-averaging procedure, this time applied to the Born-Huang formula Eq. (2.12), i.e., by taking into account that $\langle n_{ij}^x n_{ij}^y n_{ij}^x n_{ij}^y \rangle = \sum_\alpha B_{\alpha,xyxy}$, where $\langle ... \rangle = \frac{1}{4\pi} \int_\Omega ... \sin\theta_{ij} d\theta_{ij} d\phi_{ij}$ denotes the angular averaging over the solid angle Ω. Upon neglecting residual bond stresses ($t_{ij} = 0$ for all bonds), we obtain

$$G_A = \frac{1}{30}\frac{N}{V}\kappa z R_0^2, \quad (2.49)$$

where z is the coordination number, or nearest-neighbor number, which arises from the sum over all neighboring pairs of particles.

Upon putting affine and nonaffine contributions together, we finally have for the shear modulus

$$G = G_A - G_{NA} = \frac{1}{30}\frac{N}{V}\kappa R_0^2(z - 6) \quad (2.50)$$

in 3D, whereas in 2D we obtain an analogous expression, with a different numerical prefactor,

$$G = G_A - G_{NA} = \frac{1}{16}\frac{N}{V}\kappa R_0^2(z - 4) \quad (2.51)$$

and with the vanishing of rigidity occurring now at $z = 2d = 4$.

An alternative derivation, leading to exactly the same expressions, is based on approximating the eigenvectors $\mathbf{v}_p$ of the Hessian as plane waves with a random correction and then on performing the explicit calculations of G_{NA} using these forms. This alternative derivation is used in Sect. 2.4.3.

Hence, the nonaffine theory is able to recover the vanishing of material rigidity (the rigid-to-floppy transition) at the Maxwell isostatic point, i.e., $z_c = 2d$ for a generic d-dimensional space, with the scaling $G \sim (z - 2d)$. Furthermore, it also shows that the nonaffine relaxations are directly responsible for the loss of rigidity at the isostatic ("jamming") point of soft sphere packings (and analogously for random spring networks), as well as for the "softening" that disorder brings about compared to an ideal crystalline lattice. In physical terms, we can say that part of the available energy that the system can "activate" under the deformation process goes into resisting the external deformation (the affine part), while another part goes into nonaffine relaxations and rearrangements and is therefore no longer

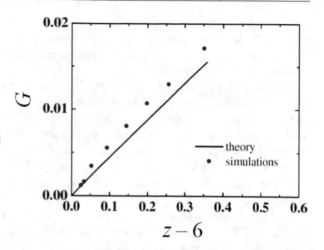

Fig. 2.7 Parameter-free quantitative comparison between the shear modulus calculated from numerical simulations of random jammed packings of harmonically repulsive spheres (data from [16]), symbols, and the theoretical prediction of Eq. (2.50), solid line

available to resist the external deformation. For non-rigid systems such as fluids, which have $G = 0$ identically, this is also in agreement with our empirical intuition, since the microscopic particles in the liquid under shear strain keep rearranging (nonaffinely) under the imposed strain, much different from the case of solids where the rearrangements are always smaller compared to the "resistive" affine motions.

The key result, Eq. (2.50), can be directly compared with numerical simulation data for the quasi-static shear modulus of jammed packings of soft repulsive harmonic spheres from [16]. The comparison is shown in Fig. 2.7.

In the next sections we will see how the above theory can be applied to more complex systems, such as networks with internal stresses, polymers, liquids, and glasses. Also, analogous expressions to Eq. (2.50) can be found for elastic moduli other than shear, such as the bulk or compression modulus, which however presents an additional difficulty due to the importance of excluded volume effects, as we shall discuss in Sect. 2.7.

2.4.3 Elasticity of Random Networks with Internal Stresses

In the above description, we always used the assumption that nearest-neighbor particles are connected via relaxed (harmonic) springs.

In soft matter systems and in glasses (with the exception of the random jammed packings discussed in the previous sections), however, the role of finite internal stresses, or bond tensions, $t_{ij} \equiv \frac{\partial U}{\partial r_{ij}} \neq 0$, cannot be neglected.

It was pointed out by S. Alexander, with the famous metaphor of the violin strings acquiring rigidity as they are stretched, that internal stresses (which cause bonds to stretch) can make underconstrained lattices (i.e., with $z < 2d$ for central-force springs) fully rigid, which would otherwise be floppy [19]. From numerical simulations, it is also known that, in disordered elastic networks, internal stresses have a profound effect on mechanical response and can indeed make underconstrained lattices become rigid [20]. This is particularly important, for example, for polymer hydrogels, where the osmotic pressure in solution generates such internal stresses on the polymer network to make it fully rigid even at very low coordination z of the network.

Alexander proposed that the standard Cauchy-Born strategy of Taylor-expanding the energy around unstressed lattice positions (used, e.g., in Sect. 2.3 to obtain the affine moduli) be generalized to stressed networks and that the net effect would be taken into account by higher-order (than linear) terms in the expression of the strain tensor as a function of the microscopic displacement field [19]. While very original and motivational for the whole field, Alexander's ideas did not concretize into a predictive framework, also because the role of nonaffinity was not explicitly taken into account.

Here we shall see how the nonaffine elasticity framework presented in Sects. 2.4.1 2.4.1 and 2.4.2 2.4.2 can be extended to include internal stresses, at the cost of modest mathematical complications and still in a fully analytical fashion.

In order to achieve this result, however, it is necessary to start from a more detailed representation of the eigenvectors of the Hessian of the amorphous solid. In an approximation (supported by simulations) suggested in [21], one can model the (normalized) eigenvectors as sinusoidal waves with wavenumber $q_p = \omega_p/v$ plus a random correction, $\epsilon_i(p)$, with zero average, and with variance $\sigma^2 = \langle \epsilon_i^2(p) \rangle$ independent of normal mode $p \in \{1, 2, ..., Nd\}$, i.e., in Cartesian components

$$v_i^\mu(p) = \hat{n}^\mu \frac{1}{\sqrt{Nd}} \left[\sqrt{2(1 - \sigma^2)} \sin(\mathbf{q}_p \cdot \mathbf{r}_i) + \epsilon_i(p) \right], \tag{2.52}$$

where $\hat{n}^\mu$ is the polarization unit vector such that $\hat{n}^\mu \hat{n}^\nu = \delta_{\mu\nu}$.

We now define the angular average as

$$\sum_{i=1}^{N} \sin^I(\mathbf{q}_p \cdot \mathbf{r}_i) \epsilon_i^J(p) = N \langle \sin^I(\mathbf{q}_p \cdot \mathbf{r}_i) \rangle \langle \epsilon_i^J(m) \rangle \tag{2.53}$$

where I and J are non-negative integer exponents.

The case of interest here is $I = 2$, which corresponds to normalization of the eigenvectors. The average can then be evaluated as follows. Assuming translation invariance (always justified for a uniform amorphous system at least in the low-q sector), there is complete freedom in choosing or shifting the origin of the reference frame, i.e., $\mathbf{r} \to \mathbf{r} + \mathbf{r}'$, where $\mathbf{r}'$ is an arbitrary shift. Hence, $\langle \sin^2(\mathbf{q} \cdot \mathbf{r}) \rangle = \langle \sin^2(\mathbf{q} \cdot (\mathbf{r} + \mathbf{r}')) \rangle = \langle \sin^2(\mathbf{q} \cdot \mathbf{r} + \mathbf{q} \cdot \mathbf{r}') \rangle$. Next we define $\vartheta \equiv \mathbf{q} \cdot \mathbf{r}'$, from which we get $\langle \sin^2(\mathbf{q} \cdot \mathbf{r}) \rangle = \langle \sin^2(\mathbf{q} \cdot \mathbf{r} + \vartheta) \rangle$. Since ϑ is an arbitrary scalar, one can choose $\vartheta = \pi/2$, without loss of generality, and the identity must hold for any values of ϑ. Then, clearly $\langle \sin^2(\mathbf{q} \cdot \mathbf{r}) \rangle = \langle \cos^2(\mathbf{q} \cdot \mathbf{r}) \rangle$, which implies

$$\langle \sin^2(\mathbf{q} \cdot \mathbf{r}) \rangle = \frac{1}{2}. \tag{2.54}$$

Using this result, it is easy to check that $v_i^\mu(p)$ is normalized,

$$\sum_{i\mu}[v_i^\mu(p)]^2 = \sum_i \frac{d}{Nd}[2(1-\sigma^2)\sin^2(\mathbf{q}_p \cdot \mathbf{r}_i)$$

$$+ 2\sqrt{2(1-\sigma^2)}\sin(\mathbf{q}_p \cdot \mathbf{r}_i)\epsilon_i(p) + \epsilon_i^2(p)]$$

$$= \frac{1}{N}\left[2(1-\sigma^2) \cdot \frac{N}{2} + N\sigma^2\right] = 1.$$

Our aim is to find the form of the eigenvalue λ of an eigenvector $\mathbf{v}$, i.e., $\mathbf{Hv} = \lambda\mathbf{v}$, that is,

$$[\mathbf{Hv}]_i^\mu = \sum_{j\nu} H_{ij}^{\mu\nu} v_j^\nu = \sum_{j\neq i}\sum_\nu H_{ij}^{\mu\nu} v_j^\nu + \sum_\nu H_{ii}^{\mu\nu} v_i^\nu$$

$$= \sum_{j\neq i}\sum_\nu \left[(s_{ij} - \frac{t_{ij}}{r_{ij}})n_{ij}^\mu n_{ij}^\nu + \frac{t_{ij}}{r_{ij}}\delta_{\mu\nu}\right](v_i^\nu - v_j^\nu)$$

$$(2.55)$$

where for the Hessian matrix we now use the following expression, which, unlike Eq. (2.40), contains also the bond tension terms t_{ij}. This is, clearly, a crucial step of the derivation, which allows us to account for internal stresses. The Hessian with bond tension terms reads as

$$H_{ij}^{\mu\nu} = \begin{cases} -(\kappa_{ij} - \frac{t_{ij}}{r_{ij}})n_{ij}^\mu n_{ij}^\nu - \frac{t_{ij}}{r_{ij}}\delta_{\mu\nu}, & i\neq j \\ \sum_{k\neq i}(\kappa_{ik} - \frac{t_{ik}}{r_{ik}})n_{ik}^\mu n_{ik}^\nu + \frac{t_{ik}}{r_{ik}}\delta_{\mu\nu}, & i=j. \end{cases} \qquad (2.56)$$

According to Ref. [10] and to what has been discussed in detail in the previous section, Sect. 2.4.2, the orientation-dependent factors $n_{ij}^\mu n_{ij}^\nu$ for a large system with uncorrelated isotropic disorder can be replaced with the isotropic (angular) average, i.e., $n_{ij}^\mu n_{ij}^\nu \to \delta_{\mu\nu}/d$, which leads to

$$[\mathbf{Hv}]_i^\mu = \sum_{j\neq i}\sum_\mu \left(\kappa_{ij} - \frac{t_{ij}}{r_{ij}}\right)n_{ij}^\mu n_{ij}^\nu(v_i^\nu - v_j^\nu) + \sum_{j\neq i}\frac{t_{ij}}{r_{ij}}(v_i^\mu - v_j^\mu)$$

$$= \sum_{j\neq i}\frac{1}{d}\left(\kappa_{ij} - \frac{t_{ij}}{r_{ij}} + d\frac{t_{ij}}{r_{ij}}\right)(v_i^\mu - v_j^\mu) = \frac{1}{d}\sum_{j\neq i}\left[\kappa_{ij} + (d-1)\frac{t_{ij}}{r_{ij}}\right]v_i^\mu$$

$$\equiv \lambda v_i^\mu. \qquad (2.57)$$

Terms proportional to v_j^μ vanish because the definition of eigenvector, obviously, requires the final result to be independent of one of its components such as v_j^μ.

Therefore, we have found an analytical expression for the eigenvectors, which contains a dependence on the bond tensions (internal stresses), t_{ij}.

With these, we are now able to write $(\Xi_{\iota\xi} \cdot \mathbf{v}_p)(\Xi_{\kappa\chi} \cdot \mathbf{v}_p)$ in an explicit analytical form:

$$(\Xi_{\iota\xi} \cdot \mathbf{v}_p)(\Xi_{\kappa\chi} \cdot \mathbf{v}_p) = \sum_{ii'}^{N} \sum_{\mu\nu}^{d} \Xi_{i,\iota\xi}^{\mu} v_i^{\mu} \Xi_{i',\kappa\chi}^{\nu} v_{i'}^{\nu}$$

$$= \sum_{ii'jj'}^{N} \sum_{\mu\nu}^{d} (r_{ij}\kappa_{ij} - t_{ij})(r_{i'j'}\kappa_{i'j'} - t_{i'j'}) n_{ij}^{\mu} n_{ij}^{\iota} n_{ij}^{\xi} n_{i'j'}^{\nu} n_{i'j'}^{\kappa} n_{i'j'}^{\chi} \cdot$$

$$\frac{1}{Nd} \hat{n}^{\mu} \hat{n}^{\nu} \left[\sqrt{2(1 - \sigma^2)} \sin(\mathbf{q}_p \cdot \mathbf{r}_i) + \epsilon_i \right] \left[\sqrt{2(1 - \sigma^2)} \sin(\mathbf{q}_p \cdot \mathbf{r}_{i'}) + \epsilon_{i'} \right].$$

$$(2.58)$$

Now, upon taking an isotropic average, the term $n_{ij}^{\mu} n_{ij}^{\iota} n_{ij}^{\xi} n_{i'j'}^{\mu} n_{i'j'}^{\kappa} n_{i'j'}^{\chi}$ may be replaced with $(\delta_{ii'}\delta_{jj'} - \delta_{ij'}\delta_{i'j})B_{\mu,\iota\xi\kappa\chi}$, where $B_{\mu,\iota\xi\kappa\chi}$ are the geometric coefficients resulting from the angular average that we have already encountered and tabulated in Sect. 2.4.2. We thus obtain [22]

$$(\Xi_{\iota\xi} \cdot \mathbf{v}_p)(\Xi_{\kappa\chi} \cdot \mathbf{v}_p) =$$

$$\frac{1}{Nd} \sum_{\mu}^{d} B_{\mu,\iota\xi\kappa\chi} \sum_{ii'jj'} (\delta_{ii'}\delta_{jj'} - \delta_{ij'}\delta_{i'j})(r_{ij}\kappa_{ij} - t_{ij})(r_{i'j'}\kappa_{i'j'} - t_{i'j'}) \cdot$$

$$[2(1 - \sigma^2)\sin(\mathbf{q}_p \cdot \mathbf{r}_i)\sin(\mathbf{q}_p \cdot \mathbf{r}_{i'}) + \epsilon_i\sqrt{2(1 - \sigma^2)}\sin(\mathbf{q}_p \cdot \mathbf{r}_{i'})$$

$$+ \epsilon_{i'}\sqrt{2(1 - \sigma^2)}\sin(\mathbf{q}_p \cdot \mathbf{r}_i) + \epsilon_i\epsilon_{i'}]$$

$$= \frac{1}{Nd} \sum_{\mu}^{d} B_{\mu,\iota\xi\kappa\chi} \sum_{ij}^{N} (r_{ij}\kappa_{ij} - t_{ij})^2 [2(1 - \sigma^2)\sin^2(\mathbf{q}_p \cdot \mathbf{r}_i)$$

$$+ 2\epsilon_i\sqrt{2(1 - \sigma^2)}\sin(\mathbf{q}_p \cdot \mathbf{r}_{i'}) + \epsilon_i^2$$

$$- 2(1 - \sigma^2)\sin(\mathbf{q}_p \cdot \mathbf{r}_i)\sin(\mathbf{q}_p \cdot \mathbf{r}_j)$$

$$- \epsilon_i\sqrt{2(1 - \sigma^2)}\sin(\mathbf{q}_p \cdot \mathbf{r}_j) - \epsilon_j\sqrt{2(1 - \sigma^2)}\sin(\mathbf{q}_p \cdot \mathbf{r}_i) - \epsilon_i\epsilon_j]$$

$$= \frac{z}{d} \sum_{\mu}^{d} B_{\mu,\iota\xi\kappa\chi} \langle (r_{ij}\kappa_{ij} - t_{ij})^2 \rangle, \qquad (2.59)$$

where z denotes, as usual, the mean coordination number of the disordered network.

The final line obtained in the above manipulation is, clearly, the sought-after generalization of Eq. (2.46) to the case of non-vanishing bond tensions $t_{ij} \neq 0$.

The presence of a factor z in front of the sum should be understood as due to the averaging $\langle...\rangle$, which implies a sum over all pairs ij.[3]

This becomes even clearer upon checking that the above expression correctly reduces to the stress-free result obtained for the elastic moduli in the previous Sect. 2.4.2. This can be shown as follows. As before, the interaction potential is a harmonic potential $U(r_{ij}) = (r_{ij} - R_0^2)/2$. Also, $\kappa \equiv \kappa_{ij}$ is the spring constant for all bonds, and R_0 is the distance between two particles in contact in the reference frame. Since the reference state is unstressed, all springs are relaxed in the minimum of the harmonic well. Hence, $t_{ij} \equiv 0$ and $r_{ij} \equiv R_0$. For the nonaffine part of the elastic stiffness tensor, recalling Eq. (2.37) in Sect. 2.4.2, we have

$$
C^{NA}_{\iota\xi\kappa\chi} = \frac{1}{V} \cdot dN \cdot \frac{zR_0^2\kappa^2\sum_\mu B_{\mu,\iota\xi\kappa\chi}}{d \cdot \frac{1}{d}z\kappa} = \frac{dNR_0^2\kappa}{V} \sum_\mu B_{\mu,\iota\xi\kappa\chi}. \tag{2.60}
$$

The affine term is, as always, given by the Born-Huang expression:

$$
C^{A}_{\iota\xi\kappa\chi} = \frac{NzR_0^2\kappa}{2V} \langle n_{ij}^\iota n_{ij}^\xi n_{ij}^\kappa n_{ij}^\chi \rangle. \tag{2.61}
$$

Here $\langle n_{ij}^\iota n_{ij}^\xi n_{ij}^\kappa n_{ij}^\chi \rangle = \sum_\mu^d B_{\mu,\iota\xi\kappa\chi}$ (cfr. Sect. 2.4.2), and we can write the elastic constant tensor as

$$
C_{\iota\xi\kappa\chi} = C^{A}_{\iota\xi\kappa\chi} - C^{NA}_{\iota\xi\kappa\chi} = \frac{NR_0^2\kappa}{2V} \sum_{\mu=1}^d B_{\mu,\iota\xi\kappa\chi}(z - 2d). \tag{2.62}
$$

For the shear modulus, $\sum_\mu^d B_{\mu,xyxy} = 1/15$ and Eq. (2.62) exactly recovers the same analytical results obtained in [10] and in Sect. 2.4.2.

We shall now, finally, relax the assumption that the interparticle distance (or bond length, in networks) R_0 coincides with the minimum of the harmonic potential, by introducing a distribution of interparticle distances peaked at an average value $R_e \neq R_0$. On average, we put $r_{ij} \equiv R_e \neq R_0$.

The fact that the actual distance between two particles in contact deviates from the minimum of the interaction automatically implies the existence of a bond tension or stress. In other words, the spring is either compressed, $R_e < R_0$, or stretched, $R_e > R_0$.

With these model assumptions, we get, for the affine part,

$$
C^{A}_{\iota\xi\kappa\chi} = \frac{NzR_0R_e\kappa}{2V} \langle n_{ij}^\iota n_{ij}^\xi n_{ij}^\kappa n_{ij}^\chi \rangle = \frac{NzR_0R_e\kappa}{2V} \sum_\mu^d B_{\mu,\iota\xi\kappa\chi}, \tag{2.63}
$$

---

[3] Also recall that there are in total $zN/2$ bonds (or node pairs ij) in the lattice, and one has to divide by this number as part of the $\langle...\rangle$ averaging process.

consistent with what we found in Sect. 2.4.2. Next, we evaluate the nonaffine contribution to the elastic moduli, $C^{NA}_{\iota\xi\kappa\chi}$. We first need to evaluate

$$(\Xi_{\iota\xi} \cdot \mathbf{v}_p)(\Xi_{\kappa\chi} \cdot \mathbf{v}_p) = \frac{1}{Nd} \sum_{\mu}^{d} B_{\mu,\iota\xi\kappa\chi} \cdot Nz \cdot (\kappa R_0)^2 = \frac{z\kappa^2 R_0^2}{d} \sum_{\mu}^{d} B_{\mu,\iota\xi\kappa\chi}. \tag{2.64}$$

Then we need to recall Eq. (2.37), which, for convenience, we rewrite here as

$$C^{NA}_{\iota\xi\kappa\chi} = \frac{1}{\mathring{V}} \sum_{p}^{Nd} \frac{(\mathbf{v}_p \cdot \Xi_{\iota\xi})(\mathbf{v}_p \cdot \Xi_{\kappa\chi})}{\lambda_p}. \tag{2.65}$$

The sum over eigenmodes is restricted to the non-zero modes only, thus excluding the trivial Goldstone zero-energy modes for rigid-body motions of the whole sample.

With a similar averaging scheme used in Sect. 2.4.2, we compute the averaged eigenvalue, which now depends also on the bond tension t_{ij},

$$\lambda = \frac{1}{d} \sum_{j \neq i} \left[\kappa_{ij} + (d-1)\frac{t_{ij}}{r_{ij}} \right] = \frac{z}{d} \left[\kappa + \kappa(d-1) \left(1 - \frac{R_0}{R_e} \right) \right] \tag{2.66}$$

and, putting things together, we obtain

$$C^{NA}_{\iota\xi\kappa\chi} = \frac{1}{V} \cdot Nd \cdot \frac{dz\kappa^2 R_0^2}{dz \left[\kappa + \kappa(d-1) \left(1 - \frac{R_0}{R_e} \right) \right]} \sum_{\mu}^{d} B_{\mu,\iota\xi\kappa\chi}$$

$$= \frac{Nd\kappa R_0^2}{V \left[1 + (d-1) \left(1 - \frac{R_0}{R_e} \right) \right]} \sum_{\mu}^{d} B_{\mu,\iota\xi\kappa\chi}, \tag{2.67}$$

where we used $t_{ij} = \kappa(R_e - R_0)$.

And, finally, upon adding together positive affine and negative nonaffine contributions as usual, we obtain, for the elastic moduli of random networks with internal stresses,

$$C_{\iota\xi\kappa\chi} = C^{A}_{\iota\xi\kappa\chi} - C^{NA}_{\iota\xi\kappa\chi} = \frac{N\kappa R_0 R_e}{2V} \left[z - \frac{2d\frac{R_0}{R_e}}{1 + (d-1) \left(1 - \frac{R_0}{R_e} \right)} \right] \sum_{\mu}^{d} B_{\mu,\iota\xi\kappa\chi}. \tag{2.68}$$

This is a key result of this section and was obtained for the first time in Ref. [22]. Clearly, if $R_e = R_0$ (i.e., the internal stress is zero), then we exactly recover Eq. (2.62) for the unstressed system (obtained in Sect. 2.4.2).

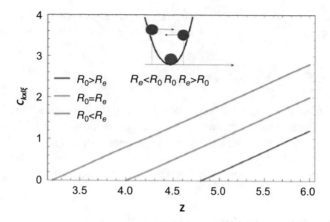

Fig. 2.8 The elastic constants $C_{\iota\xi\kappa\chi}$ as a function of coordination number z for different values of the internal stress parameter R_e/R_0, which indicates the initial particle displacement from the spring-like harmonic minimum. Results are shown for $d = 2$ systems. Reproduced from Ref. [22] with permission

Thus, we have come to the following important result: with internal stresses, the elastic constants (including the shear modulus $G \equiv C_{xyxy}$) exhibit the following modified scaling:

$$C_{\iota\xi\kappa\chi} \sim (z - 2df), \quad f = \frac{R_0/R_e}{1 + (d - 1)\left(1 - \frac{R_0}{R_e}\right)}. \tag{2.69}$$

If $R_e < R_0$, then $f > 1$, and the springs are, on average, compressed. Instead, if $R_e > R_0$, then $f < 1$, and the springs are, on average, stretched. Figure 2.8 illustrates, for the particular case of two-dimensional random networks, $d = 2$, how the ratio R_e/R_0, which controls the extent to which the springs are stretched or compressed, influences the dependence of $C_{\iota\xi\kappa\chi}$ on the mean coordination z of the network.

From a physical point of view, the behavior seen in Fig. 2.8 means that when the internal stresses are present due to initially stretched network bonds, then larger elastic constants are required to bring particles back to equilibrium positions. On the other hand, if the bonds are initially compressed, the elastic constants become smaller. This result, derived mathematically within the nonaffine elasticity theory [22], shows that networks with pre-stretched bonds lead to a larger elastic modulus. This confirms an earlier intuition of Alexander [19] and has been confirmed also in numerical simulations [23].

Finally, let us consider the case of an isotropic internal stress and, specializing on stretching, denote with $T \equiv t_{ij} = \kappa(R_e - R_0)$ the total isotropic internal stress. From a simple Taylor expansion to first order in Eq. (2.68), it is easy to obtain that, e.g., the shear modulus $G \equiv C_{xyxy}$ scales linearly with the isotropic internal stress, i.e., $G \sim T$. This mathematical prediction, based on the above theory from Ref. [22], has been recently confirmed by means of numerical simulations in Ref. [24].

2.5   The Shear Modulus of Glasses

In this section we shall apply the theory developed in the previous sections to finding closed-form expressions for the temperature-dependent shear modulus G of model thermal glasses, such as polymer glasses and Lennard-Jones glasses. We shall start with the example of glass made of flexible polymer chains, which presents a higher bonding/interaction complexity, and then consider the simpler case of LJ glasses.

2.5.1   The Shear Modulus of Polymer Glasses

Polymers are made of long molecules (in fact, *macromolecules*), which assemble together to form states of matter that may appear very different from ordinary liquids and solids. This is mainly because of the additional architectural complexity that is brought about by the ability of long molecules to intertwine, align, and fold, depending on parameters such as chain length (degree of polymerization) and polydispersity, persistence length, and, in general, the underlying inter- and intra-molecular interactions and bonding.

The molten state of polymers has posed a number of conceptual challenges before an ultimate understanding of its macroscopic properties and dynamics was reached around the end of the 1970s, thanks to mesoscopic concepts such as the tube model and reptation theory [25, 26]. The molecular dynamics in polymer melts is dominated by entanglements and by the "reptating" motion thereof inside mean-field "tubes." Based on these ideas and on allied or derived concepts (e.g., convective constraint-release [27]), quantitative theories of the viscoelastic response of polymer melts were formulated which were able to fit experimental spectra of the storage and loss viscoelastic moduli.

While the successful description of rheological and viscoelastic properties of polymer melts above the glass transition temperature T_c (also referred to as T_g) has been rewarded with a Nobel prize in physics for P.-G. de Gennes, comparatively less attention has been paid to the properties of polymers in their disordered solid (glassy) state at temperatures below T_g [28]. While "crystalline" polymers present many differences with respect to their non-polymer counterparts,[4] polymers in the glassy state present more analogies to standard atomic and molecular glasses. The reason is that below T_g the mesoscopic dynamic features and mechanisms that are peculiar of polymers in the molten state, e.g., chain entanglements and

---

[4] Most important of all the fact that polymers can never crystallize fully but always present large mesoscopic or even macroscopic amorphous regions coexisting with more ordered regions where molecules are aligned.

Fig. 2.9 Schematic
representation of a polymer
glass. Double segments
represent intra-chain covalent
bonds, whereas single
segments represent
inter-molecular van der
Waals-type interactions

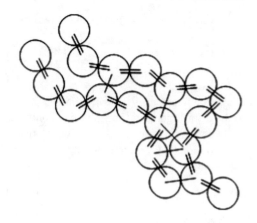

reptation, become essentially suppressed due to the low temperature and due to the impossibility of having large-scale or even mesoscopic-scale rearrangements on experimentally accessible times.

For our purpose of developing a mathematical theory with closed-form analytical results, the above considerations suggest the convenience of schematizing the polymer glass as a frozen-in amorphous lattice which retains the essential features of inter- and intra-molecular bonding and interactions typical of polymers. An illustration of this scheme is shown in Fig. 2.9. In the figure, the monomers, i.e., the smaller molecules that are glued together by covalent (intra-molecular) bonds to form the polymer chain, are represented as spheres. The covalent bonds are represented as two parallel segments (black color) and run along the chain backbone, whereas the inter-molecular non-covalent, e.g., van der Waals or Lennard-Jones (LJ) type, interactions are represented as single segments. In the following derivations, we shall assume that the chains are flexible, so we will *de facto* neglect the bond-bending component of the covalent bonds. The covalent bonds still have a stretching (central-force) spring constant which is larger than the one of LJ-type interactions. In the application of the nonaffine formulae, e.g., Eq. (2.50), this means that we will take an average spring constant $\bar{\kappa}$, which is a suitable average between the covalent and non-covalent spring constants (possibly closer in value to the covalent one), which will turn out to be the only adjustable parameter in the comparison with the experimental data.

Of course we are now considering a system where the effect of temperature is not negligible, in contrast with perfectly athermal systems considered in the previous sections. Hence the internal energy U that was sufficient for our purpose of deriving elastic moduli in an athermal setting should now be replaced by a proper free energy of deformation F [4]. Furthermore, we also need to include, besides the affine and nonaffine contributions, also the contribution to elasticity coming from thermal motions. Recalling the basic relation between the free energy of deformation F and the strain (see, e.g., [4]), we can express the shear modulus as follows:

$$G = \frac{\partial^2 (F_A + F_{NA} + F_T)}{\partial \gamma^2}, \tag{2.70}$$

where the free energy contributions $F_A = \frac{1}{2} G_A \gamma^2$ and $F_{NA} = \frac{1}{2} G_{NA} \gamma^2$ correspond to the affine and nonaffine athermal contributions to elasticity, respectively.

The thermal contribution can be expressed as follows. In general, the (negative) contribution to F due to thermal lattice vibrations is given by

$$F_T = -k_B T \ln \sum_n^\infty \exp\left(-\hbar \omega_n \left(n + \frac{1}{2} \right) / k_B T \right),$$

where n labels the eigenmodes and k_B is Boltzmann's constant. If $k_B T \gg \hbar \omega_{max}$, one can use the mean frequency $\overline{\omega}$ so that $F_T = -(3N/V)k_B T \ln(\frac{k_B}{T} \hbar \overline{\omega})$ [29]. The contribution of the elastic energy can be written as $F_T \approx -(3N/V)k_B T \theta \gamma^2$, where the non-dimensional factor $\theta = -(\partial^2/\partial\gamma^2)_{\gamma \to 0} \ln \hbar\overline{\omega}/k_B T$ has been demonstrated to be of order unity when the harmonic potential dominates the pair interaction potential [29]. This gives a good estimate:

$$F_T \approx -3 \frac{k_B T}{v} \phi \gamma^2, \tag{2.71}$$

where ϕ denotes the volume fraction occupied by the monomers (i.e., occupied by the spheres depicted in Fig. 2.9), and $v \equiv V/N$ is the volume available to each monomer in the system.

Another source[5] of temperature-dependence comes from the dependence of G on the coordination number z, already derived in the previous section for the athermal systems. Upon increasing the temperature T, while remaining always below T_g, the covalent bonds along the polymer chain remain unaffected and much higher temperatures (many hundreds of Kelvins above T_g) are needed to break the covalent bonds.

The non-covalent inter-molecular interactions (red segments in Fig. 2.9) are instead much more sensitive to temperature variations in this regime. Since these interactions can be modelled with the Lennard-Jones potential, there is a characteristic separation length scale corresponding to the minimum of the non-covalent interaction potential, r_{min}. Clearly, upon increasing the temperature, the thermal expansion of the material increases the separation between monomers belonging to nearby chain (by effectively reducing the density of the material, as verified in numerical simulations [30]), with the effect of weakening these attractive contributions to the cohesion of the solid. This reduced cohesion, as we shall

---

[5] Which will turn out to be the dominant one.

see, provides a decisive contribution to the softening of the polymer glass upon approaching the glass transition from below.

A way of accounting for this mechanism is to do a book-keeping of the *total* mean number of cohesive interactions per monomer, which we define as

$$z \equiv z_{LJ} + z_{co}, \tag{2.72}$$

where we distinguish the two contributions: z_{LJ} representing the mean number of non-covalent interactions per monomer and z_{co} the mean number of covalent interactions per monomer. Hence, z_{co} remains constant throughout the temperature window of interest for the reasons explained above, whereas z_{LJ} decreases with increasing temperature.

One can set and use different criteria to decide at which critical value of radial separation r a Lennard-Jones cohesive interaction is lost. For example, one can stipulate that the LJ cohesive interaction between two particles is lost as the two particles get farther apart than a distance corresponding to the inflection point (determined by setting $\frac{d^2 U_{LJ}}{dr^2} = 0$) of the LJ potential U_{LJ}, a criterion that has often been used to determine the dissociation of weakly bound molecular states. Alternatively, a more restrictive criterion is to declare that the cohesive interaction is lost at the distance where $\frac{dU_{LJ}}{dr} = 0$, i.e., right in the minimum of the LJ potential well. This situation is depicted in Fig. 2.10.

Having defined the total coordination number for cohesive interactions z, we now need to relate it to the temperature T. Upon introducing the thermal expansion coefficient, $\alpha_T = \frac{1}{V}(\partial V/\partial T)$, and replacing the volume V via $\phi = vN/V$, after integration, we obtain $\ln(1/\phi) = \alpha_T T + const$ (one can estimate this constant, obtaining $C \approx 0.48$ [31]). Now z can be estimated as a function of ϕ by introducing the radial distribution function (rdf) $g(r)$. Since the average connectivity due to

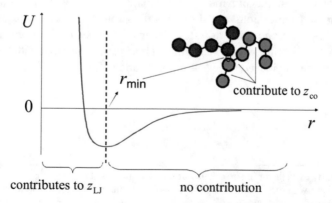

Fig. 2.10 Schematic of the criterion used to define the contribution of the contact LJ interactions and covalent bonds to the total number of mechanical bonds z. Only pairs of particles that lie within the soft repulsive part of the LJ potential contribute to the z_{LJ} counting. Adapted from Ref. [31] with permissions from the American Physical Society

covalent bonds remains fixed, only the weaker contacts contributing to z_{LJ} are changing upon increasing the packing fraction ϕ by $\delta\phi$. The increment δz can be calculated in full analogy with soft sphere systems where only the repulsive part of the potential is active. The increment δz has to be measured from the point where the system is marginally stable, i.e., $z = z_c$ at $\phi = \phi_c$, and from theory and simulations has [16, 32]: $(z - z_c) \sim \sqrt{\phi - \phi_c}$. Consistent with our modelling assumption of treating the monomers of soft spheres, we take $\phi_c \equiv \phi_{RCP} = 0.64$ [31], which, interestingly, was later confirmed quantitatively by microscopic studies on polymer glasses in [33] to a good accuracy.

In the affine approximation, the solid becomes marginally stable only in the limit $z_c \to 0$ and $\phi_c \to 0$, and hence we have $z \sim \phi^{1/2}$. By means of the relation between ϕ and α_T, we obtain $z \sim e^{-\alpha_T T/2}$. Replacing z and ϕ in $F_A + F_T$, we now can write the full expression for the temperature-dependent shear modulus in the affine approximation, $G_A = \partial^2(F_A + F_T)/\partial\gamma^2$, yielding

$$G_A(T) = \frac{2}{5\pi}\frac{1}{R_0^3}(\bar{\kappa}R_0^2 e^{-(3/2)\alpha_T T} - k_B T e^{-\alpha_T T}), \qquad (2.73)$$

where $\bar{\kappa}$ is an averaged spring constant between covalent and non-covalent interactions.[6] The Born criterion of melting [34] is given by Eq. (2.73) set to zero: $\kappa R_0^2 = k_B T e^{\alpha_T T/2}$. In most materials, one has that $\alpha_T T \ll 1$ and, remarkably, this relation reproduces the Lindemann melting criterion, which uses equipartition to state that melting occurs when the average vibrational energy of a bond equals $k_B T$. It is also known that the Lindemann criterion grossly overestimates melting temperatures for amorphous solids [19]; in the same way, Eq. (2.73) cannot capture the vanishing of rigidity as seen in the melting of glassy polymers. It turns out that to describe the melting of amorphous solids one has to account for nonaffine deformations in the lattice dynamics.

We have seen in the previous section that with purely central inter-particle interactions in d dimensions, the shear modulus vanishes at $z_c = 2d$ because the nonaffine term is proportional to the number of degrees of freedom that can be involved in the nonaffine energy relaxation. This is also consistent with the Maxwell rigidity criterion and indeed $G \sim (z - 2d)$ as found in Eq. (2.50). The latter equation

---

[6] The apparently extra factor of two in this formula, like in the subsequent expressions that will be used to compare with experimental or simulation data in the following, is due to the fact that in the above derivations we tacitly neglected the factor $\frac{1}{2}$ in the definition $G = \frac{1}{2}\frac{\partial^2 U}{\partial\gamma^2}$ (where here U is energy per unit volume) used in linear elasticity theory, where, also, the stress–strain Hooke's relation reads as $\sigma = 2\mu\gamma$, where $G \equiv \mu$ coincides with the second Lame' parameter [4]. Whence in the previous derivations $G \equiv \mu$, now for the sake of comparing with the experimental data $G \equiv 2\mu$, hence the additional factor of two.

can be slightly rewritten as follows:

$$G = G_A - G_{NA} = \frac{2}{5\pi} \frac{\bar{\kappa}}{R_0} \phi (z - z_c) \tag{2.74}$$

by noting that $\frac{N}{V} = \frac{6}{\pi} \frac{1}{R_0^3} \phi$.

Using again $\ln(1/\phi) = \alpha_T T + C$, we arrive at $\ln(\phi_c/\phi) = \alpha_T (T - T_c)$. The corresponding relation $(z - z_c) \sim \sqrt{\phi - \phi_c}$ can be manipulated into

$$\ln(\phi_c/\phi) = -\ln[1 + (z - z_c)^2/\phi_c]. \tag{2.75}$$

Combining this with the relation for $\phi(T, \alpha_T)$, we obtain $\ln[1 + (z - z_c)^2/\phi_c] = \alpha_T(T - T_c)$ and finally arrive at the condition: $z - z_c = \sqrt{\phi_c[e^{\alpha_T(T_c-T)} - 1]}$. Substituting it in Eq. (2.74), we obtain

$$G = G_A - G_{NA} = \frac{2}{5\pi} \frac{\bar{\kappa}}{R_0} \phi_c e^{\alpha_T(T_c-T)} \sqrt{\phi_c[e^{\alpha_T(T_c-T)} - 1]}. \tag{2.76}$$

According to this equation for the shear modulus $G(T)$, nonaffinity alone (induced by disorder) causes the glass melting at a critical point T_c with the scaling $\sim \sqrt{T_c - T}$, even without the effects of thermal vibrations on the rigidity. Including the effect of thermal vibrations, the full expression for $G(T)$ becomes

$$G = \frac{2}{5\pi} \left(\frac{\bar{\kappa}}{R_0} \phi_c e^{\alpha_T(T_c-T)} \sqrt{\phi_c[e^{\alpha_T(T_c-T)} - 1]} - \frac{k_B T}{R_0^3} e^{-\alpha_T T} \right). \tag{2.77}$$

One can assess the interplay and relative magnitude of nonaffinity and thermal vibrations in determining the softening of the modulus, e.g., by plotting the separate contributions for realistic values of the parameters. It turns out, as discussed in [31], that the contribution of thermal vibrations to the temperature-induced softening is much smaller compared to the effect of nonaffinity combined with thermal expansion. This is shown in Fig. 2.11 for a sample computation using the same values of the parameters used in the comparison with experimental data of polystyrene in Fig. 2.12.

This closed-form expression for the temperature-dependent shear modulus of a polymer glass can be directly compared to experiments, and the comparison is shown in Fig. 2.12. In the comparison, there is only one adjustable parameter, which is the effective spring constant for covalent bonds $\bar{\kappa}$ for which we used $\bar{\kappa} = 52$ N/m, which is on the same order of magnitude of the spring constant for $C - C$ bonds.

The theory is able to explain the dramatic loss of shear rigidity upon approaching the glass transition temperature T_c from below (or with some abuse of terminology, the "melting" of the glass). Different from perfect crystalline solids, the "melting" of the glass is caused not so much by the direct effect of thermal vibrations, which in the standard Lindemann picture bring atoms away from their equilibrium bonding

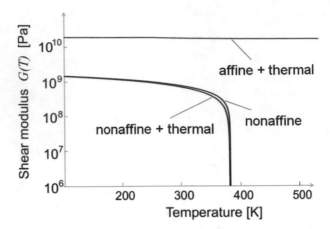

Fig. 2.11 Theoretical calculations of the various contributions to the shear modulus of polymer glass. It is seen that while the thermal vibrations can weaken the shear rigidity, the dominant contribution comes from the nonaffine relaxations, and in particular from the balance $G_A - G_{NA}$, with the first term becoming smaller as a result of thermal expansion and eventually becoming equal to the nonaffine negative term, at which point the shear rigidity is mostly lost

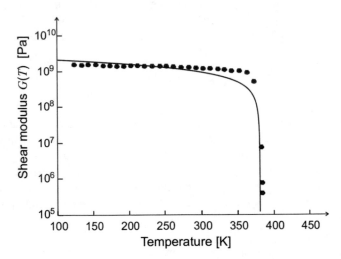

Fig. 2.12 Comparison between predictions of Eq. (2.77) (solid line) and experimental data (symbols) of [35] on polystyrene glass. With just one adjustable parameter (the average spring constant $\bar{\kappa} = 52$ N/m), the theory is able to capture the dramatic drop of the shear modulus by many orders of magnitude due to the nonaffine negative contribution which becomes equal to the affine contribution, the latter declining with increasing temperature due to thermal expansion and the concomitant decrease of density of the solid. Adapted from Ref. [31] with permission from the American Physical Society

position. Rather, it is largely caused by the (negative) nonaffine contribution which takes over with respect to the affine (cohesive) contribution, as the latter becomes weaker due to loss of interparticle interactions due to thermal expansion. In the above simplified treatment, the temperature dependence of the nonaffine term in the shear modulus is basically neglected. We shall see in Chap. 3, where we model the nonaffine contribution in frequency, that the nonaffine term also grows (in absolute value, hence becoming more negative) as the temperature is raised, an effect that is mediated by the temperature-dependent evolution of low-energy vibrational modes present in the vibrational spectrum of glasses.

Finally, we should comment on the cusp-like square-root behavior near T_c predicted by the theory. Equation (2.77) predicts a continuous transition with the following square-root cusp behavior near T_c:

$$G \sim \left(1 - \frac{T}{T_c}\right)^{1/2}. \tag{2.78}$$

It should be noted that this critical behavior has been derived under certain assumptions, in particular that of flexible chains (bond-bending terms in the interaction have been neglected). Some deviations from the exponent $1/2$ are thus expected and indeed smaller values of the exponent (down to ≈ 0.2) have been observed in numerical simulations of bead-spring polymer glasses in [36].

2.5.2 The Shear Modulus of Lennard-Jones Glasses

We can now consider a model system of spheres interacting via the Lennard-Jones potential only. This is clearly a paradigmatic system to study many of the properties and phenomena related to glassy physics and the glass transition, as it retains all the main features of structure and dynamics without unnecessary complications deriving from additional interactions such as covalent interactions. The above theory is naturally applicable to an LJ glass, where now there is a unique spring constant κ for all nearest-neighbor pairs. A variation on the theme is represented by Kob-Andersen binary mixtures where we have two types of spheres (say A and B) which differ only for the size and for the LJ interaction parameters [37]. In this case one can again assume an average spring constant $\bar{\kappa}$. We thus expect the LJ glass to follow the predictions, Eqs. (2.77) and (2.78), since those predictions were developed for the flexible chain limit, where the bond-bending role of covalent bonds is basically irrelevant.

And indeed the theoretical prediction has been confirmed in numerical simulations of Lennard-Jones and Kob-Andersen glasses, in both 2D and 3D [30], where the same square-root trend has been observed, as shown in Fig. 2.13.

The simulations fully confirm the square-root critical-like scaling $G \sim (1 - T/T_c)^{1/2}$ predicted by the theory. Interestingly, such square-root behavior is also found in seemingly unrelated systems, e.g., in a scalar Ginzburg-Landau theory

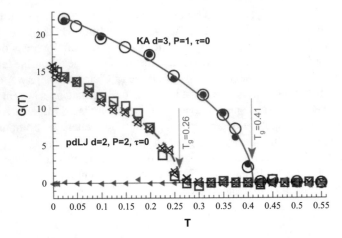

Fig. 2.13 Numerical simulations (symbols) of temperature-dependent shear modulus $G(T)$, for glasses made of Lennard-Jones (LJ) interacting soft spheres and Kob-Andersen (KA) interacting mixtures. The continuous dashed and the solid lines represent Eq. (2.78) predicted by the nonaffine theory. The arrows indicate the temperatures at which the glass transition occurs for the different systems. Adapted from Ref. [30] with permission of the American Institute of Physics

for the vanishing of rigidity of superheated spin-wave systems in the pre-critical region, although in this case there is also a finite jump typical of first-order transitions [38] (whereas no jump is predicted for the shear rigidity of glasses in the above derivation nor is observed in the numerical simulations and in the experiments). A square-root $\sim \sqrt{T_c - T}$ vanishing of the glassy cage size, which correlates with rigidity, is predicted by Mode-Coupling theory, although also in this case there is a discontinuous jump at the transition. Moreover, the transition described by the Mode-Coupling theory does not exactly coincide with the glass transition in thermal systems (with the important exception of athermal hard sphere colloids [2]), but rather with a higher-temperature crowding or caging transition in the liquid at temperatures above the glass transition [39].

2.6   The Shear Modulus of Colloidal Gels

An important class of amorphous solids is represented by colloidal systems, where the building blocks are not atoms or molecules, but colloidal (Brownian) particles suspended in a liquid solvent with sizes in the range from tens of nanometers to a few microns. Colloids can form crystals, as well as glasses, just like ordinary atomic and molecular building blocks, but they can also form new disordered states of matter, such as colloidal gels, which are unique to colloidal systems.

In this section we consider the peculiar elasticity of two main types of colloidal gels. The first type is the so-called *fractal gel*, while the second type is the *dense gel* at moderate attraction (often referred to as "depletion gel" since the interparticle interaction is given by dissolved polymer chains which induce the Asakura-Oosawa depletion attraction between particles). These are two opposite limits or regimes, in that the fractal gel forms at very low particle volume fractions $\phi \ll 0.01$, whereas the dense gel forms in the regime $\phi > 0.1$. Furthermore, the strength of attraction is also very different, as in the fractal gel the depth of interparticle interaction well is $\epsilon \gg 10\,k_BT$, whereas for the dense gels it is typically $1\,k_BT < \epsilon < 10\,k_BT$.

Also the formation mechanism is rather different. The fractal gel forms through a diffusion-limited cluster-cluster aggregation (DLCA) mechanism, where particles initially "free" in the liquid stick upon contact as they collide, due to Brownian motion, and coagulate irreversibly to form fractal clusters with a fractal dimension $d_f \approx 1.7$–1.8. These clusters eventually form a system-spanning structure with the same overall fractal dimension as the clusters, although this process is not readily described in terms of percolation-type critical phenomena. The dense gels instead form through a mechanism the origin of which is still debated. At higher ϕ, the formation mechanism is likely to occur via a glassy-like dynamical arrest inside the coarsening phase of a spinodal-type liquid–liquid phase separation [40]. At lower ϕ, recent evidence for a nonequilibrium percolation phase transition has been found [41]. At intermediate ϕ, a cluster-glass transition is also a likely route to the gel state [42,43].

2.6.1 Elasticity of Fractal Gels

The formation mechanism of fractal gels is schematically shown in Fig. 2.14a and proceeds via the growth of large fractal aggregates until the aggregates merge into a system-spanning macroscopic structure, which is also mechanically rigid.

Fractal gels are sparse, tenuous structures, with low connectivity z. Kantor and Webman, in their seminal work [44], considered that in such random elastic solids, the mesoscopic structures that are directly responsible for shear rigidity constitute only a small fraction of the network, referred to as the "stress-bearing backbone." This is because dangling ends which do not belong to the stress-bearing backbone cannot transmit stress and thus do not contribute to the elasticity. There are two modes of deformation contributing to elasticity of the stress-bearing backbone: bond-bending and bond-stretching. In two dimensions, the elastic energy of a random chain of N elastic bonds, where the bond vectors b are of length a, is

$$U = \frac{\kappa_1}{2} \sum_i \delta\Theta_i^2 + \frac{\kappa_2}{2a^2} \sum_i \delta b_i^2, \tag{2.79}$$

where κ_1 and κ_2 are local spring constants, $\delta\Theta_i$ is the change in the angle Θ_i between two adjacent bond vectors $i-1$ and i (see Fig. 2.14b), and the sum runs over bond vectors between two adjacent particles, each bond labeled by index i. Furthermore, δb is the change in length of bond i.

Thus, the first term in the above equation is related to bending energy, while the second term is related to stretching of the bond. If one applies a force to one

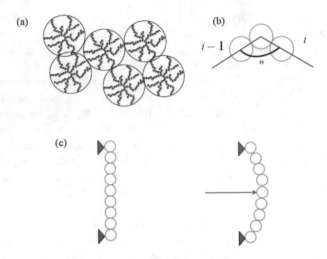

Fig. 2.14 Panel (**a**) shows the structure of a typical fractal gel with the merging of fractal clusters into a system-spanning fractal network. Panel (**b**) shows the angular bond-bending three-body interaction mediated by surface adhesion at the particle–particle contact. Panel (**c**) shows the bending rigidity of particle strands in the fractal structure that arises from the bond-bending interaction energy of panel (**b**)

end of the chain R_N with the other end R_0 held fixed, one can obtain $\delta\Theta_i$ and δb_i by minimizing the total energy Eq. (2.79), in terms of the magnitude of the applied force and the in-plane projections of the chain parallel and perpendicular to the direction of the force. In particular, $\delta\Theta_i$ depends on the bond vector component orthogonal to the force direction in the plane. Kantor and Webman therefore rewrote the bending contribution to the elastic energy as $\frac{F^2 N S_\perp^2}{2\kappa_1}$, where F is the magnitude of the applied force and $S_\perp^2$ is the squared radius of gyration of the in-plane projection of the chain along the direction orthogonal to the force. For long chains, this term dominates the elastic energy compared to the bond-stretching term. Therefore, one can obtain the bending force constant of the chain by relating the elastic energy to the squared displacement of the end of the chain as

$$\kappa_{chain} = \frac{\kappa_1}{N S_\perp}. \tag{2.80}$$

It is important to note that this effective force constant depends not only on its length but also on the direction of the applied force and the shape of the chain. DLCA fractal gels are formed by very large clusters, typically less rigid than the connections between them. Hence, the approach of Kantor and Webman is expected to give a good description of their elasticity, in terms of the elastic backbone of the cluster size R_c:

$$S_\perp \sim R_c \sim \phi^{1/(d_f - 3)}, \tag{2.81}$$

where $\phi = \frac{4}{3}a^3\frac{N}{V}$ is the volume fraction occupied by colloidal particles in the system. The scaling $R_c \sim \phi^{1/(d_f-3)}$ comes from the fractal scaling $N \sim \left(\frac{R_c}{a}\right)^{d_f}$ (cfr. Chap. 1, Sect. 1.2.3), which relates the number of colloid particles N belonging to a cluster of size (radius of gyration) R_c, to the cluster size itself and to the colloid radius a and fractal dimension d_f.

The number of particles in the stress-bearing backbone is

$$N \sim R_c^{d_B} \tag{2.82}$$

with the backbone fractal dimension $d_B \approx 1.1$, due to the nearly 1D shape of the stress-bearing backbone of colloidal gels. Upon inserting this relation, together with Eq. (2.81), into Eq. (2.80), we obtain the elastic constant of the gel as a function of the volume fraction ϕ of colloid particles:

$$\kappa_{chain} \sim \frac{\kappa_1}{\phi^{(2+d_B)/(d_f-3)}}. \tag{2.83}$$

Finally, a scaling relation for the shear modulus follows as

$$G \sim \frac{\kappa_{chain}}{R_c} \sim \phi^{\frac{3+d_B}{3+d_f}}. \tag{2.84}$$

This prediction for the exponent has been confirmed experimentally in several works, among which we refer to [45, 46].

2.6.2 Intermediate Dense Gels, and Cluster Glasses

Upon increasing the initial volume fraction, the mesoscopic organization of the gel network goes from the fractal thin structures discussed above to networks of thicker strands and to dense interconnected assemblies of bulkier flocs, typically produced by an incipient phase separation of the liquid–liquid type (with two liquid phases, one much denser in colloid, the other one much more diluted). These changes in the structure may correspond to significant changes in the elastic response of the material which make these "intermediate" concentration gels an active area of research. Upon increasing the particle volume fraction ϕ even further, one eventually hits the glass transition and a (attractive) colloidal glass is formed. In between the two limits of fractal DLCA gels and attractive glass, there exist a zoology of gels which may or not present well defined cluster structure, and where sometimes the relevant mesoscopic length scale is a correlation length of density fluctuations, rather than a well-defined cluster size. In many instances, these systems can be viewed as "cluster glasses," where clusters grow up relatively small sizes and then undergo a glass transition-type dynamical arrest into a system-spanning structure [42, 43].

Furthermore, in these experimental systems the assumption of central-force interaction (of the Lennard-Jones or depletion type, in any case with an attractive minimum) is always satisfied such that bond-bending interactions due to adhesive contact can be neglected. This of course simplifies the mathematical description and allows for more quantitative treatments.

Upon viewing these gel states from the perspective of the attractive glass, one can therefore this of starting from a good mathematical model for the shear modulus of a *homogeneous* attractive colloidal glass and then gradually introduce the structural heterogeneity upon decreasing ϕ toward less dense gel structures.

The starting point is therefore the shear modulus of a dense, structurally homogeneous colloidal attractive glass, where the relevant length scale is the nearest-neighbor separation R_0, and adopt the theory developed in Sect. 2.4.1. An appropriate starting point is thus given by Eq. (2.74). In this case, the ϕ-dependence of z is different from that that we worked out for the jammed packings, since now we are below the jamming transition, and particles are not squeezed against their neighbors, but they are at a finite separation. This means that z increases with increasing ϕ not because of the deformation of the particle surface leading to more contact points with the neighbors, but because the number of neighbors with which attractive interactions can be established increases by sheer crowding. To take this dependence into account, one can resort to liquid-state theory and integrate over the radial distribution function $g(r)$, by assuming a liquid-like structure.[7]

Also, we shall work within the affine approximation since we are going to compare predictions with rheological data where the shear modulus is measured experimentally in oscillatory shear experiments (typically in a rheometer) under substantial values of the oscillatory frequency (typically 1 Hz or larger). It is known that, under oscillatory shear at high frequency, the mechanical response is affine because the particles are given no time to reach the equilibrium nonaffine positions [47]. At intermediate frequencies of course there has to be a nonaffine component of the response, which however for simplicity we are neglecting here since it anyway does not introduce a further or different dependence on ϕ or the spring constant κ, which are the independent variables that we are going to consider in the comparisons.

For this homogeneous attractive glass, where therefore the size of the elastic building block reduces to, approximately, the particle diameter R_0, we can thus express z with the aid of liquid theory and get [43]

$$G_{hg} = (48/5)\pi^{-1}\kappa R_0^{-1}\phi \int_0^{l^\dagger} (1+l)^2 g(l; \phi)dl, \qquad (2.85)$$

---

[7] At $\phi \gtrsim 0.4$ the presence of attractive interactions does not really make the structure as encoded by $g(r)$ much different from that of a repulsive hard sphere liquid, as stipulated by the famous Weeks-Chandler-Anderson result on dominance of repulsive interactions in determining liquid structure.

where $l = (r - R_0)/R_0$ and $l^\dagger = 1/30$. Further details about the derivation of this equation can be found in Ref. [43]. For $g(r)$ near contact ($l < 0.1$), we use liquid theory valid in the dense hard sphere fluid, for which semi-analytical expressions for the $g(r)$ are available [48] and we calculate κ using the Asakura-Oosawa (AO) potential [49].

This formula applies to a homogeneous attractive glass with $\phi \gtrsim 0.4$. Upon decreasing the volume fraction, structural heterogeneity becomes important and mesoscopic structure becomes identifiable, with an average linear size R_c, that we call "cluster size" even though in certain contexts it could be more appropriate to talk of a "correlation length." The characteristic size of the elastic building blocks is now R_c because the "clusters" are internally more rigid than the tenuous points of contacts with neighboring clusters, as a result of a gel-formation process controlled by clustering kinetics [41].

This is also what one gets by considering a cluster-glass gelation mechanism that involves a double-ergodicity breaking scenario, as the one explored in [42], with local aggregation of the colloidal particles to form "renormalized beads" (clusters) which, in turn, arrest due to either caging or residual attraction (in the latter case the glass transition would be energy-driven on both levels). The major assumption is that clusters are stabilized from coalescing due to the dynamical arrest inside them [42]. They are viewed as compact (spherical or quasi-spherical) renormalized particles of diameter $\tilde{R}_0$, whose effective volume fraction may be identified with the one determined by the spheres enclosing the clusters (i.e., significantly larger than ϕ). If the cluster linear size is larger than the particle diameter by a factor say less than 10, each contact between clusters is likely to reduce to a single colloid–colloid bond. Upon neglecting, (i) the breakup probability within the cluster (an assumption both experimentally and numerically verified in [41]), and (ii) the effect of long-range repulsion, the effective interaction between clusters obviously reduces to the bare colloid–colloid interaction, in agreement with [42].

Furthermore, the mean coordination will change to $z(\phi_c)$ (where now ϕ_c is the volume fraction occupied by the clusters), but its form can be still determined as in Eq. (2.85) if ϕ_c is still in the dense glassy regime. The values of ϕ_c are identified by enforcing the reasonable requirement that the cluster volume fraction range be upper-bounded by the random close packing limit $\phi_c \equiv \phi_{\text{RCP}} = 0.64$. Hence, for the modulus of the cluster-glass, we can write

$$G_{cg} = (2/5)\pi^{-1}\phi_c z(\phi_c)\tilde{\kappa}\tilde{R}_0^{-1} \qquad (2.86)$$

with $\tilde{\kappa} \simeq \kappa$ under the assumption of small clusters, and κ denotes the spring constant between two colloids, while $\tilde{R}_0$ denotes the center-to-center separation between two nearest-neighbor clusters. The situation is schematically depicted in Fig. 2.15. Equation (2.86) gives the elastic modulus of the material, where the macroscopic elasticity is dominated by the mesoscopic structure of the gel.

The above formula can be tested against extensive experimental data of Ref. [50] for a system of colloidal silica particles with polystyrene chains as depletant agent

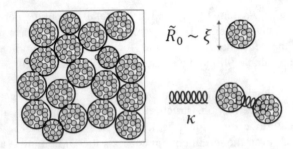

Fig. 2.15 Schematic representation of the elastic model of dense gels/cluster glasses. Clusters as renormalized elastic building blocks are highlighted as enclosed by black circles. Under certain assumptions, the spring constant between two neighboring clusters reduces to the colloid–colloid spring constant κ

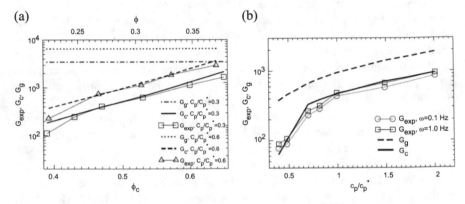

Fig. 2.16 Panel (**a**) shows the shear modulus as a function of ϕ_c (cluster volume fraction) and ϕ (colloid volume fraction) for the experimental data from [50] and for the model prediction from Eqs. (2.85) and (2.86), with the same physical parameters used in the experiments: $\tilde{R}_0/R_0 = 5.5$, polymer-to-colloid size ratio $\xi = 0.078$. Panel (**b**) shows the shear modulus of the attractive glass experimentally measured in [51] as a function of the attraction strength, expressed in terms of normalized depletant polymer concentration c_p/c_p^*, at two different frequencies (symbols), together with model predictions from Eqs. (2.85) and (2.86) with polymer-to-colloid size ratio $\xi = 0.08$ according to [51]. Reproduced from Ref. [43] with permission from the American Physical Society

in organic solvent (decalin), in the range $0.2 \lesssim \phi \lesssim 0.4$. Good agreement with Eq. (2.86) is obtained both for G as a function of colloid volume fraction and for G as a function of the normalized concentration of polymer depletant c_p/c_p^*, which sets the strength of attraction, and the spring constant κ. The comparison is shown in Fig. 2.16.

A further, in-depth, experimental verification of Eq. (2.86) was presented by Furst, Swan, Solomon, and co-workers [52] by means of experimental and simulation techniques, again on a set of depletion gels. In this study, graph-theoretic techniques were employed in order to precisely determine the cluster size R_c (or

Fig. 2.17 Experimental test of Eq. (2.86) using values of κ and $\tilde{R}_0 \equiv 2\xi$ directly accessed from experiments and simulations. The top panel shows the quantitative parameter-free comparison between the theoretically predicted (crosses) and experimentally measured (circles) values of shear modulus. The middle and bottom panels show the experimentally (filled circles) and numerically (open circles) determined values of κ and $\tilde{R}_0 \equiv 2\xi$, respectively. Adapted from Ref. [52]

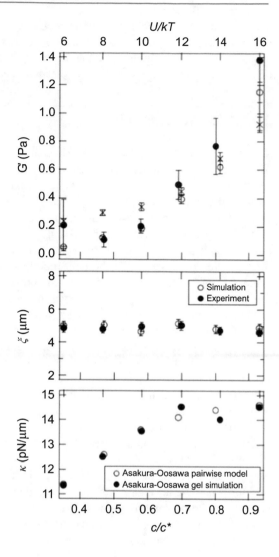

correlation length ξ) in depletion gels at $\phi = 0.20$, where the cluster identities are not readily discernible to the naked eye. The key parameters κ and $R_c \equiv \xi$ were determined experimentally and double-checked by numerical simulations and then used as input parameters in Eq. (2.86) to generate a parameter-free quantitative comparison with the experimental data, which is reported in Fig. 2.17. Also in this case $z(\phi_c)$ was estimated using liquid state theory, namely the Carnahan-Starling equation of state.

Finally, the cluster-glass model and its elastic properties described above play a role also in a different context of vortex depinning transition in type-II and type-1.5 superconductors. In those systems, 2D vortices of supercurrent are formed via the Abrikosov mechanism and often interact via effective attractive interactions mediated by the Lorentz force. Hence the vortices can form various phases, including the cluster-glass phase. Elastic models such as the one described above can be used to further understand the depinning and possibly also the vortex lattice melting transitions in these systems as discussed, e.g., in [53].

2.7   The Bulk Modulus

We have seen in Sect. 2.4.2 that the shear modulus of random jammed packings of soft spheres is quantitatively described by the same theory which describes the shear modulus of a random network of elastic (harmonic) springs. This fact has been verified also numerically by means of simulations in [14] and earlier in numerical work [17]. It was noticed already in the work of [17], however, that things are very different for the bulk modulus. While the shear modulus of random jammed packings and random spring networks are identical, their bulk modulus is quite different. In particular, while the bulk modulus of random networks displays the same scaling $\sim (z - 6)$ as the shear modulus (and quantitatively predicted by the theory that we developed at the level of Sec. 2.3.2 as shown in [10]), the bulk modulus of jammed packings scales, approximately, just as $\sim z$ and remains finite at the jamming (isostatic) transition $z = z_c = 2d$.

It was argued in [17] that this difference could be explained in terms of particle correlations which, in jammed packings, originate from self-organization of the packing under the isotropic pressure field. These correlations originate ultimately from the excluded-volume active in jammed packings, since particles have a finite volume and cannot interpenetrate or overlap upon decreasing the center-to-center separation distance, due to the increasing, high energetic cost set by the repulsive interaction. This is of course different in random spring networks where the nodes of the network have zero volume and one has much higher freedom in placing nearest-neighbors due to the absence of excluded volume. Due to the restricted freedom in finding particle arrangements set by excluded volume in jammed packings, new spatial correlations among particles arise. This situation is schematically depicted in Fig. 2.18, which also illustrates the different spatial organization of the "cage" of nearest-neighbors that surround a given particle.

The nonaffine theory thus needs to be reformulated in order to take these correlations induced by excluded volume into account properly for the case of hydrostatic compression.

Equation (2.47) was derived for random networks where bonds have randomly distributed orientations in the solid angle. In that model, any bond-orientational order parameter (which measures the "spread" in the bond orientations) is identically zero and there is full $O(3)$ symmetry for the bond unit vectors. This symmetry is broken for the more general case where correlations between bond-orientation vectors of nearest-neighbors are important.

In jammed packings, two distinct bonds (pairs of nearest-neighbor particles) ij and lm that are far apart are most likely uncorrelated, in the sense that the orientation of ij does not depend on the orientation of lm or vice versa. If, however, the two

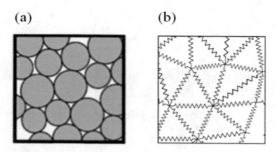

Fig. 2.18 Schematic contrast between particles packings (**a**), where excluded volume is important and there is higher likelihood of having particles that are nearly mirror-image of each other across a particle that sits in between, and (**b**) a random network where excluded volume is not active and having particles that are almost mirror-image of each other is very unlikely

bonds share a particle, such as ij and iq, they are no longer uncorrelated due to the local excluded-volume effect which forbids the overlapping of particles j and q. Therefore, the orientation of ij must depend on the orientation of iq because particles j and q cannot overlap in the excluded portion of the solid angle around i. Excluded-volume thus leads to an additional correlation term in the expression for the nonaffine part of the elastic constants, which now reads as

$$C_{\iota\xi\kappa\chi}^{NA} = 3\kappa R_0^2 \frac{N}{V} \sum_{\alpha=x,y,z} \left(A_{\alpha,\iota\xi\kappa\chi} + B_{\alpha,\iota\xi\kappa\chi} \right). \tag{2.87}$$

This equation is an extended version of Eq. (2.47), with an additional, new contribution given by the terms $A_{\alpha,\iota\xi\kappa\chi}$, which are due to excluded-volume correlations. We shall see in the following that this contribution is basically vanishing for shear deformations but is finite and provides a significant contribution in the case of hydrostatic compression, where ultimately it is responsible for the quasi-affine scaling $K \sim z$, where K is the bulk (compression) modulus.

The correlation term was derived in [54] and is given by

$$A_{\alpha,\iota\xi\kappa\chi} = \left\langle n_{ij}^{\alpha} n_{ij}^{\iota} n_{ij}^{\xi} n_{iq}^{\alpha} n_{iq}^{\kappa} n_{iq}^{\chi} \right\rangle$$

$$= \int_{\Omega} \int_{\Omega-\Omega_{cone}} \rho_{ij}(\Omega_{ij} \mid \Omega_{iq}) \rho_{iq}\left(\Omega_{iq}\right) n_{ij}^{\alpha} n_{ij}^{\iota} n_{ij}^{\xi} n_{iq}^{\alpha} n_{iq}^{\kappa} n_{iq}^{\chi} d\Omega_{ij} d\Omega_{iq},$$

$$\tag{2.88}$$

where $\langle ... \rangle$ denotes the angular averaging over bond-orientations and Ω_{cone} denotes an excluded-volume cone in the solid angle (see below).

Here we use $\rho_{ij}(\Omega_{ij} \mid \Omega_{iq})$ to denote the *conditional* probability that ij has orientation Ω_{ij} given that iq has orientation Ω_{iq}. Ω denotes the total solid angle centered on the spherical particle i, while Ω_{iq} denotes an orientation in the solid angle determined by the pair of angles (θ_{iq}, ϕ_{iq}), and, analogously, Ω_{ij} is defined

(a)   (b)

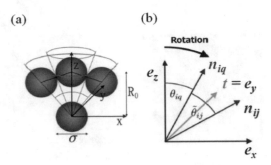

Fig. 2.19 (a) The excluded-volume cone: a bond, for example along the z-axis, leads to an excluded-cone, where no third particle can exist. R_0, as usual, denotes the average (center-to-center) distance between two nearest-neighbor particles, σ represents the diameter of the particles. (b) The frame-rotation trick to evaluate the contributions of local excluded-volume correlations to the nonaffine elastic moduli. Here, for simplicity, only the special case of $\phi_{ij} = \phi_{iq} = 0$, i.e., both ij and iq lying in the plane xz, has been shown. As we have $\phi_{iq} = 0$, the rotation axis depicted by **t** is identical to the y-axis (characterized by its unit vector $\mathbf{e}_y$). In the new coordinate frame, $\mathbf{n}_{ij}$ is easily parametrized by ϕ_{ij} (azimuthal angle) and θ_{ij} (polar angle). To get back to the non-rotated coordinate frame, one uses $n_{ij} = \mathbf{R} \cdot \mathbf{n}_{ij,rot}$ where $n_{ij,rot}$ is the bond unit vector in the rotated frame depiction. Adapted from Ref. [55]

by the pair (θ_{ij}, ϕ_{ij}). To evaluate the above integral, we first need to identify the correlation between ij and iq and then devise a strategy to evaluate the integral in the above equation.

Due to excluded-volume correlations, if $\rho_{iq}(\phi_{iq}, \theta_{iq}) = \frac{1}{4\pi}$, then the orientation of ij must depend on the orientation of iq, and the two orientations cannot be completely independent. In other words, the probability ρ_{iq} is a free probability, while ρ_{ij} is a conditional probability. The angular space available to ij, once the orientation of iq has been fixed, is restricted to $\Omega - \Omega_{cone}$, where Ω_{cone} denotes the excluded-volume cone centered on the bond iq, cfr. Fig. 2.19a. Hence, the conditional probability density for ij can be written as

$$\rho_{ij}(\Omega_{ij} \mid \Omega_{iq}) = \frac{1}{4\pi - \Omega_{cone}} = \frac{1}{4\pi - \pi \left(\frac{\sigma}{R_0}\right)^2}. \tag{2.89}$$

In the above expression, the excluded-volume cone follows from its definition in terms of the angle ψ which defines the quarter of the aperture of the cone,

$$\psi = \arcsin\left(\frac{\sigma}{2R_0}\right). \tag{2.90}$$

The solid angle sector which defines the excluded cone is given by

$$\Omega_{cone} = \int_{\theta=0}^{2\psi} \int_{\phi=0}^{2\pi} \sin\theta d\theta d\phi \qquad (2.91)$$

$$= \pi \left(\frac{\sigma}{R_0}\right)^2. \qquad (2.92)$$

For a random sphere packing where the particles are barely touching, the diameter σ is approximately equal to the equilibrium distance R_0, and $\rho_{ij}(\Omega_{ij} \mid \Omega_{iq}) \simeq 1/3\pi$.

Next, we consider the following trick, first presented in Ref. [55], to simplify the integral in Eq. (2.88). By exploiting the overall rotational invariance, the local Cartesian frame centered on the particle i is rotated such that the Cartesian z-axis (from which the azimuthal angles θ_{ij} and θ_{iq} are measured) is brought to coincide with the unit vector $\mathbf{n}_{iq}$ defining the orientation of the bond iq. Figure 2.19b provides a schematic illustration of the special case where iq and ij lie in the xz plane. This trick reduces the number of variables in the problem: instead of dealing with two sets of angles, $\{\varphi_{ij}, \theta_{ij}\}$ and $\{\varphi_{iq}, \theta_{iq}\}$, we need to consider only one set $\{\tilde{\varphi}_{ij}, \tilde{\theta}_{ij}\}$, which gives the orientation of the bond ij in the rotated frame. Upon suitably defining the rotation matrix, the above integral is much simplified.

This procedure will allow us to reduce it to a solvable integral with well-defined integration limits in the solid angle. Therefore, by exploiting the global, rigid-body, rotational invariance of the system, we change the coordinate frame for the integration over the solid angle in Eq. (2.88), to a new frame where the polar (directed along the z-axis) unit vector $\mathbf{e}_z$ coincides with the unit vector $\mathbf{n}_{iq}$. This is tantamount to rotating the coordinate frame into a new frame, where the parametrization of the allowed (not excluded) sector of the solid angle is much simpler, and one only has to exclude the polar angle from 0 to 2ψ.

The rotation is defined around an axis $\mathbf{t}$, parallel to $\mathbf{e}_y$, and perpendicular to both $\mathbf{e}_z$ and $\mathbf{n}_{iq}$, with an angle of θ_{iq} (usual convention of rotation: counterclockwise if the axis vector points in the direction of the viewer). Therefore, the unit vector $\mathbf{t}$ which defines the rotation axis is given by

$$\mathbf{t} = \frac{\mathbf{e}_z \times \mathbf{n}_{iq}}{|\mathbf{e}_z \times \mathbf{n}_{iq}|}. \qquad (2.93)$$

Here $\times$ indicates the usual cross product of vectors. Using this equation, the unit vector $\mathbf{t}$ defining the rotation axis can be calculated:

$$\mathbf{t} = \begin{pmatrix} -\sin(\phi_{iq}) \\ \cos(\phi_{iq}) \\ 0 \end{pmatrix}. \qquad (2.94)$$

Figure 2.19b schematically illustrates this rotation for the example $\phi_{iq} = 0$ and $\phi_{ij} = 0$. It can be seen from the figure that the vector $\mathbf{n}_{ij}$, i.e., the ij-bond unit vector in the non-rotated frame, is given through the ij-bond unit vector in the rotated frame $\mathbf{n}_{ij,rot}$ via

$$\mathbf{n}_{ij} = \mathbf{R} \cdot \mathbf{n}_{ij,rot}. \tag{2.95}$$

The rotation matrix $\mathbf{R}$ is defined by Rodrigues' rotation formula [56]:

$$\mathbf{R} = \cos\left(\theta_{iq}\right) \mathbf{1} + \sin\left(\theta_{iq}\right) [\mathbf{t}]_\times + \left(1 - \cos\left(\theta_{iq}\right)\right) \mathbf{t} \otimes \mathbf{t}. \tag{2.96}$$

Additionally, the following definitions are used, where t_x, t_y, and t_z represent the components of the x-axis, y-axis, and z-axis, respectively,

$$\mathbf{t} \otimes \mathbf{t} = \begin{pmatrix} t_x^2 & t_x t_y & t_x t_z \\ t_x t_y & t_y^2 & t_y t_z \\ t_x t_z & t_y t_z & t_z^2 \end{pmatrix} \tag{2.97}$$

$$= \begin{pmatrix} \sin^2\left(\phi_{iq}\right) & -\sin\left(\phi_{iq}\right)\cos\left(\phi_{iq}\right) & 0 \\ -\sin\left(\phi_{iq}\right)\cos\left(\phi_{iq}\right) & \cos^2\left(\phi_{iq}\right) & 0 \\ 0 & 0 & 0 \end{pmatrix}$$

$$[\mathbf{t}]_\times = \begin{pmatrix} 0 & -t_z & t_y \\ t_z & 0 & -t_x \\ -t_y & t_x & 0 \end{pmatrix} \tag{2.98}$$

$$= \begin{pmatrix} 0 & 0 & \cos\left(\phi_{iq}\right) \\ 0 & 0 & \sin\left(\phi_{iq}\right) \\ -\cos\left(\phi_{iq}\right) & -\sin\left(\phi_{iq}\right) & 0 \end{pmatrix}.$$

Next, we shall consider the integral I defined by

$$I_{\alpha\iota\xi} = \int_{\Omega - \Omega_{cone}} n_{ij}^\alpha n_{ij}^\iota n_{ij}^\xi \sin\left(\theta_{ij}\right) d\theta_{ij} d\phi_{ij}. \tag{2.99}$$

This integral shows up in the expression for $A_{\alpha,\iota\xi\kappa\chi}$, Eq. (2.88). Upon noticing that $\rho_{ij}(\Omega_{ij} \mid \Omega_{iq}) = const$ in the allowed solid angle $\Omega - \Omega_{cone}$ for ij, we can factor $\rho_{ij}(\Omega_{ij} \mid \Omega_{iq}) = const$ out of the ij integral leaving a product between $I_{\alpha\iota\xi}$ and $\rho_{ij}(\Omega_{ij} \mid \Omega_{iq})$ inside the integral for $A_{\alpha,\iota\xi\kappa\chi}$,

$$A_{\alpha,\iota\xi\kappa\chi} = \int_\Omega \int_{\Omega - \Omega_{cone}} \rho_{ij}(\Omega_{ij} \mid \Omega_{iq}) \rho_{iq}\left(\Omega_{iq}\right) n_{iq}^\alpha n_{iq}^\iota n_{iq}^\xi n_{ij}^\alpha n_{ij}^\kappa n_{ij}^\chi d\Omega_{ij} d\Omega_{iq}$$

$$= \int_\Omega I_{\alpha\iota\xi} \rho_{ij}(\Omega_{ij} \mid \Omega_{iq}) \rho_{iq}\left(\Omega_{iq}\right) n_{iq}^\alpha n_{iq}^\kappa n_{iq}^\chi d\Omega_{iq}. \tag{2.100}$$

As the next step, one can rewrite the $I_{\alpha\iota\xi}$ integral by transforming from spherical coordinates to Cartesian coordinates:

$$
\begin{aligned}
I_{\alpha\iota\xi} &= \int_{\Omega-\Omega_{cone}} n_{ij}^{\alpha} n_{ij}^{\iota} n_{ij}^{\xi} \sin\left(\theta_{ij}\right) d\theta_{ij} d\phi_{ij} \\
&= \int_{\Omega-\Omega_{cone}} \int_{r_{ij}=0}^{\infty} n_{ij}^{\alpha} n_{ij}^{\iota} n_{ij}^{\xi} \delta\left(r_{ij}-R_0\right) \sin\left(\theta_{ij}\right) d\theta_{ij} d\phi_{ij} dr_{ij} \\
&= \int_{V-V_{cone}} n_{ij}^{\alpha} n_{ij}^{\iota} n_{ij}^{\xi} \delta\left(r_{ij}-R_0\right) dx_{ij} dy_{ij} dz_{ij}.
\end{aligned}
\tag{2.101}
$$

We now define the transformation of rotated coordinates $\mathbf{r}_{ij,rot}$ to non-rotated coordinates $\mathbf{r}_{ij}$ as

$$
\mathbf{r}_{ij} = \Phi\left(\mathbf{r}_{ij,rot}\right) \equiv \mathbf{R} \cdot \mathbf{r}_{ij,rot},
\tag{2.102}
$$

where $\mathbf{R}$ is the rotation matrix defined in Eq. (2.96). In the following, $D\Phi$ denotes the Jacobian matrix of the function Φ and $\det(D\Phi)$ is the determinant of the Jacobian. Using the transformation theorem in three dimensions [57], one obtains:

$$
\begin{aligned}
I &= \int_{[V-V_{cone}]_{rot}} n_{ij}^{\alpha} n_{ij}^{\iota} n_{ij}^{\xi} \delta\left(r_{ij,rot}-R_0\right) \left|\det(D\Phi)\right| dx_{ij,rot} dy_{ij,rot} dz_{ij,rot} \\
&= \int_{[V-V_{cone}]_{rot}} n_{ij}^{\alpha} n_{ij}^{\iota} n_{ij}^{\xi} \delta\left(r_{ij,rot}-R_0\right) dx_{ij,rot} dy_{ij,rot} dz_{ij,rot} \\
&= \int_{[\Omega-\Omega_{cone}]_{rot}} \int_{\tilde{r}_{ij}=0}^{\infty} n_{ij}^{\alpha} n_{ij}^{\iota} n_{ij}^{\xi} \delta\left(\tilde{r}_{ij}-R_0\right) \sin\left(\tilde{\theta}_{ij}\right) d\tilde{\theta}_{ij} d\tilde{\phi}_{ij} d\tilde{r}_{ij} \\
&= \int_{[\Omega-\Omega_{cone}]_{rot}} n_{ij}^{\alpha} n_{ij}^{\iota} n_{ij}^{\xi} \sin\left(\tilde{\theta}_{ij}\right) d\tilde{\theta}_{ij} d\tilde{\phi}_{ij} \\
&= \int_{\tilde{\theta}_{ij}=\theta_{\min}}^{\pi} \int_{\tilde{\phi}_{ij}=0}^{2\pi} n_{ij}^{\alpha} n_{ij}^{\iota} n_{ij}^{\xi} \sin\left(\tilde{\theta}_{ij}\right) d\tilde{\theta}_{ij} d\tilde{\phi}_{ij}.
\end{aligned}
\tag{2.103}
$$

Here we used $D\Phi = \mathbf{R}$ and $|\det(\mathbf{R})| = 1$ for rotation matrices $\mathbf{R}$ [57]. The angle $\theta_{\min}$ is determined by the excluded volume cone [54] as

$$
\theta_{\min} = 2\psi = 2\arcsin\left(\frac{\sigma}{2R_0}\right).
\tag{2.104}
$$

We recall that n_{ij}^{α} is defined as the α-th Cartesian component of the bond unit vector $\mathbf{n}_{ij}$ and is related to the bond unit vector of the rotated frame $\mathbf{n}_{ij,rot}$ via $\mathbf{n}_{ij} = \mathbf{R} \cdot \mathbf{n}_{ij,rot}$. In the rotated frame, $\mathbf{n}_{ij,rot}$ is given by

$$\mathbf{n}_{ij,rot} = \begin{pmatrix} \sin\left(\tilde{\theta}_{ij}\right)\cos\left(\tilde{\phi}_{ij}\right) \\ \sin\left(\tilde{\theta}_{ij}\right)\sin\left(\tilde{\phi}_{ij}\right) \\ \cos\left(\tilde{\theta}_{ij}\right) \end{pmatrix}. \tag{2.105}$$

Therefore, we can now use Eq. (2.103) together with Eq. (2.100) to arrive at the following expression for the correlation term $A_{\alpha,\iota\xi\kappa\chi}$:

$$\begin{aligned} A_{\alpha,\iota\xi\kappa\chi} &= \left\langle n_{ij}^{\alpha} n_{ij}^{\iota} n_{ij}^{\xi} n_{iq}^{\alpha} n_{iq}^{\kappa} n_{iq}^{\chi} \right\rangle \\ &= \int_{\Omega}\int_{\Omega-\Omega_{cone}} \rho_{ij}\rho_{iq} n_{iq}^{\alpha} n_{iq}^{\iota} n_{iq}^{\xi} n_{ij}^{\alpha} n_{ij}^{\kappa} n_{ij}^{\chi} d\Omega_{ij} d\Omega_{iq} \\ &= \int_{\theta_{iq}=0}^{\pi}\int_{\phi_{iq}=0}^{2\pi}\int_{\tilde{\theta}_{ij}=2\psi}^{\pi}\int_{\tilde{\phi}_{ij}=0}^{2\pi} \rho_{ij}\rho_{iq} n_{iq}^{\alpha} n_{iq}^{\iota} n_{iq}^{\xi} \\ &\quad \times n_{ij}^{\alpha} n_{ij}^{\kappa} n_{ij}^{\chi} \sin\left(\tilde{\theta}_{ij}\right)\sin\left(\theta_{iq}\right) d\tilde{\theta}_{ij} d\tilde{\phi}_{ij} d\theta_{iq} d\phi_{iq}. \end{aligned} \tag{2.106}$$

With the last identity, Eq. (2.106), we have reduced the original integral for $A_{\alpha,\iota\xi\kappa\chi}$ to a much simpler, manageable integral with well-defined integration limits in the solid angle.

Also, we have now reached an expression for $A_{\alpha,\iota\xi\kappa\chi}$ as an integral of two independent sets of variables, namely $\{\tilde{\theta}_{ij}, \tilde{\phi}_{ij}\}$ and $\{\theta_{iq}, \phi_{iq}\}$. The integration can be easily done numerically or analytically and yields the following values for the excluded-volume correlation coefficients $A_{\alpha,\iota\xi\kappa\chi}$:

$$\begin{array}{c|c|c|c} \alpha & x & y & z \\ \hline A_{\alpha,xxxx} & -0.0304 & -0.00357 & -0.00357 \\ A_{\alpha,xyxy} & -0.00357 & -0.00357 & -0.000149 \\ A_{\alpha,xxyy} & -0.00982 & -0.00982 & -0.00327 \end{array} \tag{2.107}$$

These coefficients can now be implemented in Eq. (2.87) to determine the nonaffine contributions to the elastic constants in the presence of excluded-volume correlations.

We shall also note that the affine contribution to the elastic tensor, $C_{\iota\xi\kappa\chi}^{A}$, in

$$C_{\iota\xi\kappa\chi} = C_{\iota\xi\kappa\chi}^{A} - C_{\iota\xi\kappa\chi}^{NA} \tag{2.108}$$

is not affected by the excluded-volume correlations. Hence, we just need to evaluate $C_{\iota\xi\kappa\chi}^{NA}$ by means of Eq. (2.87) with the correlation coefficients of Eq. (2.107). For a shear deformation, i.e., $\iota\xi\kappa\chi = xyxy$, it turns out from the calculation that the excluded-volume correction is one order of magnitude smaller than the standard nonaffine term of Eq. (2.48). Physically, this is a manifestation of the anisotropy of the shear field, which leaves a small projection of the interparticle forces in the direction of two diametrically opposed bonds. Therefore, there cannot be an efficient "cancellation" of the affine forces under shear, which, in turn, leaves nonaffinity, and the $(z - 2d)$ scaling basically intact.

The non-zero, though small, correction predicted by the analytical theory to the shear modulus scaling might be an artifact due to model approximations used in the above derivation. For example, we always overestimate the excluded-volume cone by not considering the deformability of the soft particles in jammed packings. If this was properly taken into account, it would lead to a smaller excluded-volume cone and weaker correlations, hence to a higher nonaffinity than predicted in this approximation. In turn, that would yield an even smaller, practically negligible correction due to excluded volume in the case of shear. Another, though related, source of inaccuracy is the neglect of deviations from the average nearest-neighbor distance R_0.

In the case of hydrostatic compression, the isotropic character of the external field preserves the abovementioned cancellation of affine forces Ξ on a given particle and leads to a significant correction to the nonaffine component, as we are going to show next.

Upon recalling that the bulk modulus is defined in terms of components of the rank-4 elastic tensor as $K \equiv \frac{1}{3}(C_{xxxx} + 2C_{xxyy})$, for the bulk modulus, we obtain [55]

$$K = \frac{1}{18}\kappa\sigma^2\frac{N}{V}(z - 6) + K_{\mathrm{corr}}, \qquad (2.109)$$

with $K_{\mathrm{corr}} = 0.087$, in units of $\kappa R_0^2(N/V)$. This is now a significantly larger correction, basically of the same order of magnitude as the other contributions to the modulus. The reason again lies in the fact that the forces transmitted by neighbors are on average cancelling each other effectively under isotropic compression, thus leaving smaller values of the affine forces acting on a given particle.

These theoretical predictions recover, and explain, the known effect of vanishing of the ratio G/K at the rigidity transition of jammed packings [17].

Predictions of the above theory for the bulk modulus are shown in Fig. 2.20.

The above theory provides a quantitative prediction of moduli and of the discontinuous jump of the bulk modulus at the jamming transition, quantified by K_{corr}. We introduce the shorthand $\beta = (1/30)\kappa\sigma^2(N/V)$ and $\alpha = (1/18)\kappa\sigma^2(N/V)$ for the prefactors of G and K, respectively, for convenience of notation. Recalling that κ has units of [N/m], σ is a length, and N/V is in units of [m^{-3}], it is clear that α and β are measured in units of Pascal, although here we discuss their calculated values in units of $\kappa\sigma^2(N/V)$. Calculating the slope in $G \approx \beta(z - z_c)$, we find

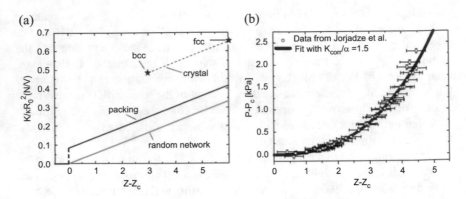

Fig. 2.20 (a) Comparison between bulk modulus for random sphere packings and for random networks. For comparison, also the values of simple crystals (fcc and bcc) are shown. (b) Comparison between the theory and the experimental data of compressed emulsions of [58]. Only one fitting parameter (the prefactor) is used in the comparison. Adapted from Ref. [55]

$\beta \approx 0.60$ for the shear modulus, in good agreement with the value $\beta \approx 0.75$ found in the simulations of [59]. For the jump in the bulk modulus at jamming, using the short-hand $K \approx \alpha(z - z_c) + K_{corr}$, our theory gives $K_{corr}/\alpha = 1.50$, which is of the right order of magnitude but smaller than the value $K_{corr}/\alpha = 4.50$ given in [59]. This discrepancy might be due to the obviously different approximations and assumptions done in numerical simulation protocols, which were discussed at the beginning of this section.

One can compare the theoretical prediction for the jump of the bulk modulus at the jamming transition with experimental data on compressed emulsions [58]. In the experiment, different values of pressure applied to the packing were recorded, and the values of z corresponding to the different pressure values P were measured using a fluorescent dye in the interparticle contacts between emulsion droplets. The output of this measurement is a curve relating $\delta P = P - P_c$ to $\delta z = z - z_c$, where we have to interpret z_c as the limit of isostaticity. The bulk modulus is defined in terms of pressure and coordination z via $K = -V(dP/dV) = -V(\frac{dP}{d\delta z})\frac{d\delta z}{dV}$. There is a one-to-one mapping between the volume fraction occupied by the drops, ϕ, and the contact number, z, in compressed emulsions, which was determined empirically in [58] to be $\delta z = z_0\sqrt{\delta\phi}$, with $z_0 = 10.6$, for that system. Using this relation and the definition of volume fraction $\phi = V_{drops}/V$, one obtains: $d\delta z/dV = -z_0 V_{drops}/2\sqrt{\delta\phi}V^2 = -z_0^2\phi/2\delta z V$. Upon replacing in the formula for K, we finally have a relationship between K, δz, and δP, given by $K = \frac{\phi z_0^2}{2\delta z}(\frac{dP}{d\delta z})$. We can thus replace the theoretical expression for $K = \alpha\phi\delta z + K_{corr}$, where α is the only fitting parameter containing the spring constant and integrate the differential equation to get

$$\delta P = P - P_c = \frac{K_{corr}}{z_0^2}(\delta z)^2 + \frac{2\alpha}{3z_0^2}(\delta z)^3. \tag{2.110}$$

The one-parameter fit comparison between the analytical theory, given by Eq. (2.110) and the experimental data of [58], is shown in Fig. 2.20b. The only fitting parameter is $\alpha \propto \kappa/R_0$ which is directly proportional to the spring constant of the drop–drop interaction, hence contains the dependence on the chemistry of the emulsion and on the local deformability of the emulsion drops, and is inversely proportional to the drop diameter. The theory accounts for the creation of excess contacts with pressure, and for the nonaffine particle rearrangements, and is able to provide a one-parameter fit of the data. In [58], the same data were modelled by accounting for the creation of excess contacts only, and neglecting rearrangements, which requires two adjustable parameters.

The above derivation provides a minimal model to account for geometric particle correlations in granular media, induced by the excluded volume between particles. More refined descriptions are no longer analytically solvable and must rely on master kinetic equations for the formation of particle contacts as the granular system evolves from a "floppy" structure to a denser one with a higher average coordination number z. As discussed in [60], such master kinetic equations for the time-evolution number of particles that surround a void can be solved numerically to yield a wealth of information about the structure evolution of granular packings, including the distribution of particles around a given particle.

2.8 Stress-Fluctuation Formalism for the Elastic Moduli of Thermal Systems

In Sect. 2.5.1 of this chapter we presented a schematic, approximate, approach to incorporating the effects of thermal fluctuations on the elastic modulus. We have done so by schematizing the intermolecular interactions as harmonic oscillators whose eigenfrequency gets renormalized by the applied shear strain, following an original derivation of [29]. We shall now present a more first-principles and systematic approach to the effect of thermal fluctuations on the nonaffine elasticity based on statistical mechanics, known as the stress-fluctuation formalism.

We start from the derivation valid for equilibrium phases of matter (e.g., crystals at finite temperature), and we shall then extend the results to nonequilibrium amorphous solids such as glasses.

2.8.1   General Formalism

We consider the scalar distance between two particles i and j in the deformed lattice, r_{ij}, and we write it in terms of the Lagrangian strain tensor defined in Eq. (2.8),

$$r_{ij} = \left[(r_i^\alpha - r_j^\alpha)(r_i^\beta - r_j^\beta) \, R_0^2 \, (2\eta_{\alpha\beta} + \delta_{\alpha\beta}) \right]^{1/2}. \tag{2.111}$$

Under the usual assumption that particles in the lattice interact in a pairwise fashion and that the interaction depends only on the distance between them and denoting $U(r_{ij})$ the potential energy of the interaction, then the canonical partition function is given by

$$Z = \frac{(2\pi m k_B T)^{3N/2}}{h^{3N}} \int \cdots \int d\mathbf{r}_m \dots d\mathbf{r}_n \exp\left(- \sum_{<m,n>} U(r_{mn})/k_B T \right), \tag{2.112}$$

where, as usual, m is the particle mass, N is the number of particles in the system, k_B is Boltzmann's constant, h is Planck's constant, and T is the temperature. If the phase is a crystal, the integrals are restricted to a small region around the lattice site. Furthermore, the summation runs over all pairs of interacting particles, as in previous derivations.

The Helmholtz free energy A is related to the partition function Z via the standard relation:

$$A = -k_B T \ln Z. \tag{2.113}$$

The above relations, Eqs. (2.111)–(2.112)–(2.113), can now be combined to determine the derivatives of the free energy, based on which one can evaluate the elastic constants. It should be noted that the Helmholtz free energy is indeed a free energy of deformation since the energy in Eq. (2.112) is evaluated corresponding to *deformed* distances r_{ij} expressed by Eq. (2.111). The result of the computation for the first derivatives is as follows:

$$\frac{1}{V}\left(\frac{\partial A}{\partial \eta_{\alpha\beta}} \right)_T = \left\langle \frac{R_0^2}{V} \sum_{\langle i,j \rangle} \frac{U'}{r_{ij}} (r_i^\alpha - r_j^\alpha)(r_i^\beta - r_j^\beta) \right\rangle \tag{2.114}$$

and, for the second derivatives,

$$
C_{\iota\xi\kappa\chi} = \frac{1}{V}\left(\frac{\partial^2 A}{\partial \eta_{\iota\xi}\partial \eta_{\kappa\chi}}\right)_T = \frac{R_0^4}{Vk_BT}\left\{\left\langle \sum_{\langle i,j\rangle}\frac{U'}{r_{ij}}(r_i^\iota - r_j^\iota)(r_i^\xi - r_j^\xi)\right\rangle\right.
$$

$$
\cdot \left\langle \sum_{\langle i,j\rangle}\frac{U'}{r_{ij}}(r_i^\kappa - r_j^\kappa)(r_i^\chi - r_j^\chi)\right\rangle -
$$

$$
-\left\langle\left[\sum_{\langle i,j\rangle}\frac{U'}{r_{ij}}(r_i^\iota - r_j^\iota)(r_i^\xi - r_j^\xi)\right]\cdot\left[\sum_{\langle i,j\rangle}\frac{U'}{r_{ij}}(r_i^\kappa - r_j^\kappa)(r_i^\chi - r_j^\chi)\right]\right\rangle\right\}+
$$

$$
+\frac{R_0^4}{V}\left\langle \sum_{\langle i,j\rangle}\frac{U''}{r_{ij}^2}(r_i^\iota - r_j^\iota)(r_i^\xi - r_j^\xi)(r_i^\kappa - r_j^\kappa)(r_i^\chi - r_j^\chi)\right\rangle -
$$

$$
-\frac{R_0^4}{V}\left\langle \sum_{\langle i,j\rangle}\frac{U'}{r_{ij}^3}(r_i^\iota - r_j^\iota)(r_i^\xi - r_j^\xi)(r_i^\kappa - r_j^\kappa)(r_i^\chi - r_j^\chi)\right\rangle,
$$

$$
(2.115)
$$

where $\langle\ldots\rangle$ now has a different meaning compared to the previous subsections and is used here to denote the canonical ensemble averaging:

$$
\langle X\rangle = Z = \frac{(2\pi mk_BT)^{3N/2}}{h^{3N}}\int\ldots\int d\mathbf{r}_m\ldots d\mathbf{r}_n X\exp\left(-\sum_{<m,n>}U(r_{mn})/k_BT\right)
$$

$$
(2.116)
$$

for a generic quantity X.

In addition to the terms in the above expressions, there are also kinetic terms that arise from the terms $-Nk_BT\ln V$ contained in the Helmholtz free energy, which have been omitted for brevity in the above equation. These contributions, for the stress tensor read as $-Nk_BT\delta_{\alpha\beta}$, and for the elastic constant tensor as $Nk_BT\delta_{\iota\xi}\delta_{\kappa\chi}$. Typically these contributions are much smaller than the contributions listed in Eq. (2.115), since they are proportional to k_BT, which is an energy scale in most solids much smaller than the bonding energy (with the important exception of weakly interacting colloids).

Upon including also the kinetic term and upon setting $U'' = \kappa_{ij}$ and $U' \equiv t_{ij}$, consistent with our previous notation, the expression for the shear modulus reads as

$$G = \frac{Nk_BT}{V} + \frac{1}{V}\left\langle \sum_{\langle i,j \rangle} \frac{\kappa_{ij}}{r_{ij}^2}(r_{ij}^x r_{ij}^y)^2 \right\rangle - \frac{1}{V}\left\langle \sum_{\langle i,j \rangle} \frac{t_{ij}}{r_{ij}^3}(r_{ij}^x r_{ij}^y)^2 \right\rangle -$$

$$- \frac{1}{Vk_BT}\left\{ \left\langle \left(\sum_{\langle i,j \rangle} \frac{t_{ij}}{r_{ij}} r_{ij}^x r_{ij}^y \right)^2 \right\rangle - \left\langle \left[\sum_{\langle i,j \rangle} \frac{t_{ij}}{r_{ij}} r_{ij}^x r_{ij}^y \right]^2 \right\rangle \right\} \tag{2.117}$$

where we can identify terms that we have already encountered. First of all, the kinetic term $\frac{Nk_BT}{V}$ is small for all atomic and molecular systems, as well as for colloids with binding energy $\gg k_BT$ and therefore can be often neglected. The other two terms on the r.h.s. in the first line are easily recognized to represent the affine Born-Huang modulus, given by Eq. (2.12), in our previous derivation in Sect. 2.2, i.e., what we called G_A. The identification is exact in the athermal limit where the ensemble averaging $\langle ... \rangle$ reduces to the orientational averaging over bond orientations.

The contributions on the second line of Eq. (2.117), instead, represent the nonaffine contribution to the modulus, what we used to call G_{NA}. In the above formula, this term can also be rewritten (using the covariance formula well known in statistics) as

$$G_{NA} = \frac{1}{Vk_BT}\left\langle \left(\sum_{\langle i,j \rangle} \frac{t_{ij}}{r_{ij}} r_{ij}^x r_{ij}^y - \left\langle \sum_{\langle i,j \rangle} \frac{t_{ij}}{r_{ij}} r_{ij}^x r_{ij}^y \right\rangle \right)^2 \right\rangle. \tag{2.118}$$

Upon using the Irving-Kirkwood formula for the stress

$$\sigma_{\alpha\beta} = \frac{1}{V} \sum_{\langle i,j \rangle} \frac{t_{ij}}{r_{ij}} r_{ij}^\alpha r_{ij}^\beta, \tag{2.119}$$

we thus arrive at the following expression for the nonaffine contribution to the shear modulus:

$$G_{NA} = \frac{V}{k_BT}\langle \sigma_{xy} - \langle \sigma_{xy} \rangle \rangle^2 \tag{2.120}$$

which is a simple, yet very useful formula for computing the nonaffine correction to the shear modulus in glasses, liquids and also in crystals, by means of molecular simulations.

The above expression was derived early on in [61] in a seminal contribution as an extension of the Born-Huang formalism to crystals at finite temperature (see the discussion below) and only later rediscovered in the context of amorphous solids.

2.8.2   The Temperature-Dependent Shear Modulus of Perfect Crystals

The fact that there could be a negative nonaffine contribution to the shear modulus also in perfect centrosymmetric crystals at finite temperatures is not in contradiction with our discussion about the microscopic origin of nonaffinity in the previous sections. The only way that $\langle \sigma_{xy} - \langle \sigma_{xy} \rangle \rangle^2$ does not vanish in a crystal is due to thermal fluctuations of the particles at finite temperature $T > 0$. The thermal fluctuations bring the particles instantaneously away from the lattice points in a random fashion, such that the randomness induced by thermal agitation breaks the inversion symmetry in any instantaneous snapshot. Therefore, the thermally induced breaking of inversion symmetry generates forces in the affine positions that need to be re-equilibrated by means of nonaffine motions. This explains, for example, the decrease of elastic constants of crystals with increasing temperature up to melting, where a melting criterion for crystals is indeed the vanishing of the shear modulus (the so-called Born melting criterion that we discussed also for glasses) [62]. Sample calculations that illustrate this effect for the case of fcc crystals of Lennard-Jones interacting particles (a good model for Argon crystals) are shown in Fig. 2.21.

2.8.3   The Case of Liquids

The above expressions for the elastic constants are very general for equilibrium states of matter and can be applied also to liquids. In this case, our daily life intuition tells us that the zero-frequency shear modulus of a simple liquid must be identically zero since liquids can flow under infinitesimal perturbations without opposing any rigid resistance. Within the above framework, this empirical fact must imply that the nonaffine, and stress-fluctuation term G_{NA} must be equal to the affine term G_A. This can be shown rigorously using the tools of equilibrium statistical mechanics (since

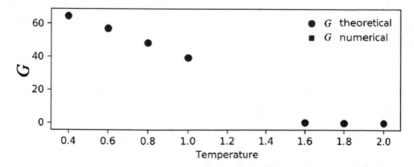

Fig. 2.21 Shear modulus G of fcc crystals of LJ interacting particles as a function of temperature T, computed from MD simulations. Numerical results refer to the direct measurement of G from box deformation in the simulations, whereas theoretical data points refer to the use of Eq. (2.117) using MD simulation data as input. Reproduced with permission from [63]

liquids are equilibrium phases of matter) together with the virial representation of stress valid for liquids. Let us redefine the *instantaneous* shear stress as

$$\hat{\sigma}_{xy} \equiv \frac{1}{V} \sum_{\langle i,j \rangle} \frac{t_{ij}}{r_{ij}} r_{ij}^x r_{ij}^y \tag{2.121}$$

and the ensemble-averaged shear stress as

$$\sigma_{xy} \equiv \frac{1}{V} \left\langle \sum_{\langle i,j \rangle} \frac{t_{ij}}{r_{ij}} r_{ij}^x r_{ij}^y \right\rangle. \tag{2.122}$$

Upon further noticing that $\frac{t_{ij}}{r_{ij}} = f_{ij}$ is a force, the following generic mathematical identity applies

$$\langle f(x) A(x) \rangle = -k_B T \langle A'(x) \rangle, \tag{2.123}$$

where x is some unconstrained coordinate, $A(x)$ some property, $f(x) = -U'(x)$ a force with respect to some energy $U(x)$, and the average is as usual Boltzmann weighted like in Eq. (2.116). Following [47], the instantaneous shear stress $\hat{\sigma}_{xy}$ can be expressed by means of the alternative virial representation valid for liquids:

$$\hat{\sigma}_{xy} = \sigma_{xy} - \frac{1}{V} \sum_{ij} r_{ij}^y f_{ij}^x. \tag{2.124}$$

The second term is sometimes called the "inner virial" and vanishes on average as can be seen using Eq. (2.123):

$$\langle r_{ij}^y f_{ij}^x \rangle = -k_B T \left\langle \frac{\partial r_{ij}^y}{\partial x} \right\rangle = 0, \tag{2.125}$$

which therefore leaves

$$\langle \hat{\sigma}_{xy} \rangle = \sigma_{xy} \tag{2.126}$$

as it should. Upon going back to the stress fluctuation formula Eq. (2.118) with the above results, one can perform integration by parts with respect to the Boltzmann integral and arrive at [30]

$$G_{NA} = -\frac{1}{V} \left\langle \sum_{\langle i,j \rangle} (r_{ij}^y)^2 \frac{\partial f_{ij}^x}{\partial r_{ij}^x} \right\rangle \tag{2.127}$$

and upon further recalling that

$$f_{ij}^x(r_{ij}) = -\frac{U'(r_{ij})r_{ij}^x}{r_{ij}},$$
(2.128)

one can then immediately retrieve that Eq. (2.127) reduces identically to the same expression that defines the affine shear modulus G_A and hence $G = G_A - G_{NA} = 0$ for a liquid. In the above steps we used the virial expression for the stress valid for a liquid [47]. Furthermore, in Eq. (2.127), the stress fluctuations stemming from different interactions are decoupled, which is obviously a valid assumption only for disordered systems and in particular for liquids. Finally, when we performed the integration by parts, we also had to require that all particle positions are unconstrained and independent (generalized) coordinates, which is possible at imposed stress but cannot hold at fixed strain. This implies that the above cancellation leading to $G = 0$ is also present in thermodynamic ensembles where the strain is kept fixed as discussed in [30] (see also below, Fig. 2.23 and the corresponding discussion, in the next section).

2.8.4 The Temperature-Dependent Shear Modulus of Glasses Revisited

In view of its derivation, and formulation, in terms of equilibrium statistical mechanics, the stress-fluctuation formalism that we presented above is therefore naturally and immediately applicable to equilibrium condensed states of matter, such as crystals and liquids. Its application to glasses, which are instead *nonequilibrium* states of matter (and typically quite far away from thermodynamic equilibrium [64]), presents some nontrivial challenges. These arise mostly from the need to suitably redefine the ensemble averaging for systems out of equilibrium. A possible pragmatic strategy has been developed in [65] and will be discussed in the following.

If the supercooled liquid is further cooled below the glass transition temperature, a full relaxation of time correlation functions, such as the intermediate scattering function, cannot be observed anymore within the experimentally available time window. This implies that the system is non-ergodic and cannot sample the entire phase space. To remedy this situation, the phase space of the system is divided into N_D subsystems or domains. Every domain is assumed to be in equilibrium, but between the domains, the system is out of equilibrium.

One then defines the probability density of the domain a as given by

$$f_a(\Gamma) = s_a(\Gamma)\frac{\exp(-\beta H(\Gamma))}{Z_a}$$
(2.129)

with the *restricted* partition function Z_a given by

$$Z_a = \int d\Gamma s_a(\Gamma) \exp(-\beta H(\Gamma)), \tag{2.130}$$

where H is the Hamiltonian of the system, $\beta = \frac{1}{k_B T}$ the Boltzmann factor, Γ the set of phase space coordinates, and s_a a switching function which is equal to one as long as Γ belongs to the domain (subsystem) a and zero otherwise. The probability distribution of the entire system is given by a composition of the single-domain distribution weighted with a nonequilibrium weight, as follows:

$$f(\Gamma) = \sum_{a=1}^{N_D} w_a f_a(\Gamma), \tag{2.131}$$

where w_a represents the nonequilibrium weight, and $\sum_{a=1}^{N_D} w_a = 1$.

With this reformulation for the phase-space averaging, the stress-fluctuation formalism for the zero-frequency shear modulus can be written as [65]

$$G = G_A - \frac{V}{k_B T} \sum_{a=1}^{N_D} w_a \left(\langle \sigma_{xy}^2 \rangle_a - \langle \sigma_{xy} \rangle_a^2 \right), \tag{2.132}$$

where G_A is the usual affine modulus given by the Born-Huang theory. The above formula can be used with numerical MD simulations to provide estimates of the zero-frequency shear modulus of glasses, by performing a set of simulations on different realizations of the glass. Under the assumption that every simulation is sampling a single domain and that the set of prepared samples is representative of the distribution of weights w_a, one can then evaluate Eq. (2.132) by simple time averages, by taking advantage of the fact that each subsystem is locally at equilibrium and hence locally ergodic (which implies that, therefore, the local ensemble average is locally equivalent to time-averaging). See also [66, 67].

As pointed out in [68], this method is somewhat sensitive to the choice of time window over which the time-averaging is taken. As an alternative to the above approach, one can resort to taking the stress-fluctuation expression Eq. (2.118) minus the stress–stress autocorrelation function $C(t) = \langle \sigma_{xy}(t)\sigma_{xy}(0) \rangle$ at $t = 0$, which decays to a non-zero plateau value for a broad class of glasses. This is equivalent to considering the stress fluctuations relative to the frozen-in stresses in the particular domain. The autocorrelation is calculated up to a cutoff time t_c which is much larger than the relaxation time of the fast processes in the sample. The physical intuition behind this correction is as follows. Each individual glassy realization contains frozen-in stresses, which are induced due to the initial conditions and cooling protocol of the sample. For instance, a fast cooling brings a liquid out of equilibrium very quickly. This does not leave enough time for stress relaxation processes to occur, leading to significant frozen-in stresses in the glass sample which might not be present if a slower cooling rate would have been used. While these residual stresses are sometimes deliberately introduced during the manufacturing process and influence mechanical/rheological measurements on individual glassy materials, they are not a characteristic property of the material but a by-product of its manufacturing process.

Here we should focus on the method embodied by Eq. (2.132) and we shall now schematically illustrate the calculation procedure.

For a domain a which is assumed to be in local thermodynamic equilibrium, we apply the stress-fluctuation formalism protocol to the subsystem and compute the shear stress fluctuation in the *restricted* canonical ensemble (recalling Eq. (2.116)):

$$\langle \sigma_{xy}^2 \rangle_a = \frac{1}{Z_a} \int \cdots \int d\mathbf{r}_m d\mathbf{r}_n s_a \sigma_{xy}^2 \exp\left(- \sum_{<m,n>} \beta U(r_{mn}) \right), \qquad (2.133)$$

where the restricted partition function is given by

$$Z_a = \int \cdots \int d\mathbf{r}_m d\mathbf{r}_n s_a \exp\left(- \sum_{<m,n>} \beta U(r_{mn}) \right). \qquad (2.134)$$

Now, both U and σ_{xy} can be Taylor-expanded around the so-called inherent state of each simulated sample. The inherent state or inherent structure (IS) is the local (deepest) minimum in the energy landscape of the system. The IS can be easily determined from the MD simulations since it is nothing but a set of coordinates $\{\mathbf{r}_i^{IS}\}$ for which the total energy U attains a local minimum.

Upon further inserting the Taylor expansions around $\{\mathbf{r}_i^{IS}\}$ back into Eq. (2.133) and manipulating the resulting Gaussian integrals, one arrives eventually at the following expression [68]:

$$\frac{V}{k_B T} \langle \sigma_{xy}^2 \rangle_f \approx V \sum_a w_a \left(\frac{(\sigma_{xy}^{IS,a})^2}{k_B T} + A + B k_B T + O(k_B T^2) \right), \qquad (2.135)$$

where A and B are coefficients that can be easily evaluated from the knowledge of the particle positions and interactions in the IS configurations [68]. Once again, the set of subsystems $\{a\}$ can be taken to coincide with individual realizations of the glassy system. The above formula can be used as a starting point for perturbative calculations of $G(T)$ valid in the low-temperature glassy regime. In Fig. 2.22, calculations are shown for the shear modulus of LJ glass as a function of T, which are based on the formulae, Eq. (2.132), together with perturbative calculations (red line) at low T using the method of Eq. (2.135).

The above expression, Eq. (2.135), suggests that the temperature dependence of the nonaffine stress-fluctuation term G_{NA} is rather nontrivial, with some terms that decrease with decreasing T and other terms which, instead, increase with decreasing T. Not only, but the temperature dependence of G_{NA} may differ depending on

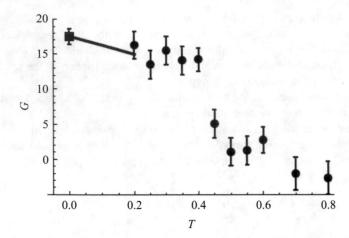

Fig. 2.22 Zero-frequency shear modulus G of the LJ glass computed using the nonequilibrium stress-fluctuation formula Eq. (2.132) for the nonaffine part, black symbols. The red line represents perturbative calculations at low T performed using Eq. (2.135). The glass transition temperature for this system is $T_c \approx 0.4$, and the shear modulus is basically zero in the equilibrium liquid at $T > 0.4$ in agreement with the discussion in Sect. 2.8.3. Adapted from Ref. [68] with permission of the American Institute of Physics

the interparticle interactions and even on the choice of the ensemble for the MD simulation.

In [30], the temperature dependence of the nonaffine term within the stress-fluctuation formalism has been studied in detail. Results are shown in Fig. 2.23 for LJ-type glasses.

It is seen in Fig. 2.23 that the temperature dependence of G_{NA} is indeed quite non-trivial and in particular is non-monotonic with a maximum occurring at the glass transition temperature T_g. The non-monotonicity could be explained with the competing contributions in Eq. (2.135), although more computational work is required to fully elucidate this feature. It is also seen that the affine component G_A decreases monotonically with increasing temperature, which is mostly controlled by thermal expansion and the associated decrease of density [30], thus confirming the picture of Sect. 2.4.3.

From the perspective of the analytical theory, not only the analytical prediction about G_A declining with temperature is confirmed but also the insensitivity of G_{NA} with temperature assumed in the analytical theory presented in Sect. 2.5.1 is also reasonably confirmed since G_{NA} in Fig. 2.23 is seen to vary very little with temperature up to very close to the glass transition from $T = 0$ all the way up to $T = 0.75T_g$. Also, the shear modulus becomes identically zero at the glass transition due to the exact balancing $G = G_A - G_{NA} = 0$ of affine and nonaffine contributions, as theorized in [10, 31].

The procedure discussed above for the evaluation of the partition function of a glassy system out of equilibrium is somewhat heuristic. A more formal approach

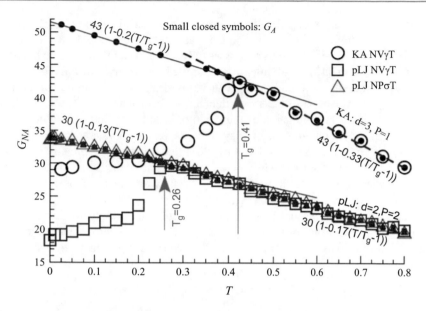

Fig. 2.23 The nonaffine part of the zero-frequency shear modulus, G_{NA} as a function of temperature T for three different LJ-type glass formers and ensembles. The corresponding affine parts G_A are represented as closed small symbols. Circles are simulation data for a Kob-Andersen (KA) binary mixture in the ensemble $NV\gamma T$ where the strain is fixed. Squares are polydisperse LJ particles in the same ensemble. Triangles are polydisperse LJ particles in the stress-controlled ensemble $NV\sigma T$. As discussed in [30], the latter ensemble always has $G = 0$ due to a cancellation similar to the one which occurs in the equilibrium liquid discussed in Sect. 2.8.3. Adapted from [30] with permission from the American Institute of Physics

is the one known as "cloned-liquid theory," which combines the replica method, to evaluate quenched-disorder averages, with standard liquid state theory. Also, in this case, one considers a cloned liquid with $a = 1, .., N_D$ replicas. In a nutshell, in order to evaluate elastic constants, one has to take ensemble averages of quantities derived from the free energy $A = -k_B T \ln Z$. Now, since the system is out of equilibrium and non-ergodic, the ensemble average can only be performed over different realizations (replicas) of the disorder and not as the standard canonical ensemble. In order to average the logarithm over many replicas, one resorts to the so-called replica trick:

$$\ln Z = \lim_{n \to 0} \frac{Z^n - 1}{n}, \tag{2.136}$$

which simplifies the calculation of $\overline{\ln Z}$ and reduces it to the much more manageable problem of calculating the disorder average $\overline{Z^n}$, where n is an integer which labels the replicas. A key assumption, which is also implicit in our presentation of Eq. (2.132), is that of *self-averaging*, whereby the average over a large realization of the system is indistinguishable from the average over many (smaller) realizations (replicas) of the disorder. Introducing replicas allows one to perform this macroscopic average over different disorder realizations. For example, the nonaffine contribution to the shear modulus of glasses turns out to be a self-averaging quantity as shown, e.g., in [13]. Building on the replica symmetry breaking theory of spin glasses, the cloned liquid theory can be used to provide a more formal and first-principles evaluation of the stress-fluctuation formalism to compute the shear modulus of glasses as a function of temperature. The theory [69, 70], which turns out be exact for infinite-dimensional systems, is very elaborate and goes beyond the scope of this book. We should also mention that the replica symmetry breaking approach to evaluate the free energy of deformation has been also fruitfully used in the mathematical theory of elasticity of polymer networks (e.g., rubber), for which the interested reader is referred to [71].

Finally, we should discuss how the stress-fluctuation formalism reduces to the nonaffine response result for G_{NA}, Eq. (2.87), in the limit of zero temperature. This can be readily seen as follows. First we shall rewrite the stress-fluctuation as

$$G_{NA} = \frac{V}{k_B T}\langle \sigma_{xy}^2 \rangle - \langle \sigma_{xy} \rangle^2 = \frac{V}{k_B T}\langle \delta\sigma_{xy}\delta\sigma_{xy}\rangle, \tag{2.137}$$

where $\delta\sigma_{xy} = \sigma_{xy} - \langle \sigma_{xy}\rangle$. In the athermal limit, the nonaffine contribution can be written as [72]

$$G_{NA} = V \sum_{p=1}^{3N-3} \frac{1}{\lambda_p} \left(\sum_i^N \mathbf{v}_i^p \cdot \frac{\partial \sigma_{xy}}{\partial \mathbf{r}_i} \right) \left(\sum_j^N \mathbf{v}_j^p \cdot \frac{\partial \sigma_{xy}}{\partial \mathbf{r}_j} \right) = \sum_{p=1}^{3N-3} \frac{V}{\lambda_p} \delta\sigma_{xy}^p \delta\sigma_{xy}^p, \tag{2.138}$$

where $\delta\sigma_{xy}^p = \sum_i^N \mathbf{v}_i^p \cdot \frac{\partial \sigma_{xy}}{\partial \mathbf{r}_i}$ is the fluctuation of the stress induced by the p-th eigenmode. One should note that the above equation is obtained by taking the limit $T \to 0$ in the finite $T > 0$ formulation given by Eq. (2.137), which involves performing this limit in the Boltzmann integral.

References

1. M. Born, K. Huang, *Dynamical Theory of Crystal Lattices* (Clarendon Press, Oxford, 1954)
2. K. Binder, W. Kob, *Glassy Materials and Disordered Solids* (World Scientific, Singapore, 2011)
3. F. Leonforte, R. Boissière, A. Tanguy, J.P. Wittmer, J.L. Barrat, Phys. Rev. B **72**, 224206 (2005)
4. L. Landau, E. Lifshitz, *Theory of Elasticity: Volume 6* (Pergamon Press, Oxford, 1986)
5. D.C. Wallace, *Thermodynamics of Crystals* (Wiley, New York, 1972)
6. A. Acharya, J. Bassani, J. Mech. Phys. Solids **48**(8), 1565 (2000)
7. J.A. Zimmerman, D.J. Bammann, H. Gao, Int. J. Solids Struct. **46**(2), 238 (2009)
8. A. Zaccone, J. Phys. Condens. Matter **21**(28), 285103 (2009)
9. C.E. Maloney, A. Lemaître, Phys. Rev. E **74**, 016118 (2006)
10. A. Zaccone, E. Scossa-Romano, Phys. Rev. B **83**, 184205 (2011)
11. B.A. DiDonna, T.C. Lubensky, Phys. Rev. E **72**, 066619 (2005)

12. G.R. Huang, B. Wu, Y. Wang, W.R. Chen, Phys. Rev. E **97**, 012605 (2018)
13. A. Lemaître, C. Maloney, J. Stat. Phys. **123**(2), 415 (2006)
14. R. Milkus, A. Zaccone, Phys. Rev. B **93**, 094204 (2016)
15. M. van Hecke, J. Phys. Condens. Matter **22**(3), 033101 (2009)
16. C.S. O'Hern, L.E. Silbert, A.J. Liu, S.R. Nagel, Phys. Rev. E **68**, 011306 (2003)
17. W.G. Ellenbroek, Z. Zeravcic, W. van Saarloos, M. van Hecke, EPL (Europhys. Lett.) **87**(3), 34004 (2009)
18. G.K. Batchelor, R.W. O'Brien, Proc. R. Soc. Lond. A. Math. Phys. Sci. **355**(1682), 313 (1977)
19. S. Alexander, Phys. Rep. **296**(2), 65 (1998)
20. E.M. Huisman, T.C. Lubensky, Phys. Rev. Lett. **106**, 088301 (2011)
21. V. Mazzacurati, G. Ruocco, M. Sampoli, Europhys. Lett. **34**(9), 681 (1996)
22. B. Cui, G. Ruocco, A. Zaccone, Granul. Matter **21**(3), 69 (2019)
23. S. Zhang, E. Stanifer, V. Vasisht, L. Zhang, E. Del Gado, X. Mao. Prestressed elasticity of amorphous solids, Phys. Rev. Research **4**, 043181 (2022)
24. C.T. Lee, M. Merkel, Soft Matter **18**, 5410 (2022)
25. M. Doi, S.F. Edwards, *The Theory of Polymer Dynamics* (Oxford University Press, Oxford, 1988)
26. P. de Gennes, *Scaling Concepts in Polymer Physics* (Cornell University Press, Ithaca, NY, 1979)
27. A.E. Likhtman, T.C.B. McLeish, Macromolecules **35**(16), 6332 (2002)
28. P.G. De Gennes, J. Polym. Sci. B Polym. Phys. **43**(23), 3365 (2005)
29. J. Frenkel, *Kinetic Theory of Liquids* (Oxford University Press, Oxford, 1955)
30. J.P. Wittmer, H. Xu, P. Polińska, F. Weysser, J. Baschnagel, J. Chem. Phys. **138**(12), 12A533 (2013)
31. A. Zaccone, E.M. Terentjev, Phys. Rev. Lett. **110**, 178002 (2013)
32. M. Wyart, Ann. Phys. Fr. **30**(3), 1 (2005)
33. X.Y. Cui, S.P. Ringer, G. Wang, Z.H. Stachurski, J. Chem. Phys. **151**(19), 194506 (2019)
34. M. Born, J. Chem. Phys. **7**(8), 591 (1939)
35. K. Schmieder, K. Wolf, Kolloid Z. **134**, 149 (1953)
36. I. Kriuchevskyi, J.P. Wittmer, H. Meyer, O. Benzerara, J. Baschnagel, Phys. Rev. E **97**, 012502 (2018)
37. W. Kob, H.C. Andersen, Phys. Rev. Lett. **73**, 1376 (1994)
38. K. Binder, Rep. Prog. Phys. **50**(7), 783 (1987)
39. W. Goetze, *Complex Dynamics of Glass-Forming Liquids: A Mode-Coupling Theory* (Oxford University Press, Oxford, 2009)
40. P.J. Lu, E. Zaccarelli, F. Ciulla, A.B. Schofield, F. Sciortino, D.A. Weitz, Nature **453**(7194), 499 (2008)
41. J. Rouwhorst, C. Ness, S. Stoyanov, A. Zaccone, P. Schall, Nat. Commun. **11**(1), 3558 (2020)
42. K. Kroy, M.E. Cates, W.C.K. Poon, Phys. Rev. Lett. **92**, 148302 (2004)
43. A. Zaccone, H. Wu, E. Del Gado, Phys. Rev. Lett. **103**, 208301 (2009)
44. Y. Kantor, I. Webman, Phys. Rev. Lett. **52**, 1891 (1984)
45. A.H. Krall, D.A. Weitz, Phys. Rev. Lett. **80**, 778 (1998)
46. W.H. Shih, W.Y. Shih, S.I. Kim, J. Liu, I.A. Aksay, Phys. Rev. A **42**, 4772 (1990)
47. R. Zwanzig, R.D. Mountain, J. Chem. Phys. **43**(12), 4464 (1965)
48. D. Henderson, E.W. Grundke, J. Chem. Phys. **63**(2), 601 (1975)
49. S. Asakura, F. Oosawa, J. Chem. Phys. **22**(7), 1255 (1954)
50. S. Ramakrishnan, Y.L. Chen, K.S. Schweizer, C.F. Zukoski, Phys. Rev. E **70**, 040401 (2004)
51. M. Laurati, G. Petekidis, N. Koumakis, F. Cardinaux, A.B. Schofield, J.M. Brader, M. Fuchs, S.U. Egelhaaf, J. Chem. Phys. **130**(13), 134907 (2009)
52. J.P. Pantina, E.M. Furst, Phys. Rev. Lett. **94**, 138301 (2005)
53. W. Wang, R. Díaz-Méndez, M. Wallin, J. Lidmar, E. Babaev, Phys. Rev. B **104**, 144206 (2021)
54. A. Zaccone, E.M. Terentjev, J. Appl. Phys. **115**(3), 033510 (2014)
55. M. Schlegel, J. Brujic, E.M. Terentjev, A. Zaccone, Sci. Rep. **6**(1), 18724 (2016)
56. D. Koks, *Explorations in Mathematical Physics* (Springer, New York, 2006)

57. K. Rektorys, *Survey of Applicable Mathematics* (Springer Netherlands, New York, 1994)
58. I. Jorjadze, L.L. Pontani, J. Brujic, Phys. Rev. Lett. **110**, 048302 (2013)
59. C.P. Goodrich, A.J. Liu, S.R. Nagel, Nat. Phys. **10**(8), 578 (2014)
60. C.C. Wanjura, P. Gago, T. Matsushima, R. Blumenfeld, Granul. Matter **22**(4), 91 (2020)
61. D. Squire, A. Holt, W. Hoover, Physica **42**(3), 388 (1969)
62. J. Wang, J. Li, S. Yip, D. Wolf, S. Phillpot, Physica A Stat. Mech. Appl. **240**(1), 396 (1997)
63. M. Xiong, X. Zhao, N. Li, H. Xu, Comput. Phys. Commun. **247**, 106940 (2020)
64. H.C. Oettinger, *Beyond Equilibrium Thermodynamics* (Wiley, New York, 2005)
65. S.R. Williams, D.J. Evans, J. Chem. Phys. **132**(18), 184105 (2010)
66. S. Abraham, P. Harrowell, J. Chem. Phys. **137**(1), 014506 (2012)
67. R. Dasgupta, H.G.E. Hentschel, I. Procaccia, Phys. Rev. Lett. **109**, 255502 (2012)
68. I. Fuereder, P. Ilg, J. Chem. Phys. **142**(14), 144505 (2015)
69. H. Yoshino, M. Mézard, Phys. Rev. Lett. **105**, 015504 (2010)
70. H. Yoshino, J. Chem. Phys. **136**(21), 214108 (2012)
71. N. Goldenfeld, *Lectures on Phase Transitions and the Renormalization Group* (CRC Press, Boca Raton, FL, 2018)
72. H. Mizuno, L.E. Silbert, M. Sperl, S. Mossa, J.L. Barrat, Phys. Rev. E **93**, 043314 (2016)

Viscoelasticity

3

Abstract

In this chapter, we will include viscous dissipation into the microscopic description of the mechanical response of disordered solids and arrive at a self-contained microscopic theory of the frequency-dependent viscoelastic moduli that can be compared with simulation and experimental data of real materials. Polymer glasses and metallic glasses will be considered as case studies. The same theory can also provide results for the relaxation creep modulus to rationalize the characteristic power-law creep behavior observed experimentally in disordered solids. In this chapter, the internal eigenfrequency of vibrational modes of atoms/molecules/grains in the solid will be denoted as ω_p to distinguish it from ω used for the external oscillatory frequency of the applied deformation.

3.1 Fundamentals of Linear Viscoelasticity

An ideal elastic solid responds to an applied strain γ with a stress σ, which is proportional to γ according to Hooke's law, $\sigma = G\gamma$, where G is the elastic modulus, and is independent of the rate at which the strain is applied. Classical elasticity theory provides a mathematical formulation of such behavior, cfr. Appendix A. At the opposite end of material's response, a viscous fluid responds with a stress that is proportional to the applied strain rate, $\dot{\gamma}$, through Newton's law of viscous liquids, $\sigma = \eta\dot{\gamma}$, with η the viscosity of the fluid. In the broad range comprised between these two limits, all materials that surround us in our daily life can be found, and, in fact, anything which exists in the Universe, with the exception of gases. These real materials therefore share features of both ideal fluids and ideal solids. Depending on the relative extent of their fluid rather than solid response, they can be classified as liquidlike or solidlike. In the first case, they typically lack shear rigidity on a long time scale of response, but they exhibit solidlike response at higher

© The Author(s), under exclusive license to Springer Nature Switzerland AG 2023
A. Zaccone, *Theory of Disordered Solids*, Lecture Notes in Physics 1015,
https://doi.org/10.1007/978-3-031-24706-4_3

rates of strain, whereas in the latter case, they possess shear rigidity on longer time scales, but they also exhibit significant viscous dissipation.

For an ideal elastic solid with no dissipation, the applied stress, as a function of time, and the resulting strain, as a function of time, are perfectly in phase. For a purely viscous fluid, instead, there is a 90-degree phase lag of the stress behind the applied strain. This is simply because if one takes the applied strain $\gamma(t)$ to be a sinusoidal function with frequency ω, $\gamma \sim \sin \omega t$, then according to the above quoted Newton's law, $\sigma \sim \dot{\gamma} = \frac{d}{dt} \sin \omega t = \cos \omega t$. Therefore, since $\cos \theta = \sin(\theta + \frac{\pi}{2})$, it is clear that in this case, the stress is not in phase with the applied strain but is instead lagging behind by 90°. Hence, all real materials will fall within these two limits and will be characterized by their own lag phase δ, which is an intrinsic material property, such that (for the case of applied strain):

$$\gamma = \gamma_0 \sin \omega t,$$
$$\sigma = \sigma_0 \sin(\omega t + \delta) \tag{3.1}$$

with $0 \leq \delta \leq \frac{\pi}{2}$.

According to Boltzmann's superposition principle in linear response theory, the stress at time t resulting from the application of strain at previous times t' is given by:

$$\sigma(t) = \int_{-\infty}^{t} G(t - t')\dot{\gamma}(t')dt \tag{3.2}$$

where $G(t)$ is the time-dependent elastic modulus, also called the "relaxation" modulus.

Hence, with the strain rate given by $\dot{\gamma}(t) = \omega \gamma_0 \cos(\omega t)$ and defining $t - t' = s$, one obtains:

$$\sigma(t) = \gamma_0 \int_{0}^{\infty} \omega G(s) \cos[\omega(t - s)]ds \tag{3.3}$$

Using the trigonometric identity $\cos(\theta \pm \phi) = \cos(\theta)\cos(\phi) \mp \sin(\theta)\sin(\phi)$, we arrive at the expression:

$$\frac{\sigma(t)}{\gamma_0} = \left[\omega \int_{0}^{\infty} G(s) \sin(\omega s)ds \right] \sin(\omega t) + \left[\omega \int_{0}^{\infty} G(s) \cos(\omega s)ds \right] \cos(\omega t), \tag{3.4}$$

which thus identifies:

$$G' = \omega \int_{0}^{\infty} G(s) \sin(\omega s)ds$$
$$G'' = \omega \int_{0}^{\infty} G(s) \cos(\omega s)ds. \tag{3.5}$$

The real and imaginary part, G' and G'', corresponds to the dissipationless and to the dissipative part of the response, respectively, and they are also known as the storage (elastic) modulus and the loss (viscous) modulus. While G' is by definition in phase with the stress, G'', instead, lags behind by 90°. For a generic stress wave $\sigma_0 e^{i\omega t}$, the strain wave lagging behind would be $\gamma_0 e^{(i\omega t + \frac{\pi}{2})}$, hence the factor $i \equiv e^{\frac{\pi}{2}}$ in front of G'', leading to the complex shear modulus defined as:

$$G^* = G' + iG''. \tag{3.6}$$

In the linear response regime, with intrinsic material properties that do not vary with time and with causality being obeyed, G' and G'' are related to each other by the Kramers-Kronig relations.

Applying the trigonometric identity $\sin(\theta \pm \phi) = \sin(\theta)\cos(\phi) \pm \cos(\theta)\sin(\phi)$ to $\sigma(t) = \sigma_0 \sin(\omega t + \delta)$ leads to:

$$\frac{\sigma(t)}{\gamma_0} = \frac{\sigma_0}{\gamma_0} \cos(\delta)\sin(\omega t) + \frac{\sigma_0}{\gamma_0} \sin(\delta)\cos(\omega t). \tag{3.7}$$

from which, by direct comparison with Eq. (3.4), we get that $G' \propto \cos\delta$ and $G'' \propto \sin\delta$, and:

$$\delta = \arctan\frac{G''}{G'}. \tag{3.8}$$

Furthermore, from Eqs. (3.4) and (3.5), we obtain:

$$\sigma(t) = G'\gamma(t) + \frac{G''}{\omega}\dot{\gamma}(t) \tag{3.9}$$

which, by comparison with Newton's law of viscous fluids, $\sigma = \eta\dot{\gamma}$, yields the following identification:

$$\eta = \frac{G''}{\omega} \tag{3.10}$$

between the viscosity η of the system and the loss modulus G''. The first part of Eq. (3.9) is nothing but Hooke's law and thus represents the elastic part of the response, with $G' \rightarrow G$ in the limit $G'' \rightarrow 0$, where G is the elastic shear modulus. Also, for an impulse Dirac-delta applied strain $\dot{\gamma} = \gamma_0 \delta(t)$ in Eq. (3.2) $G(t) = \frac{\sigma(t>0)}{\gamma_0}$ provides the response to a mechanical *creep* test.

Dynamical mechanical analysis (DMA) of a material is most conveniently performed in the frequency domain, by experimentally performing a sweep of frequencies ω in a certain range. This gives access to the viscoelastic moduli as functions of frequency, $G'(\omega)$ and $G''(\omega)$. Their typical behavior for glassy material of variable "hardness" is shown in Fig. 3.1.

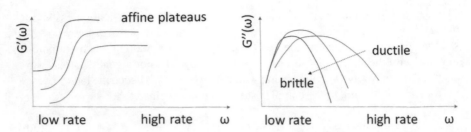

Fig. 3.1 Schematic representation of the viscoelastic moduli $G'(\omega)$ and $G''(\omega)$ for typical glassy disordered materials of varying rigidity/hardness. G' typically exhibits a plateau at low frequency, which is strongly affected by nonaffine motions, since at these low rates of oscillation/deformation, there is plenty of time for nonaffine displacements to take place and $G' \approx G_A - G_{NA}$; conversely, at high frequencies/rates, there is a higher plateau where $G' \approx G_A$ is basically affine, hence larger than the low-frequency plateau. The loss modulus G'' provides a picture of the dissipated energy in the oscillatory deformation process. The characteristic peak in G'' corresponds to a "resonance" where the applied frequency ω matches the inverse of some fundamental relaxation time scale of the system

Until recently, most viscoelastic models have been largely phenomenological, i.e., without any connection to the microscopic physics (and chemistry) of atoms, molecules, and interactions, bonding, and structuring thereof. The most important, historically, of such models has been the Maxwell model for viscoelastic liquids. This is a mathematical interpolation between the two limits of Hookean elastic solid and viscous Newtonian fluid, where an elastic Hookean element (spring) is connected in series with a dissipative element (a damper or dashpot). In such a series connection, the stress on each element is the same and equal to the total (imposed) stress, while the total strain γ is the sum of the strain in each element:

$$\sigma = \sigma_D = \sigma_E$$
$$\gamma = \gamma_D + \gamma_E. \tag{3.11}$$

Then, upon taking the time derivative of the strain equation and using the Hookean relation and the Newtownian one for the elastic (E) and dissipative (D) element, respectively, we get:

$$\frac{d\gamma}{dt} = \frac{d\gamma_D}{dt} + \frac{d\gamma_E}{dt} = \frac{\sigma}{\eta} + \frac{1}{G}\frac{d\sigma}{dt}, \tag{3.12}$$

which leads to the following (constitutive) relation:

$$\frac{\dot{\sigma}}{G} + \frac{\sigma}{\eta} = \dot{\gamma} \tag{3.13}$$

between stress, strain, elastic modulus, viscosity, and strain rate. Upon elementary integration, one has a characteristic exponential decay of the stress with time, with a characteristic (Maxwell) time scale τ_M:

$$\sigma(t) = \sigma_0 \exp(-t/\tau_M) \qquad (3.14)$$

where σ_0 denotes the initial stress at time $t = 0$. The relaxation modulus $G(t)$ can be obtained from this relation, by noting that at $t = 0$, only the spring will deform, while the initial stress and strain are related by $\sigma_0 = G(0)\gamma_0$. Therefore:

$$G(t) = \frac{\sigma(t)}{\gamma_0} = \frac{\sigma_0}{\gamma_0} \exp(-t/\tau_M) \qquad (3.15)$$

with a characteristic single-exponential decay in time identified as:

$$\tau_M \equiv \frac{\eta}{G}. \qquad (3.16)$$

The Maxwell time τ_M can be interpreted [1] as the time scale of an Arrhenius, thermally activated jumping over a barrier by which a molecule or atom jumps out of the transient "cage" formed by its nearest neighbors in the liquid.

At lower temperatures, more generically, one can describe the energy landscape of a supercooled liquid as a rugged function with many minima separated from each other by barriers (saddles) of variable height. This naturally introduces a distribution of barrier heights. If the main transport mechanism is diffusive, relaxation from a local potential energy well into another across the barrier is a simple exponential process $\sim \exp(-t/\tau)$ and $\tau \approx \tau_M$. At lower temperatures, upon approaching the glassy regime, this is no longer true, and there is, rather, a distribution of relaxation times $\rho(\tau)$ (induced by the distribution of barrier heights). Hence, in this case, the relaxation modulus will be a weighted average of the form [2]:

$$G(t) \approx \int_0^\infty \rho(\tau) \exp(-t/\tau) d\tau \approx \exp(-t/\bar{\tau})^\beta. \qquad (3.17)$$

If $\rho(\tau)$ is Gaussian distributed, it is very easy to see that $G(t)$ starts off as a simple exponential decay and then crosses over into a stretched-exponential with $\beta \approx 2/3$. This is a typical value of the stretching exponent that is observed in the α-relaxation of structural glasses. An assumption underlying Eq. (3.17) is that the relaxation spectrum $\rho(\tau)$ exists a priori and does not evolve with time on the observation time scale. In general, a stretched-exponential relaxation will result from the integral in Eq. (3.17) if the distribution $\rho(\tau)$ satisfies certain mathematical properties, which can be, more or less directly, related to the value of β as discussed, e.g., in Ref. [3]. While stretched-exponential relaxation is ubiquitous in the glassy and supercooled liquid states, it has not been possible to trace it back to a single well-defined

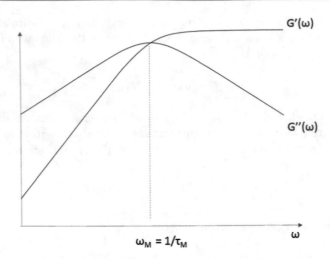

Fig. 3.2 The storage modulus, G', and the loss modulus, G'', as predicted by the Maxwell model, Eq. (3.19), plotted as a function of the oscillation frequency ω, for oscillatory shear deformations $\gamma = \gamma_0 \sin \omega t$. The storage modulus G' reaches an infinite-frequency plateau $G' \to G_\infty$, where the elasticity is basically affine, such that $G_\infty = G_A$. This fact was emphasized already by Y. Frenkel [1] and, at the microscopic level, e.g., in Ref. [5]

mechanism in the many-body dynamics or to a well-defined microscopic descriptor of the dynamics, and it remains a very active field of research.[1]

Upon Fourier transforming Eq. (3.2), we obtain:

$$G \, i\omega \, \gamma_0 = \left(i\omega + \frac{1}{\tau_M} \right) \sigma_0 \tag{3.18}$$

from which the complex shear modulus is identified explicitly as:

$$G^*(\omega) \equiv \frac{\sigma_0}{\gamma_0} = G' + iG'' = \frac{G_\infty \tau_M^2 \omega^2}{\tau_M^2 \omega^2 + 1} + i\frac{G_\infty \tau_M \omega}{\tau_M^2 \omega^2 + 1}. \tag{3.19}$$

Here we have denoted $G \equiv G_\infty$ to indicate that, in this case, G is what is left in the response in the infinite frequency limit, $\omega \to \infty$, where, thus, the response approaches the purely elastic limit.

The trends of G' and G'' predicted by the Maxwell model are sketched in Fig. 3.2.

From a more microscopic point of view (which of course is not present in the Maxwell model, which knows nothing about the molecular structure and dynamics of matter), this is again consistent with the fact that, in the infinite frequency limit,

---

[1] On a more fundamental level, the stretched-exponential relaxation in supercooled liquid is predicted by Mode-Coupling theory [4].

the response of a liquid is solidlike, and this plateau, shown in Fig. 3.19 in G', coincides with the affine plateau, as we shall see below when we will develop the microscopic theory. The fact that simple liquids respond to deformations like solids in the high frequency limit is a well-known fact to anyone who dives into water from an elevated height. This fact was emphasized, in the early days of liquid state theory, by Y. Frenkel in his early monograph on liquids and amorphous solids [1]. We also note that, again consistent with the phenomenology of liquids, the Maxwell model predicts $G' \to 0$ at $\omega \to 0$, i.e., the material is not rigid since it has a vanishing shear modulus at zero frequency/rate of deformation.

These facts are recovered by the Maxwell model, quite amazingly, without anything in the model which connects to the physics of real liquids and to their microscopic physics. The above facts bear also important consequences for a broader understanding of liquids. In particular, although solids, liquids, and glasses feature propagating longitudinal sound waves (though with different velocities), the dynamics of transverse (shear) waves in the three states of matter are very different. Solids exhibit propagating shear waves down to arbitrarily low momentum k, and their velocity is set by the shear elastic modulus G. Liquids, on the contrary, do not display propagating shear waves at low momenta, thus consistent with the fact that $G' = 0$ at ω and k going to zero. Nevertheless, propagating shear waves appear above a certain critical momentum, again consistent with the above picture provided by the Maxwell model, which has historically been called the Frenkel theory of liquids [1], or more recently, the k-gap theory developed by Trachenko and coworkers [6].

The starting point of the Frenkel or k-gap theory is the simple equation:

$$\omega(k)^2 + i\,\omega(k)\,\frac{1}{\tau} = c_s^2 k^2 , \tag{3.20}$$

which is known as *telegraph equation*, originally introduced by Heaviside in a different context. Here c_s is the speed of sound, and $\omega(k)$ is the frequency of sound waves (including, e.g., phonons) propagating through the material (not to be confused with the frequency ω of externally applied sinusoidal deformations considered in the previous paragraphs). This equation originates from the Fourier transform of the following one-dimensional manipulation of the Navier-Stokes-like equation originally developed by Y. Frenkel [1] to accommodate viscoelasticity à la Maxwell:

$$\eta\,\frac{\partial^2 v}{\partial x^2} = \rho\,\tau_M\,\frac{\partial^2 v}{\partial t^2} + \rho\,\frac{\partial v}{\partial t},$$

where v is the fluid velocity component in either the y or z direction, $\tau_M = \eta/G_\infty$ is the Maxwell viscoelastic relaxation time introduced above, and ρ is the density. Contrary to the standard Navier-Stokes equation, the above equation contains a second derivative in time, thus allowing for sound propagation with speed $c_s = \sqrt{G_\infty/\rho}$. This sound mode results from implementing Maxwell's so-called viscoelastic interpolation inside the Navier-Stokes equation. In other words, the above equation is a wave equation with viscous damping. The above Navier-Stokes-like equation was proposed by Frenkel without the support of a formal underlying theory and rather as an *ad hoc* phenomenological procedure. More precisely, it is not even clear if Frenkel's derivation of Eq. (3.20) is compatible with the usual conservation laws of standard hydrodynamics.

The same equation can be derived in the context of *generalized hydrodynamics* [7] but, again, in a phenomenological way, i.e., by postulating the Maxwell relaxation mechanism without a formal justification. Recently, a more formal derivation of the k-gap equation based on fundamental symmetries and on the topological defects induced by nonaffine motions in liquids was provided in Ref. [8].

Finally, other possibilities in terms of combining elastic (spring) and dissipative (dashpot) elements have been explored, with various predictions about different viscoelastic behaviors. For example, the Kelvin-Voigt model implies that a spring and a dashpot are added in parallel. By performing the same type of analysis that we did above for the Maxwell model, one arrives at a constitutive equation:

$$\sigma = G\gamma + \eta\dot{\gamma} \tag{3.21}$$

which clearly predicts a solidlike behavior, with fully developed shear rigidity at zero deformation rates/frequencies. This is a reasonably good model to describe a solid undergoing reversible, viscoelastic strain. In particular it is applicable to modelling of creep in polymer materials where, after application of a constant stress, the material slowly relaxes to its undeformed state.

The most important among these models for solids is, however, the Zener model, also referred to as the standard linear solid (SLS) model. Here, the original Maxwell model (i.e., a spring and a dashpot added in series) is added in parallel with a spring element. This effectively "corrects" the Maxwell model by imposing a fully rigid behavior at vanishing deformation rate/frequency, thanks to the additional spring added in parallel. The predicted behavior for G' and G'' as a function of the applied deformation frequency ω resembles that of the top (red) curves in Fig. 3.1. In the next section, we shall see how these behaviors can be obtained on a fundamental and microscopic level based on the nonaffine response theory of disordered solids, by extending the theory of Chap. 2 to include viscous dissipation at the microscopic level.

3.2   Microscopic Nonaffine Theory of Viscoelasticity

In this section, we shall extend the nonaffine elasticity theory of Chap. 2 to include dissipative effects at the microscopic level. The starting point is the writing of a suitable nonequilibrium (generalized Langevin) equation of motion for a tagged particle (atom, molecule, grain), which interacts with many other particles, the latter schematized as harmonic oscillators. This dynamical coupling, which, at the microscopic level, is mediated by the long-range part of the interatomic or intermolecular interaction forces, gives rise to a viscous-type friction term in the equation of motion and to a stochastic thermal noise.

3.2.1 Nonequilibrium Dissipative Equation of Motion

Our aim is to derive a suitable equation of motion for a tagged atom (or ion, molecule, particle) in a glass in response to an applied strain. This can be done by extending the Zwanzig-Caldeira-Leggett (ZCL) [9–11] approach to microscopic dynamics in disordered materials. In the construction of this approach, the dynamical coupling between the tagged particle and many other particles is an effective way of describing anharmonicity, and a mapping between the ZCL model Hamiltonian and real molecular systems can be demonstrated, although this mapping is not believed to be one to one [12].

In the ZCL approach, the Hamiltonian of a tagged particle (mass M, position Q, and momentum P) coupled to all other particles (treated as harmonic oscillators with mass M_m, position X_m and momentum P_m) in the material can be written as (cfr. Ref. [10]):

$$\mathcal{H} = \frac{P^2}{2M} + U(Q) + \frac{1}{2}\sum_m \left[\frac{P_m^2}{M_m} + M_m \omega_m^2 \left(X_m - \frac{\gamma_m Q}{M_m \omega_m^2} \right)^2 \right]. \tag{3.22}$$

This formalism can be extended to include the presence of externally applied field. The term describing the dynamical coupling between the tagged particle and the mth-oscillator is defined as γ_m. Introducing the mass-scaled tagged-particle displacement $s = Q\sqrt{M}$, the resulting generalized Langevin equation of motion for the displacement of the tagged particle becomes (cfr. [10] for the full derivation):

$$\ddot{s} = -U'(s) - \int_{-\infty}^{t} \nu(t - t')\frac{ds}{dt'}dt' + F_p(t), \tag{3.23}$$

where $F_p(t)$ is the thermal stochastic noise with zero average, U is a local interaction potential (e.g., with the nearest neighbors), and ν is the friction resulting from many long-range interactions with all other particles in the system, imposed by the dynamical bilinear coupling. For dynamical response to an oscillatory strain, one can average the dynamical equation over many cycles, which amounts to a time average. Since the noise F_p is defined to have zero mean [10], an average over many cycles leaves $\langle F_p \rangle = 0$ in the above equation.[2]

---

[2] According to Ref. [13], when the system is non-ergodic below T_g, nothing guarantees this is true a priori, but there is evidence that this approximation might be reasonable in the linear regime where the response converges to a reproducible noise-free average stress.

3.2.2   Derivation of Microscopic Viscoelastic Moduli

We rewrite Eq. (3.23) for a tagged atom in d-dimensions, which moves with an affine velocity prescribed by the strain-rate tensor $\dot{\mathbf{F}}$ (where the dot indicates a time derivative):

$$\ddot{\mathbf{r}}_i^{\mu} = \mathbf{f}_i^{\mu} - \int_{-\infty}^{t} \nu(t - t') \left(\dot{\mathbf{r}}_i^{\mu} - \mathbf{u}^{\mu} \right) dt' \tag{3.24}$$

where $\mathbf{f}_i^{\mu} = -\partial U / \partial \mathbf{r}_i^{\mu}$ generalizes the $-U'(s)$ to the tagged atom. Furthermore, we used the Galilean transformations to express the particle velocity in the moving frame: $\dot{\mathbf{r}}_i = \dot{\mathring{\mathbf{r}}}_i - \mathbf{u}$ where $\mathbf{u} = \dot{\mathbf{F}}\mathring{\mathbf{r}}_i$ represents the local velocity of the moving frame. This notation is consistent with the use of the circle on the particle position variables introduced in Chap. 2 to signify that they are measured with respect to the reference rest frame. In terms of the original rest frame $\{\mathring{\mathbf{r}}_i\}$, the equation of motion can be written, for the particle position averaged over several oscillations, as:

$$\mathbf{F}\ddot{\mathring{\mathbf{r}}}_i = \mathbf{f}_i - \int_{-\infty}^{t} \nu(t - t') \cdot \frac{d\mathring{\mathbf{r}}_i}{dt'} dt', \tag{3.25}$$

where $\langle F_p \rangle = 0$ was dropped for the reasons explained above.[3]

We work in the linear regime of small strain $\| \mathbf{F} - \mathbf{1} \| \ll 1$ by making a perturbative expansion in the small displacement $\{\mathbf{s}_i(t) = \mathring{\mathbf{r}}_i(t) - \mathring{\mathbf{r}}_i\}$ around a known rest frame $\mathring{\mathbf{r}}_i$. That is, we take $\mathbf{F} = \mathbf{1} + \delta\mathbf{F} + \dots$ where $\delta\mathbf{F} \approx \mathbf{F} - \mathbf{1}$ is the small parameter. Replacing this back into Eq. (3.25) gives:

$$(\mathbf{1} + \delta\mathbf{F} + \dots) \frac{d^2 \mathbf{s}_i}{dt^2} = \delta\mathbf{f}_i - (\mathbf{1} + \delta\mathbf{F} + \dots) \int_{-\infty}^{t} \nu(t - t') \cdot \frac{d\mathbf{s}_i}{dt'} dt'. \tag{3.26}$$

For the term $\delta\mathbf{f}_i$, imposing mechanical equilibrium again, which is $\mathbf{f}_i = 0$, implies:

$$\delta\mathbf{f}_i = \frac{\partial \mathbf{f}_i}{\partial \mathring{\mathbf{r}}_j} \delta\mathring{\mathbf{r}}_j + \frac{\partial \mathbf{f}_i}{\partial \eta} : \delta\eta \tag{3.27}$$

where in the first term we recognize:

$$\frac{\partial \mathbf{f}_i}{\partial \mathring{\mathbf{r}}_j} \delta\mathring{\mathbf{r}}_j = -\mathbf{H}_{ij}\mathbf{s}_j \tag{3.28}$$

[3] Terms $\ddot{\mathbf{F}}\mathring{\mathbf{r}}_i$ and $\int_{-\infty}^{t} \nu(t - t')\dot{\mathbf{F}} \cdot \mathring{\mathbf{r}}_i dt'$ are not allowed to enter the equation of motion because they depend on the position of the particle and therefore have to vanish for a system with translational invariance, as noted already in [14] and in [15].

while for the second term we have:

$$\mathbf{\Xi}_{i,\kappa\chi} = \frac{\partial \mathbf{f}_i}{\partial \eta_{\kappa\chi}} \tag{3.29}$$

and the limit $\eta \to 0$ is implied.

With these identifications, we can write Eq. (3.26), to first order:

$$\frac{d^2\mathbf{s}_i}{dt^2} + \int_{-\infty}^{t} \nu(t - t')\frac{d\mathbf{s}_i}{dt'}dt' + \mathbf{H}_{ij}\mathbf{s}_j = \mathbf{\Xi}_{i,\kappa\chi}\eta_{\kappa\chi}, \tag{3.30}$$

where $\eta_{\kappa\chi}$ are the components of the Cauchy-Green strain tensor defined as $\eta = \frac{1}{2}\left(\mathbf{F}^T\mathbf{F} - \mathbf{1}\right)$. This is a second-rank tensor, and should not be confused with the fluid viscosity η (a scalar), discussed in the previous sections. The above equation can be solved by Fourier transformation followed by a normal mode decomposition, as we shall see next. If we specialize on time-dependent uniaxial strain along the x direction, $\eta_{xx}(t)$, then the vector $\mathbf{\Xi}_{i,xx}\eta_{xx}(t)$ represents the force acting on particle i due to the motion of its nearest-neighbors, which are moving toward their respective affine positions (see e.g. [16] for a more detailed discussion). Hence, all terms in Eq. (3.30) are vectors in $\mathbb{R}^3$, and the equation is in manifestly covariant form. In metallic glasses, this "drag force" also includes electronic effects taken into account semi-empirically, e.g., via the embedded-atom model (EAM) presented in Chap. 1.

To make it convenient for further manipulation, we extend all matrices and vectors to $Nd \times Nd$ and Nd-dimensional, respectively, and we will select $d = 3$. After applying Fourier transformation to Eq. (3.30), we obtain:

$$-\omega^2\tilde{\mathbf{s}} + i\tilde{\nu}(\omega)\omega\tilde{\mathbf{s}} + \mathbf{H}\tilde{\mathbf{s}} = \mathbf{\Xi}_{\kappa\chi}\tilde{\eta}_{\kappa\chi}, \tag{3.31}$$

where $\tilde{\nu}(\omega)$ is the Fourier transform of $\nu(t)$, etc. (we use the tilde consistently throughout to denote Fourier-transformed quantities). In the above equation, all the terms are now vectors in $\mathbb{R}^{3N}$ space. Next, we apply normal mode decomposition in $\mathbb{R}^{3N}$ using the $3N$-dimensional eigenvectors of the Hessian as the basis set for the decomposition. This is equivalent to diagonalize the Hessian matrix $\mathbf{H}$. Proceeding in the same way as in the static elasticity case discussed in Chap. 2, we have that the m-th mode of displacement can be written as:

$$-\omega^2\hat{\tilde{s}}_m(\omega) + i\tilde{\nu}(\omega)\omega\,\hat{\tilde{s}}_m(\omega) + \omega_m^2\hat{\tilde{s}}_m(\omega) = \hat{\Xi}_{m,\kappa\chi}(\omega)\tilde{\eta}_{\kappa\chi}. \tag{3.32}$$

It was shown [16], by means of MD simulations for a LJ glass, that $\hat{\Xi}_{m,\kappa\chi} = \mathbf{v}_m \cdot \mathbf{\Xi}_{\kappa\chi}$ is self-averaging, and one might therefore introduce the smooth correlator function on eigenfrequency shells

$$\Gamma_{\mu\nu\kappa\chi}(\omega) = \langle\hat{\Xi}_{m,\mu\nu}\hat{\Xi}_{m,\kappa\chi}\rangle_{\omega_m \in \{\omega, \omega + d\omega\}} \tag{3.33}$$

on frequency shells. Following the general procedure of [16] to find the oscillatory stress for a dynamic nonaffine deformation, the stress is obtained to first order in strain amplitude as a function of ω (note that the summation convention is active for repeated indices):

$$
\begin{aligned}
\tilde{\sigma}_{\mu\nu}(\omega) &= C^A_{\mu\nu\kappa\chi}\,\tilde{\eta}_{\kappa\chi}(\omega) - \frac{1}{\mathring{V}}\sum_m \hat{\Xi}_{m,\mu\nu}\hat{\tilde{s}}_m(\omega) \\
&= C^A_{\mu\nu\kappa\chi}\,\tilde{\eta}_{\kappa\chi}(\omega) - \frac{1}{\mathring{V}}\sum_m \frac{\hat{\Xi}_{m,\mu\nu}\hat{\Xi}_{m,\kappa\chi}}{\omega_m^2 - \omega^2 + i\tilde{\nu}(\omega)\omega}\tilde{\eta}_{\kappa\chi}(\omega) \\
&\equiv C_{\mu\nu\kappa\chi}(\omega)\tilde{\eta}_{\kappa\chi}(\omega).
\end{aligned}
\tag{3.34}
$$

In the thermodynamic limit and assuming a continuous vibrational spectrum, we can replace the discrete sum over $3N$ degrees of freedom with an integral over vibrational frequencies up to the Debye (cutoff) frequency ω_D. In this case, we need to replace the discrete sum over the $3N$ degrees of freedom (eigenmodes) with an integral, $\sum_{m=1}^{3N}\ldots \rightarrow \int_0^{\omega_D} g(\omega_p)\ldots d\omega_p$, where $g(\omega_p)$ is the vibrational density of states (VDOS).[4]

Then, the complex elastic constants can be obtained as:

$$
C_{\mu\nu\kappa\chi}(\omega) = C^A_{\mu\nu\kappa\chi} - 3\rho \int_0^{\omega_D} \frac{g(\omega_p)\Gamma_{\mu\nu\kappa\chi}(\omega_p)}{\omega_p^2 - \omega^2 + i\tilde{\nu}(\omega)\omega}d\omega_p
\tag{3.35}
$$

where $\rho = N/\mathring{V}$ denotes the density of the solid in the initial state. This is a crucial result obtained in [17], which differs from a previous result obtained in [16] because the friction is non-Markovian, hence frequency or time/history-dependent, whereas in [16] it is just a constant, corresponding to Markovian dynamics. This turns out to be a fundamental difference, because as we shall see later, experimental data of real materials cannot be described by a constant friction coefficient. In the above we used mass-rescaled variables throughout, so that the atomic/molecular mass is not present explicitly in the final expression. If one, instead, uses non-rescaled variables and specializing to shear deformations, $\mu\nu\kappa\chi = xyxy$, we obtain the following expressions for the complex shear modulus G^*:

$$
G^*(\omega) = G_A - 3\rho \int_0^{\omega_D} \frac{g(\omega_p)\Gamma_{xyxy}(\omega_p)}{m\omega_p^2 - m\omega^2 + i\tilde{\nu}(\omega)\omega}d\omega_p.
\tag{3.36}
$$

In the above expression, the frequencies are now in physical units of Hertz. The first term on the r.h.s. is the affine shear modulus G_A, introduced in Chap. 2, which is independent of ω. The low-frequency behavior of $\Gamma(\omega_p)$ can be estimated analytically using Eq. (2.46) derived in Chap. 2 (and originally in Ref. [18]), which

---

[4] See Chap. 5 for discussion of the VDOS and related quantities.

gives $\hat{\Xi}_{p,xy}^2 \propto \lambda_p$, thus implying (from its definition above): $\Gamma(\omega_p) \propto \omega_p^2$. This analytical estimate appears to work reasonably well in the low-eigenfrequency part of the $\Gamma(\omega_p)$ spectrum [16, 19, 20].

By separating real and imaginary part of the above expression, we then get to the storage and loss moduli as [20]:

$$G'(\omega) = G_A - 3\rho \int_0^{\omega_D} \frac{m\, g(\omega_p)\, \Gamma(\omega_p)\, (\omega_p^2 - \omega^2)}{m^2(\omega_p^2 - \omega^2)^2 + \tilde{v}(\omega)^2 \omega^2} d\omega_p$$

$$G''(\omega) = 3\rho \int_0^{\omega_D} \frac{g(\omega_p)\, \Gamma(\omega_p)\, \tilde{v}(\omega)\, \omega}{m^2(\omega_p^2 - \omega^2)^2 + \tilde{v}(\omega)^2 \omega^2} d\omega_p.$$

(3.37)

It is easy to check that the storage modulus $G'(\omega)$ reduces to $G_A \equiv G_\infty$ in the infinite-frequency limit, $\omega \to 0$. From the point of view of practical computation, the VDOS $g(\omega_p)$ can be obtained numerically via direct diagonalization of the Hessian matrix $\mathbf{H}_{ij}$, since its eigenvalues are related to the eigenfrequencies via $\lambda_p = m\omega_p^2$. Similarly, the affine-force correlator $\Gamma(\omega_p)$ can also be computed from its definition by knowing the positions of all the particles, their interactions, and forces (so that the affine force fields Ξ can be computed) as well as the eigenvectors of the Hessian $\mathbf{v}_p$. Calculating eigenvalues and eigenvectors of the Hessian matrix is a computationally demanding task, especially in terms of RAM. Direct diagonalization (DD) of the Hessian is feasible on standard computers only up to $N \sim 10^4$ particles/atoms. As shown in Ref. [21], the RAM usage for direct diagonalization of the Hessian scales with the number of atoms as $\sim N^2$, i.e., very unfavorably. The computational time scales also very unfavorably, as $\sim N^{2.71}$. As a way to obviate this problem, a computational protocol based on the Kernel polynomial method (KPM) has been developed in [21], which is based on approximating eigenvector-based quantities of the Hessian with Chebyshev polynomials. With this methodology, it is possible to have a much more favorable scaling, i.e., linear in N, for both the RAM and the computational time. This makes it possible to compute $g(\omega_p)$ and $\Gamma(\omega_p)$ for much larger systems, i.e., $N > 10^5$, which would otherwise be impossible with direct diagonalization.

Illustrative calculations based on Eq. (3.37) for defective fcc lattices are shown in Fig. 3.3, for the simplified case of Markovian friction $v = const$ and purely harmonic bonds, with $\kappa_{ij} = 1$ (cfr. Eq. (2.2)), and $m = 1$. The VDOS $g(\omega_p)$ and the affine-force correlator $\Gamma(\omega_p)$ are computed from direct diagonalization of the Hessian for $N = 10^4$ atoms.

The calculations recover the expected behavior, with a nonaffine-to-affine transition being apparent in $G'(\omega)$, i.e., a crossover from a low-frequency (dominated by nonaffinity) plateau to a high-frequency plateau where $G' = G_A = G_\infty$ and the response is fully affine. This crossover corresponds to the maximum absorption peak in $G''(\omega)$. These features reproduce the generic trends predicted by the phenomenological Zener (standard linear solid) mentioned in the previous section.

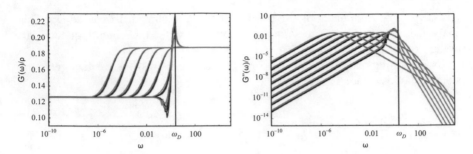

Fig. 3.3 The storage modulus, $G'(\omega)$ (left), and the loss modulus, $G''(\omega)$ (right), computed using Eq. (3.37) for fcc lattices with regularly depleted bonds (red), randomly depleted bonds (blue), and vacancies (green). All lattices have $z = 9$ on average: in the regularly depleted lattice, bonds are removed such that each atom has $z = 9$ exactly, while in the randomly depleted lattice, they are randomly cut according to a binomial distribution. The friction coefficient ν ranges from 10^2 to 10^5 with stepsize equal to one order of magnitude. The leftmost curve has $\nu = 10^5$, while the rightmost curve with the overshoot has the lowest friction, $\nu = 10^{-2}$. The friction has no impact on the shape of the curves but shifts the nonaffine-to-affine transition in G' to lower strain oscillation frequencies ω. An overshoot and undershoot arise when $\nu \sim \omega_D$. Courtesy of Dr. Rico Milkus

In the next sections, we shall apply the above theory to models of real materials such as polymer glasses and metallic glasses. In particular, we shall start from validating the above theory on a simulated amorphous material, for which one has full control over the interparticle interactions, and all the particle positions and dynamics, from MD simulations. Based on this input, the physical quantities entering Eq. (3.37) can then be taken from the simulations, and the calculated $G'(\omega)$ and $G''(\omega)$ based on Eq. (3.37) can then be compared, in a *parameter free* way, with the same quantities computed by the MD simulation. This will provide a stringent test of the theory and of the assumptions used in its derivation.

3.3   Case Study: Polymer Glasses

As discussed in Chap. 2, polymer glasses are complex systems due to the interplay between the random-coil structure of the closed-packing polymer chains with their covalent bonding along the chain [22] and weaker Lennard-Jones-type interactions between monomers belonging to different chains, which impart cohesion and rigidity to the material. Below the glass transition temperature T_g chain (Rouse), degrees of freedom and entanglements can be safely considered to be frozen-in, whereas they play a crucial role for the elastic properties at temperatures comparable to or larger than T_g. Nonetheless, chain constraints still have a certain importance in that, for example, the shear modulus may depend on the average chain length (i.e., the degree of polymerization or molecular weight), which controls the relative "concentration" of covalent bonds per monomer subunit.

While making quantitative predictions at the atomistic level for the viscoelastic moduli of polymer materials is still out of reach even for atomistic simulations in high-performance computing,[5] coarse-grained models have been very useful to scale down the mechanical response and to make sense out of phenomenological observations.

The most schematic of such coarse-grained models, and perhaps most useful, therefore, to arrive at more general physical principles, is offered by the so-called Kremer-Grest model [23]. In this framework, the subunits (monomers) are schematically modelled as spherical Brownian beads interacting with their neighbors along the chain via the finitely extensible nonlinear elastic (FENE) bond. It basically simplifies the chain of monomers by connecting a sequence of beads with nonlinear springs, while non-bonding interactions are represented by a shifted Lennard-Jones (LJ) pair potential. With reference to Fig. 2.9 in Chap. 2, the FENE bonds would correspond to the "double segments" along the chain, whereas the LJ bonds correspond to the red segments between monomers on nearby chains.

For the FENE potential $U_{\text{FENE}} = -0.5 K R_0^2 \ln \left[1 - \left(\frac{r}{R_0} \right)^2 \right]$, where the parameters used in the comparison below were set as $K = 30$, $R_0 = 1.5$. For LJ potential (cfr. Sect. 1.1.1 in Chap. 1), $U_{\text{LJ}} = 4\epsilon \left[\left(\frac{\sigma}{r} \right)^{12} - \left(\frac{\sigma}{r} \right)^6 - \left(\left(\frac{\sigma}{r_c} \right)^{12} - \left(\frac{\sigma}{r_c} \right)^6 \right) \right]$, where the constants were chosen to be $\epsilon = 1$, $\sigma = 1$ and the cutoff radius $r_c = 2.5$. This is enough to simulate flexible polymers, while for taking into full account the bending rigidity, potential terms like that in Eq. (1.2) must be added.

In a typical simulation, bead trajectories are updated according to simple Langevin dynamics:

$$\dot{\mathbf{p}}_i = \mathbf{f}_i - \nu \, \mathbf{p}_i + \mathbf{F}_p \qquad (3.38)$$

with $\mathbf{p}_i = \dot{\mathbf{r}}_i$ the momentum of particle i, $\mathbf{f}_i = -\sum_j \nabla U_{ij}$ the total force originating from interactions with other beads, and $\mathbf{F}_p$ the stochastic force, the time correlation of which obeys the standard Markovian fluctuation-dissipation theorem [10], by construction. This is basically Eq. (3.23) in the Markovian limit with $\nu(t-t') = const$. As usual, ϵ sets the LJ energy scale, and K is the bond energy scale, where $K/\epsilon = 30$. With reference to fundamental units of mass M, length d, and energy $\mathcal{E}$, we set $\sigma = 1$ and $m = 1$, giving a time unit of $\tau = \sqrt{m\sigma^2/\varepsilon}$. The system is equilibrated in a melted state at $T = 1.0$ (LJ units are used throughout), maintaining zero external pressure using a Nose-Hoover barostat. It is subsequently cooled down by decreasing T at rate $\tau_c \sim O(10^5)\tau$.

In simulations, one then obtains the viscoelastic moduli by mechanical spectroscopy or DMA, i.e., applying small amplitude oscillatory simple shear strain to the simulated box, as in Refs. [13, 24]. From the stress-strain curves, one can then

---

[5] Due to the shortness of time step, on the order of femtoseconds, only deformation rates/frequencies on the order of $\omega \gg 10^{10}\text{s}^{-1}$ can be accessed in atomistic MD simulations.

compute the storage G' and loss G'' moduli (cfr. Eqs. (3.7) and (3.8)):

$$G' = \frac{\sigma_0}{\gamma_0} \cos \delta, \quad G'' = \frac{\sigma_0}{\gamma_0} \sin \delta, \tag{3.39}$$

where σ_0 is the average amplitude of stress, γ_0 is the amplitude of strain (fixed at 2%), and δ is the phase shift between the two, introduced earlier in Sect. 3.1. All the dynamics in the simulations can be computed using the LAMMPS (large-scale atomic/molecular massively parallel simulator) code.

Static simulation snapshots of the Kremer-Grest polymer glass described above can be used as starting point to evaluate the viscoelastic moduli based on the microscopic nonaffine theory presented in the previous section. An important point, however, is to properly take into account the effects of temperature. The theory presented in the previous section is, in its essence, an athermal theory, since temperature T does not explicitly appear in the equations, whereas temperature is, of course, an essential physical parameter in the MD simulations.

From the point of view of lattice dynamics, the main effect of temperature is to displace (by random "thermal kicks") each particle (atom, molecule, grain) from the interaction attractive minimum with its nearest neighbors.

Schematically, let us consider two bonded particles, as depicted in Fig. 3.4.

When the particle is in the bonding minimum with the particle at the center (not shown in the cartoon), $dU/dr = 0$, and there is no net force, or in other words, the "bonding tension" is zero. The effect of temperature is to displace the particle either to the left or the right of the bonding minimum, toward regions where $dU/dr \neq 0$, thus giving rise to internal bond tensions, also referred to as "internal stresses", which are nothing but the t_{ij} terms discussed in Chap. 2 (cfr. Section 2.4.3). These forces clearly have to be relaxed or re-equilibrated to

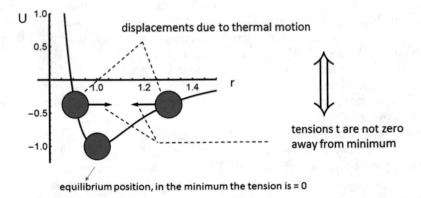

Fig. 3.4 Schematic illustration of the effect of temperature T on the lattice dynamics of glasses. Thermal kicks push the particle away from the interaction energy minimum, which leads to local mechanically unstable states associated with negative eigenvalues of the Hessian matrix. Increasing T leads to an increase in the population of these unstable modes, also referred to as instantaneous normal modes (INMs)

ensure mechanical equilibrium, which occurs via relaxation processes by which the particle tends to move closer to the minimum. In other words, temperature generates locally unstable states that need to be relaxed. Such local mechanically unstable states correspond to negative eigenvalues of the Hessian matrix. Recall that $\lambda_p = m\omega_p^2$, where m is mass and ω_p is the vibrational frequency. A negative eigenvalue, $\lambda_p < 0$, implies that the corresponding vibrational frequency ω_p is purely imaginary, $\omega_p = 0 + i\,\mathrm{Im}\omega_p$. Upon inserting this into a plane wave solution to the dynamical problem, we get:

$$\exp(i\omega_p t) \sim \exp(-\mathrm{Im}\omega_p t), \tag{3.40}$$

which, in turn, implies that the excitation decays in time exponentially, i.e., it has the character of an overdamped *relaxation* rather than a wave. Although somewhat simplistic, this argument gives an intuition about the fundamentally different nature of unstable modes compared with standard plane wave-like normal modes. These unstable modes associated with exponential relaxation are, clearly, spatially localized and are referred to in the liquid physics literature as instantaneous normal modes (INMs). Indeed, these modes have been intensively studied in the context of dynamics in the liquid state, where the atoms/molecules are much more mobile and perform much larger departures from local minima [25, 26].

In MD simulations of liquids (and glasses), the INMs are obtained as negative eigenvalues from direct diagonalization of the *instantaneous* Hessian matrix, hence the name. In practice, the longer one runs an MD simulation, the more likely the system will be in a local minimum of the potential energy landscape (PEL). Deep minima in the PEL of supercooled liquids and glasses are called "inherent structures". Based on the above discussion, it is clear that by driving the system into an inherent structure (typically by steepest descent energy minimization), i.e., through well-equilibrated simulations, the INMs are effectively washed out as the particles, on average, tend to be localized in the energy minima. If, instead, one performs shorter or not fully equilibrated simulations and then averages over many such simulations or snapshots, the INMs become clearly visible upon diagonalizing the Hessian matrix constructed based on the MD snapshots.[6] As we shall see next, retaining the INMs in the vibrational density of states (VDOS) $g(\omega_p)$ obtained from diagonalization of the Hessian is crucial in order to correctly describe the effect of temperature at the level of Eq. (3.37).

By applying this procedure, i.e., upon averaging over at least 10 MD snapshots that are not fully equilibrated in the inherent structures, for the KG polymer glass, one obtains the vibrational spectrum shown in Fig. 3.5 for different temperatures.

The VDOS is obtained based on input from MD of 50 bead-spring polymer chains with 100 monomer subunits per chain simulated in LAMMPS using a

---

[6] Here one should recall that the Hessian matrix $\mathbf{H}_{ij} = \frac{\partial^2 U}{\partial \mathbf{r}_i \partial \mathbf{r}_j}$ can be directly computed from a snapshot of an MD simulation, i.e., by taking as input the set of positions of all the particles in the snapshot $\{\mathbf{r}_i\}$ and the energy of the system U.

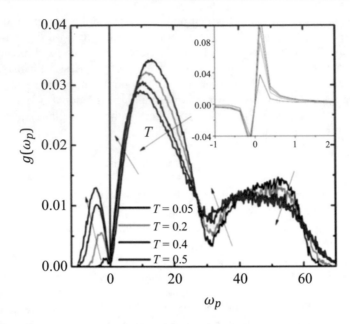

Fig. 3.5 Vibrational density of states (VDOS) of the Kremer-Grest model polymer glass. 50 bead-spring polymer chains with 100 monomer subunits per chain were simulated with MD in LAMMPS using a Langevin thermostat at different temperatures as specified in the legend. The glass transition for the system is $T_g \approx 0.4$, in Lennard-Jones units where $\sigma = 1$ and $\epsilon = 1$ (cfr. Chap. 1, Eq. (2.10)) [23]. The inset shows the VDOS normalized by the Debye law of solids (see Chap. 5), $g(\omega_p)/\omega_p^2$, with the so-called boson peak (see Chap. 5), i.e., an excess of low-energy modes over the Debye law, which increases with temperature. The INMs are plotted, as conventionally done, on the negative frequency axis [26]. Adapted, with modifications, from Ref. [19]

Langevin thermostat at different temperatures. The glass transition for the KG system is $T_g \approx 0.4$, in Lennard-Jones reduced units where $\sigma = 1$, $m = 1$, and $\epsilon = 1$ [23]. Two fairly distinct vibrational bands, at low frequency and high frequency, are visible, which correspond to large-scale phononic and other vibrational modes involving intermolecular LJ interactions (low-energy part of the spectrum) and intramolecular modes involving the FENE bonds (high-energy part of the spectrum), respectively. As per the standard convention used in the liquid physics literature, the INMs are plotted on the negative frequency axis. The population of INMs is clearly seen to increase with temperature as the glass transition is approached and crossed from below (i.e., from solid to liquid). As the temperature increases, also an excess of low energy modes in the low-energy part of the spectrum appears, which is known as the boson peak (this phenomenon, ubiquitous in glasses and also present in many crystals, will be discussed in more detail in Chap. 5).

Crucially, the frequency integrals for the viscoelastic moduli in Eq. (3.37) have to be evaluated by also including the INMs. This can be done either by rewriting the integral as a discrete sum over the set of frequencies ω_p (this is anyway a discrete

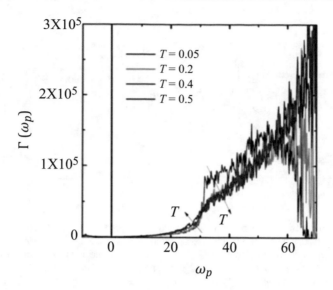

Fig. 3.6 The affine-force correlator, as defined by Eq. (3.33), and computed numerically based on averaging over MD snapshots of the Kremer-Grest polymer glass at temperatures (in Lennard-Jones units) specified in the legend. Adapted from Ref. [19]

set since they are obtained by numerical diagonalization of the Hessian matrix of a finite system) as in Ref. [19] or, alternatively, by using an integral in the complex plane so as to include the purely imaginary frequencies of the INMs, which lie on the imaginary axis [21].

In order to evaluate the integral for the viscoelastic moduli, one still has to compute also the affine-force correlator Γ_p. Recalling its definition from Eq. (3.33), it can be computed again based on MD snapshots which provide all necessary input to compute the $3N$-dimensional affine force vectors Ξ and, again by direct diagonalization of the Hessian, the eigenvectors required for the normal mode decomposition of the Ξ vectors. The result for the correlator $\Gamma(\omega_p)$ is shown in Fig. 3.6.

In the low-frequency part of the spectrum, the $\Gamma(\omega_p)$ approximately follows the quadratic $\sim\omega_p^2$ law predicted analytically in [18], with no significant variation with temperature.

Upon further recalling that the friction ν in Eq. (3.37) is actually a constant (Markovian) with frequency due to the (Markovian) Langevin thermostat used in the simulations, we now have all the ingredients to compute the viscoelastic moduli for the KG polymer glass using Eq. (3.37). All the quantities are therefore inferred from static MD simulation snapshots and from the Langevin thermostat (ν), and no free or adjustable parameter is present, which therefore provides a stringent test of the theory.

The comparison between the predictions of viscoelastic moduli from Eq. (3.37) and MD simulations of the oscillatory shear strain (using the "wiggle" command

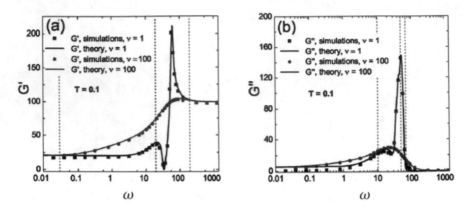

Fig. 3.7 Parameter-free comparison between theoretical predictions (continuous lines) of non-affine response theory based on Eq. 3.37 (using input from static MD simulations as discussed in the text) of the Kremer-Grest polymer glass (100 polymer chains each with 100 monomer subunits) and MD simulations (symbols) of the oscillatory shear strain. Panel (**a**) shows the comparison for the storage shear modulus G', panel (**b**) for the loss viscous modulus G''. Adapted with modifications from Ref. [19]

in LAMMPS) is shown in Fig. 3.7 as a function of frequency for a temperature ($T = 0.1$) well below the glass transition temperature ($T_g \approx 0.4$).

A perfect agreement, with no adjustable parameters, is obtained. Two different values of the friction constant ν of the Langevin thermostat used in the MD simulations are used to show the effect of friction on the DMA curves. The plot of G' clearly displays the crossover between the low-frequency plateau, dominated by nonaffine contributions to the elasticity (the negative term in the first of Eq. (3.37)), and the essentially affine high-frequency plateau where only G_A in the first of Eq. (3.37) survives, while the nonaffine integral vanishes in the limit $\omega \to \infty$. The plots of G'' exhibit the characteristic "absorption" peak in correspondence to the nonaffine-to-affine crossover in G'.

The theoretical calculations thus make use of the INMs in the vibrational spectrum as a successful way of reproducing thermal effects on the viscoelastic moduli. To further test the ability of this scheme, based on INMs, to capture the variation of the viscoelastic moduli with temperature, calculations of G' and G'' as a function of T at selected oscillation frequencies ω are shown in Fig. 3.8.

It is clear that the theory is able to correctly describe the T-dependence of the shear modulus G', thanks to the INMs in the VDOS $g(\omega_p)$, which enters the formulae Eq. (3.37); in turn, the INMs are directly related to thermal effects as depicted schematically in Fig. 3.4. The agreement is good at all three oscillation frequencies considered although with more scatter at the lowest frequency ($\omega = 0.03$) where the nonaffine contributions are dominant and which can be attributed to finite-size effects in the VDOS since it was computed for finite systems of 10^3 particles due to the size limitations of direct diagonalization of the Hessian discussed previously. It is also observed that G' decreases upon decreasing the frequency

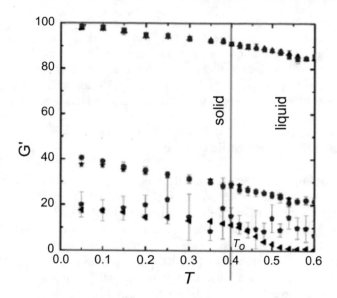

Fig. 3.8 Comparison between G' computed theoretically with the first of Eq. (3.37), red symbols, and G' from MD simulations, black symbols. The oscillation frequency of the external shear strain is varied from bottom to top as follows: $\omega = 0.03$ (bottom), $\omega = 20$ (central), $\omega = 200$ (top), in MD units. All other parameters are the same as in the previous figure, and the friction constant is set to $\nu = 1$. Adapted with modifications from Ref. [19]

ω, consistent with Fig. 3.7a, and becomes very small at the lowest frequency ($\omega = 0.03$) near the glass transition temperature T_g. At $T = T_g \approx 0.4$ (for the KG polymer glass) it is expected, indeed, that $G' \to 0$ in the limit $\omega \to 0$, although, for obvious reasons, this limit cannot be accessed in simulations or experiments, in practice. Furthermore, at the lowest frequency, $\omega = 0.03$, there is somewhat of a crossover taking place near $T_g \approx 0.4$, after which, in the liquid-like melt, the theory predictions become noisy. Conversely, no crossover is observed at higher frequencies where the affine contribution is comparatively more important.

The above comparison with a model polymer glass allows for a total control over the physical parameters, especially given the fact that the friction constant ν is specified by the thermostat of the Langevin MD simulations. In a real atomistic system, there is no such friction parameter, since friction is rather an emergent quantity arising from long-range anharmonic dynamic couplings between atoms (where long-range electrostatics usually plays a dominant role, at least in organic materials). In this case, ν must be taken as a free parameter in the comparison between Eq. (3.37) and the atomistic simulation data. Results have been shown in [27] with a satisfactory one-parameter comparison between predictions from Eq. (3.36) for the complex modulus and atomistic simulations data for glassy thermosets dicyclopentadiene (DCPD) and 5-norbornene-2-methanol (NBOH). The theoretical framework, combined with machine learning, has proved useful to identify specific atomic-level vibrational contributions to the nonaffine softening contribution to the modulus, as a route to optimize the mechanical performance for material design. A huge limitation remains due to the extremely high oscillation frequencies accessible in atomistic MD simulations, much larger than 10^{10} Hz. It is a formidable challenge for future research to crack this time scale bridging problem and produce quantitative predictions that can be directly compared to experiments.

3.4   Case Study: Metallic Glasses

Compared with traditional solid materials, metallic glasses (MGs) exhibit extraordinary physical properties, in terms of their ability to sustain large mechanical loads prior to yielding and in terms of their ductility and lightness [28]. They are solid metallic materials, usually alloys of a few metals, with a non-crystalline, disordered structure at the atomic level. While early atomic-scale theories based on defect physics and lattice dynamics, like [29, 30], have provided a good understanding of mechanical relaxation and internal friction in crystalline metals, revealing from the same microscopic scale, the relation between viscoelasticity and dynamical heterogeneity in MGs has been, instead, a long-term challenge.

With the advent of MGs as the next-generation metallic materials for mechanical and civil engineering applications, extensive experimental investigations like stress relaxation techniques and dynamic mechanical analysis (DMA) have brought a wealth of observations about the viscoelasticty and anelasticity of these materials. Intensive research, like [31, 32], has focused on the stress relaxation of various MGs, showing that localized plastic flow could be activated during viscoelastic and plastic deformation. As is often the case with glassy materials (e.g., like in the dielectric relaxation of glasses and supercooled liquids), the whole DMA spectrum of viscoelastic materials is usually fitted by an empirical (Kohlrausch) stretched-exponential function, cfr. Eq. (3.17) introduced earlier in this chapter. Linking the stretched-exponential relaxation to atomistic dynamics is an open challenge for our understanding of glassy materials.

Here, in an attempt to answer these questions by taking a microscopic approach, we develop the nonaffine atomic-scale theory of viscoelastic response and relaxation for metallic glasses, in a bottom-up way, starting from the microscopic particle-bath Hamiltonian introduced in Sect. 3.2.1 and suitably adapted to metals. The resulting Langevin equation for the atomic displacements, in this case, turns out to be a generalized Langevin equation (GLE) with a memory kernel for the friction. Although it is currently not possible to specify the functional form of the time dependence of the friction kernel from first principles, approximated stretched-exponential forms for the microscopic friction derived, e.g., in [33], based on many-body kinetic theory and mode-coupling approximations, can be used. Since in this case the memory kernel for the friction contains more adjustable parameters, a direct comparison with data from experimental measurements can be done. Finally, the atomistic description of the ion-ion interaction in metals via the embedded atom method (EAM) introduced in Chap. 1 (Sect. 1.1.3) requires a careful extension of

the nonaffine lattice dynamics framework to correctly evaluate the Hessian matrix
and the affine-force vectors, in this case.

3.4.1 Application to $Cu_{50}Zr_{50}$ Metallic Glass

In order to test the nonaffine response theory for metals, stress-relaxation experiments on $Cu_{50}Zr_{50}$ glassy system are used. As before, the VDOS is needed as an
input to calculate the viscoelastic response. To this aim, we used numerical MD
simulations of the same MGs, which takes also electronic structure effects into
account at the level of the embedded atom method (EAM).

3.4.2 Experiments

Experimentally, metallic glass ribbons made up of $Cu_{50}Zr_{50}$ with length over
7 mm are processed by the melt-spinning technique in an inert argon atmosphere.
Differential scanning calorimetry (DSC) is used to determine the thermal properties
of the samples, which have a glass transition temperature $T_g \approx 670$ K at a heating
rate of 20 K/min. The tensile stress relaxation experiments are performed at a
constant strain of 0.4% (for more details about the experimental procedure, see
[17]). The resulting stress relaxation in the form of tensile stress σ as a function
of time is fitted by the Kohlrausch function $\sigma(t) = \sigma_0 \exp[-(t/\tau)^\beta]$ with σ_0
being stress relaxation at $t = 0$, shown in Fig. 3.9, at three different temperatures:
$T = T_g(670\,\text{K})$, $T = 0.9T_g(603\,\text{K})$, and $T = 0.8T_g$ (536 K). It should be noted that
the stress at sufficiently long time, σ_∞, which is $\sigma(t)$ at $t = \infty$, has been estimated
to be zero for the three temperatures.

In Fig. 3.9, the Kohlrausch stretched-exponential (empirical) fitting is excellent
apart from deviations at long time, which are due to processes other than the
α-relaxation (e.g., other long-time or low-frequency relaxation processes). In the
following, we use the fitted Kohlrausch function to obtain the dynamic viscoelastic
Young moduli $E'(\omega)$ and $E''(\omega)$ in the frequency domain.

3.4.3 MD Simulations with EAM Potentials

In the molecular dynamics (MD) simulations, the Finnis-Sinclair-type EAM potentials optimized for realistic amorphous Cu-Zr structures are used, whose expression
is shown in Eq. (B.1) [34]. Seven independent $Cu_{50}Zr_{50}$ MG models were obtained
by quenching the system at cooling rate 10^{10} K/s from a liquid state equilibrated
at 2000 K with different initial positions and velocity distributions. Each model
was composed of 8192 atoms, and the external pressure was held at zero during
the quenching process using a Parrinello-Rahman barostat [35]. Periodic boundary

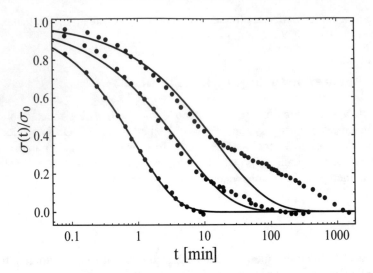

Fig. 3.9 Kohlrausch (stretched-exponential) empirical fits (solid lines) of experimental data (symbols). Top to bottom corresponds to temperatures in the following order: 536 K, 603 K, 670 K (the latter equals the T_g). Solid curves are Kohlrausch $\sigma(t) \sim \exp[-(t/\tau)^\beta]$ empirical fittings used to calibrate results, where the two parameters β and τ were chosen to be 0.69, 0.87 (mins); 0.55, 4.03 (mins); 0.55, 14.87 (mins) for $T = T_g$, $T = 0.9T_g$ and $T = 0.8T_g$, respectively. Reproduced from Ref. [17] with permission from the American Physical Society

conditions were imposed in the MD simulations. The resulting VDOS for different temperatures and averaged over $N = 7$ independent replicas are shown in Fig. 3.10.[7]

The VDOS is calculated as usual by diagonalizing the Hessian matrix based on static MD configurations of the CuZr alloys in mass-rescaled coordinates, which are also used to calculate the Ξ_i vectors and hence the affine-force correlator $\Gamma(\omega_p)$. Analytical expressions for the Hessian matrix, for the force acting on a tagged atom and for the Ξ_i vectors using the EAM interaction for atomic dynamics in metallic glasses, can be found in Appendix B.

3.4.4   Memory Kernel for the Friction

The friction kernel $\nu(t)$ that appears in Eqs. (3.23) and (3.30) can be rewritten in the continuum limit (in units where mass is equal to 1) as:

$$\nu(t) = \int_0^\infty \frac{\gamma^2(\omega_p)}{\omega_p^2} \cos(\omega_p t) g(\omega_p) d\omega_p, \qquad (3.41)$$

---

[7] It can be easily seen that the eigenfrequency spectrum is not particularly sensitive to temperature; therefore, INMs have not been considered in this case [17]. Nonetheless, it would be necessary to properly compute INMs and take them into account for describing the temperature dependence of the moduli also in the present case if one were to attempt a parameter-free comparison.

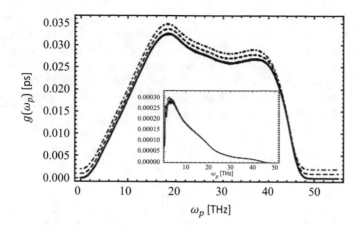

Fig. 3.10 VDOS of the simulated $Cu_{50}Zr_{50}$ systems. Solid, dashed, and dotted lines correspond to VDOS at 670 K, 603 K, and 536 K, respectively. The curves have been lifted upward in order to be distinguishable for the reader. The inset shows the VDOS normalized by the Debye law ω_p^2, which shows clear evidence of a strong boson peak. Reproduced from Ref. [17] with permission from the American Physical Society

where $\gamma(\omega_p)$ represents the continuous limit (spectrum) of the discrete set of dynamic coupling constants $\{\gamma_p\}$ in the ZCL Hamiltonian Eq. (3.22). For any given (well-behaved) VDOS function $g(\omega_p)$, the existence of a well-behaved function $\gamma(\omega_p)$ that satisfies Eq. (3.41) is guaranteed by the fact that we can always decompose $v(t)$ onto a basis set of $\{\cos(\omega_p t)\}$ functions, i.e., by taking a cosine (Fourier) transform. The inverse cosine transform, in turn, returns the spectrum of coupling constants $\gamma(\omega_p)$ as a function of the memory kernel $v(t)$:

$$\gamma^2(\omega_p) = \frac{2\omega_p^2}{\pi g(\omega_p)} \int_0^\infty v(t) \cos(\omega_p t) dt. \tag{3.42}$$

This spectrum of coupling terms $\gamma(\omega_p)$ contains information on how strongly the atom's motion is dynamically coupled to the motion of other atoms in a mode with vibrational frequency ω_p. This is an important information, because it tells us about the degree of long-range anharmonic couplings in the atomic motions.

Looking at Eq. (3.41) again, it is evident that the ZCL Hamiltonian Eq. (3.22) does not provide any prescription for the form of the memory function $v(t)$, which can take any (sufficiently smooth) functional form depending on the values of the coefficients γ_p, as discussed in [9]. Hence, a shortcoming of Caldeira-Leggett-type models, including ZCL, is that the functional form of $v(t)$ cannot be derived a priori for a given system, because while the VDOS is certainly an easily accessible quantity from simulations of a physical system, the spectrum of coupling constants $\{\gamma_p\}$ is basically a phenomenological parameter.

In general, the determination of the memory kernel is an open problem for which several approaches have been proposed very recently, most of which have been tested only on model systems so far; some examples can be found in [36–40]. However, for a supercooled liquid, the time-dependent friction, which is dominated by slow collective dynamics, has been derived within many-body kinetic theory (Boltzmann equation) using a mode-coupling-type approximation in Ref. [33] and is given by the following elegant expression:

$$v(t) = \frac{\rho k_B T}{6\pi^2 M} \int_0^\infty dq q^4 F_s(q, t)[c(q)]^2 F(q, t) \tag{3.43}$$

where $c(q)$ is the direct correlation function of liquid-state theory; $F_s(q, t)$ is the self-part of the intermediate scattering function (ISF) $F(q, t)$, the latter defined in Chap. 1, Sect. 1.3, (cfr. Eq. (1.59)); and ρ is the density (cfr., e.g., Ref. [41] for more details on the definition of these physical quantities). All of these quantities are functions of the wavenumber q. Clearly, the integral over q leaves a time dependence of $v(t)$, which is controlled by the product $F_s(q, t)S(q, t)$. For a relatively homogeneous system, $F_s(q, t)S(q, t) \sim F(q, t)^2$, especially in the long-time regime. From mode-coupling theories [4] and simulations in supercooled liquids [41], it is typically found $F(q, t) \sim \exp[-(t/\tau)^\xi]$, with $\xi < 1$, typically in the range $0.3 - 0.8$. Hence, given the stretched-exponential behavior of $F(q, t)$ in Eq. (3.43), the above arguments imply that $v(t)$ must also follow a stretched-exponential of time (with a different exponent),

$$v(t) = v_0 \exp(-a\, t^b) \tag{3.44}$$

a result which is, therefore, motivated by Sjogren and Sjolander [33], where v_0, a, and b are adjustable parameters to be determined in the comparison with experimental data.

3.4.5   Dynamic Viscoelastic Young's Moduli of $Cu_{50}Zr_{50}$

Before presenting a comparison between the theory and the experimental data on $Cu_{50}Zr_{50}$, we first convert the experimentally measured linear response in terms of stress relaxation $\sigma(t)$ in the time domain to the frequency-dependent dynamic viscoelastic Young's moduli $E'(\omega)$ and $E''(\omega)$, for a uniaxial strain of amplitude ϵ_0:

$$E'(\omega) = \frac{\sigma_\infty}{\epsilon_0} + \frac{\sigma_0\,\omega}{\epsilon_0} \int_0^\infty e^{-(t/\tau)^\beta} \sin(\omega t)dt, \tag{3.45}$$

$$E''(\omega) = \frac{\sigma_0\,\omega}{\epsilon_0} \int_0^\infty e^{-(t/\tau)^\beta} \cos(\omega t)dt. \tag{3.46}$$

A more detailed derivation of this result can be found in the Appendix of Ref. [17].

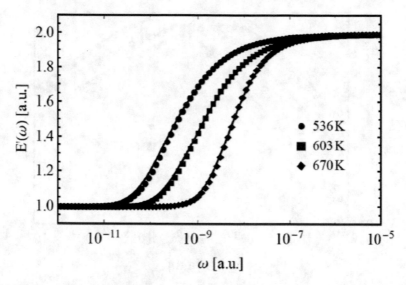

Fig. 3.11 Real part of the complex viscoelastic Young's modulus. From right to left, solid lines represent E' for $T = T_g$, $T_g = 0.9T_g$, and $T = 0.8T_g$ respectively, from the Kohlrausch best fitting of the experimental stress-relaxation data of Fig. 3.9. Symbols are calculated based on the nonaffine theory Eq. (3.35) with $\mu\nu\kappa\chi = xxxx$ for tensile deformation. For $T = T_g$, $T = 0.9T_g$ and $T = 0.8T_g$, b was chosen to be 0.72, 0.58, and 0.58, respectively; a was taken to be 1.2×10^{-6}, 7×10^{-6}, and 3.4×10^{-6}. $\nu_0 = 0.137$ is the same for all temperatures. From Ref. [17], reprinted with permission from the American Physical Society

   In Fig. 3.11, the comparison between nonaffine response theory given by Eq. (3.35) (with $\mu\nu\kappa\chi = xxxx$ for tensile deformation) for $E'(\omega)$, and the experimental data at $T_g = 670$ K, i.e., exactly at T_g, and at two other temperatures below the glass transition, is shown. It is clear that the theoretical model is in excellent agreement with the experimental data and is also exhibiting the typical trend of α-relaxation phenomena. This shows how crucial soft modes in the VDOS are, as well as the memory effects embodied in the non-Markovian friction kernel, for the understanding of the viscoelastic response and of α-relaxation below the glass transition. In Fig. 3.12, fittings of the loss modulus, $E''(\omega)$, are shown. Also in this case, it is seen that the theory provides an excellent description of the experimental data and is able to reproduce the characteristic α-wing asymmetry of the absorption peak, typical of α-relaxation. Note that, for clarity of presentation, we have changed the unit of time to shift curves horizontally. This means that we have arbitrary units on both the x and y axes.
   The theoretical framework thus provides a direct connection between the excess of low-energy (boson-peak) modes of the VDOS near T_g and in the glassy state, the non-Markovian memory effects in the dynamics, and the corresponding features of the mechanical response such as the α-wing asymmetry in $E''(\omega)$. It is in fact impossible to obtain a successful fitting of the data using a simple Debye model (cfr. Chap. 5) for the VDOS that has no excess of soft modes (i.e., with no such

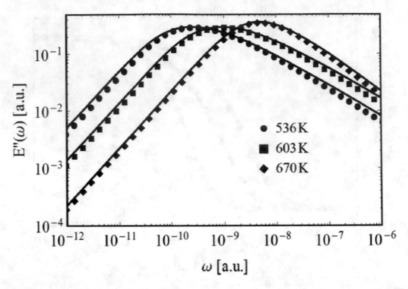

Fig. 3.12 Imaginary part of the complex viscoelastic Young's modulus. From right to left, solid lines represent E'' for $T = T_g$, $T_g = 0.9T_g$, and $T = 0.8T_g$ respectively, from the Kohlrausch best fitting of the experimental stress-relaxation data of Fig. 3.9. Symbols are calculated based on the nonaffine theory Eq. (3.35) with $\mu\nu\kappa\chi = xxxx$ for tensile deformation. For $T = T_g$, $T = 0.9T_g$, and $T = 0.8T_g$, b in the memory kernel was chosen to be 0.72, 0.58, and 0.58, respectively; a was taken to be $1.2 \times 10^{-6}, 7 \times 10^{-6}$ and 3.4×10^{-6}. $\nu_0 = 0.137$ is the same for all temperatures. From Ref. [17], reprinted with permission from the American Physical Society

peak as the one in the inset of Fig. 3.10). The link between boson peak in the VDOS and the absorption peak in G'' of metallic glasses has been discussed also based on atomistic MD simulations data in Ref. [42].

Importantly, the theory demonstrates that non-Markovianity of the atomic dynamics is as important as the boson peak in the VDOS in order to describe the experimental data. It was indeed checked that using a constant (Markovian) friction $\nu = const$, or even a simple-exponential time dependence for $\nu(t)$, it is not possible to describe the experimental data. Only a stretched-exponential form of $\nu(t)$ with a value of the stretching exponent in the range 0.58–0.72, which decreases upon decreasing T further down from T_g, allows one to describe the data. Since ν in the theory physically arises from spectrum of dynamic coupling terms between an atom and all other atoms in the material (cfr. Eq. (3.41)), this result implies that every atom is in long-range dynamic coupling to many other atoms beyond the nearest-neighbor shell. In practice, this is the result of the anharmonicity of the ion-ion interaction and of the non-locality of the electronic contributions to the interatomic interaction (cfr. Chap. 1, Sect. 1.1.3).

Also, this theoretical analysis shows that the time scale over which atoms retain memory of their previous collision history, $\tau \equiv a^{-1/b}$, also increases upon decreasing the temperature, by more than a factor two overall, even though this increase appears to be somewhat non-monotonic, from $\tau \approx 1.14$ at $T = T_g$ to $\tau \approx 3.1$ at $T = 0.9T_g$ to $\tau \approx 2.0$ at $T = 0.8T_g$.

The above approach has been successfully extended to ternary alloys such as $La_{60}Ni_{15}Al_{25}$ in [43]. By adding extra complexity, ternary alloys often feature a larger separation of time scales in the dynamics, often due to the presence of light and heavy atoms in the alloy, in this case Al and La, respectively. This separation of characteristic time scales is such that a single relaxation time τ or a single stretched-exponential relaxation process no longer suffices to provide a description of the data. Hence, a second, additional, stretched-exponential process has to be added to the memory kernel with a different relaxation time. The dynamic of lightest atom thus gives rise to a secondary relaxation, which resembles the so-called β-relaxation process. Providing a microscopic foundation to these secondary relaxation processes is a very active field of research in terms of experimental and computational efforts.

3.5   Microscopic Theory of Relaxation Modulus and Creep

The relaxation modulus $G(t)$ was introduced in Sect. 3.1 in the context of creep, i.e., when the system responds to a step input in either stress or strain. Recalling the Boltzmann superposition integral (cfr. Eq. (3.2)):

$$\sigma(t) = \int_{-\infty}^{t} G(t - t')\dot{\gamma}(t')dt \tag{3.47}$$

which defines $G(t)$ as the characteristic linear response function of the material, often referred to as the relaxation function or creep modulus. We recall that $G(t)$ is measured via the system response to a step strain of amplitude γ_0: $\dot{\gamma} = \gamma_0\delta(t')$ from which the above equation becomes:

$$\sigma(t) = G(t)\gamma_0. \tag{3.48}$$

Upon inverting the Hooke's relation, one can make it explicit for the strain instead of the stress. The above Boltzmann relation can be inverted:

$$\gamma(t) = \int_{-\infty}^{t} J(t - t')\dot{\sigma}(t')dt \tag{3.49}$$

to obtain:

$$\gamma(t) = J(t)\sigma_0 \tag{3.50}$$

which is the strain response to a step in the stress and where $J(t)$ is the compliance or compliant modulus, i.e., the ratio of strain to stress or just the inverse of the elastic modulus $G(t)$.[8]

In ordinary solids such as crystalline metals, the creep behavior was observed already in 1910 by Andrade who reported what is now referred to as the Andrade creep law:

$$\gamma(t) \sim t^{1/3} \tag{3.51}$$

which thus implies:

$$G(t) \sim t^{-1/3}. \tag{3.52}$$

The $1/3$ exponent in the Andrade creep can be explained in terms of thermally activated motion of dislocations, thanks to pioneering work and ideas by N. F. Mott and F. R. N. Nabarro, among others [29].

Power-law creep, also referred to as power-law rheology or scale-free rheology (due to the absence of a characteristic time scale), is ubiquitous in disordered materials and in particular in soft matter systems. In soft materials, creep is typically measured in terms of:

$$\dot{\gamma}(t) \sim t^{-x} \tag{3.53}$$

with a variable exponent x in the range from $1/3$ to 1 [44–46]. The latter limit corresponds to logarithmic creep, $\gamma(t) \sim \ln t$, which, typically, represents an upper bound. In terms of $\dot{\gamma}(t)$, Andrade creep is retrieved with $x = 2/3$.

For polymers, Rouse chain dynamics predicts $G(t) \sim t^{-1/2}$ [47], although this is unlikely to be the main mechanism in the solid glassy state below T_g where chain degrees of freedom are frozen-in. An exponent $x \approx 1/2$ in Eq. (3.53), which corresponds to $G(t) \sim t^{-1/2}$, has been reported in the Kremer-Grest polymer glass (cfr. Sect. 3.3) in Ref. [47], colloidal glasses [45] including colloidal glasses near yielding [48], and granular packings near the jamming (isostatic) transition [49].

Unlike the case of crystalline metals, there are currently no microscopic predictions of creep in disordered solids across the broad range of exponents reported experimentally. Among the first-principles theories, mode-coupling theory predicts Andrade creep, with $G(t) \sim t^{-1/3}$, for hard-sphere colloidal glasses [4]. The nonaffine response theory, as we shall see below in detail, predicts the $t^{-1/2}$ creep near the vanishing of rigidity of disordered solids, in agreement with simulations

---

[8] This setup, where the time variation of the strain γ in response to step stress load is recorded, is traditionally the most common in the practice of material testing.

and experiments of various systems mentioned above. Besides these microscopic approaches, we should mention phenomenological models based either on thermally activated local flowing regions [50] or fractional Maxwell-type models (cfr. Sect. 3.1 for the basic Maxwell model) obtained by adding in series a spring and a fractional dashpot. The latter is a dashpot with fractional dynamics: $\sigma = \eta^\beta \frac{d^\beta \gamma}{dt^\beta}$, where β is a fractional number. The fractional derivative d^β/dt^β is a formal generalization of a derivative to noninteger order and mathematically gives rise to a variety of different creep scenarios as discussed in [51, 52].

In the next section, we shall derive the power-law creep of disordered solids from first principles, making use of the nonaffine response theory developed earlier in this chapter, together with previous results obtained in Chap. 2.

3.5.1 Theory of Power-Law Creep in Disordered Solids

We consider the time-dependent, relaxation shear modulus $G(t)$. This can be calculated by taking the inverse Fourier transform of the frequency-dependent complex shear modulus $G^*(\omega)$ in Eq. (3.36), which gives:

$$G(t) = G - \frac{3}{2\pi}\rho \int_{-\infty}^{\infty} \int_0^{\infty} \frac{g(\omega_p)\Gamma(\omega_p)\exp(i\omega t)}{\omega_p^2 - \omega^2 + i\nu\omega} d\omega_p d\omega$$

$$= G - 3\rho\, t e^{-\frac{\nu}{2}t} \int_0^{\infty} g(\omega_p)\Gamma(\omega_p)\, \text{sinc}\left(\frac{1}{2}\sqrt{4\omega_p^2 - \nu^2 t}\right) d\omega_p. \tag{3.54}$$

Here G is the quasistatic (infinite-time or zero-frequency) shear modulus, which is dominated by nonaffinity, and $\text{sinc}(x) = \sin(x)/x$ denotes the cardinal sine function. Furthermore, we used the assumption of Markovian constant friction, $\nu = const$, for simplicity.

Numerical evaluations of Eq. (3.54) for the three lattices (random network, randomly depleted fcc and fcc with random vacancies) discussed in Chap. 1, Sect. 1.5.2, at two representative values of z, are shown in Fig. 3.13.

At short times, a plateau that corresponds to the high-frequency affine response is observed, after which a power-law decay is observed with an exponent in the range between $-1/2$ and $-3/4$. This power law can be understood mechanistically as follows.

We focus on the limit overdamped systems, which is both important for various colloidal systems and turns out to be amenable to analytic simplifications. For large ν and large times, we can simplify the expression in Eq. (3.54). First we take $\sqrt{\nu^2 - 4\omega_p^2} \approx \nu - 2\frac{\omega_p^2}{\nu}$, where we use $\omega_p \ll \nu$, thanks to the overdamped dynamics

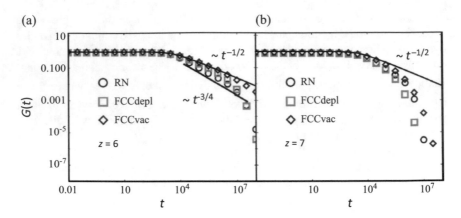

Fig. 3.13 Creep relaxation modulus $G(t)$ of the three lattice model systems depicted in Fig. 1.13 (random network, RN) and Fig. 1.14 (fcc randomly depleted, FFCdepl, and fcc with vacancies, FFCvac) with $z = 6$ in (**a**) and $z = 7$ in (**b**). For all these systems, the infinite-time shear modulus goes G to zero at the isostatic point $z = 6$. The scaling of $G(t)$ changes significantly upon going from the marginal-rigidity state with $z = 6$ to the fully rigid state with $z = 7$. This is due to due to the Debye $\sim \omega^2$ regime in the VDOS, which is present for $z = 7$ and is absent, replaced by constant plateau down to $\omega_p \to 0$ for $z = 6$. Reproduced from Ref. [20] with permission from the American Physical Society

assumption. We replace this into Eq. (3.54) and use the definition of $\sinh(x)$ to get:

$$
G(t) \approx 6\rho\, e^{-\frac{v}{2}t} \int_0^\infty \frac{g(\omega_p)\Gamma(\omega_p) \sinh\left(\frac{v}{2}t - \frac{\omega_p^2}{v}t\right)}{\omega_p^2\left(v - 2\frac{\omega_p^2}{v}\right)}\, d\omega_p
$$

$$
= 3\rho \int_0^\infty \frac{g(\omega_p)\Gamma(\omega_p)\left(e^{-\frac{\omega_p^2}{v}t} - e^{-vt+\frac{\omega_p^2}{v}t}\right)}{\omega_p^2\left(v - 2\frac{\omega_p^2}{v}\right)}\, d\omega_p
$$

$$
\approx 3\rho\, \frac{1}{v} \int_0^\infty \frac{g(\omega_p)\Gamma(\omega_p)}{\omega_p^2} e^{-\frac{\omega_p^2}{v}t} d\omega_p. \tag{3.55}
$$

In the last step, we used $v \gg 2\omega_p^2/v$ and $vt - \omega_p^2 t/v \gg 1$. This corresponds to a set of Maxwell elements with relaxation times $\tau_p = v/\omega_p^2$. We now recall the standard relationship between the vibrational density of states (VDOS) and the eigenvalue spectrum $\rho(\lambda)$ of the Hessian matrix, $g(\omega_p)d\omega_p = \rho(\lambda)d\lambda$, with $\omega_p^2 = \lambda$. At the isostatic point of disordered solids, $z = 6$, the VDOS, instead of the Debye law $g(\omega_p) \sim \omega_p^2$, develops a plateau of soft modes [53], $g(\omega_p) \sim const$. This limit corresponds to the scaling $\rho(\lambda) \sim \lambda^{-1/2}$ in the eigenvalue distribution. This scaling can be shown to arise from the random-matrix character of the spectrum, and this can be derived, e.g., from the Marchenko-Pastur distribution of random matrix theory, as discussed recently in [54] and as we shall see in more detail in Chap. 5. In the VDOS obtained numerically by digonalization of the theree lattices, a scaling $\rho(\lambda) \sim a + \lambda^{-1/2}$, where a is a constant, is actually more appropriate since we are in fact slightly above $z = 6$ due to numerical precision.[9] However, we will stick to the simple $\rho(\lambda) \sim \lambda^{-1/2}$ for the asymptotic analysis. Recall now that $\Gamma(\omega_p) \sim \omega_p^2$, from the analytical theory of nonaffine elasticity (cfr. Eq. (2.46) in Chap. 2) [18], which implies $\tilde{\Gamma}(\lambda) \sim \lambda$. Inserting these results in the last line of Eq. (3.55), we obtain the following Laplace transform, which can be easily evaluated to give:

$$G(t) \sim \int_0^\infty \frac{\rho(\lambda)\tilde{\Gamma}(\lambda)}{\lambda} e^{-\lambda t} d\lambda \sim \int_0^\infty \frac{\lambda^{-1/2}\lambda}{\lambda} e^{-\lambda t} d\lambda$$
$$\sim t^{-1/2}, \tag{3.56}$$

where the last line arises upon performing the integral. This result was derived in Ref. [20].

This derivation shows that the power-law $G(t) \sim t^{-1/2}$ often observed in many soft materials, particularly near marginal rigidity, can be rationalized in terms of (i) the development of soft mode instability in the VDOS and (ii) the internal nonaffine particle dynamics under deformation.

Away from marginal rigidity, the VDOS exhibits Debye behavior at low frequency, $g(\omega_p) \sim \omega_p^2$, which translates to $\rho(\lambda) \sim \lambda^{1/2}$. Inserting this in the above Eq. (3.56) leads to larger creep exponents; in particular, one has $G(t) \sim t^{-3/2}$, which is unrealistic. In reality, if the full VDOS is used, the boson peak in the VDOS is going to produce lower exponents. Real materials will thus be comprised between these two limits. In future work, the nonaffine framework can be extended to make more precise estimates of creep exponents for different material classes, based on the knowledge of $\Gamma(\lambda)$ and of $\rho(\lambda)$ from MD simulations.

References

1. J. Frenkel, *Kinetic Theory of Liquids* (Oxford University Press, Oxford, 1955)
2. R.G. Palmer, D.L. Stein, E. Abrahams, P.W. Anderson, Phys. Rev. Lett. **53**, 958 (1984)
3. D.C. Johnston, Phys. Rev. B **74**, 184430 (2006)
4. W. Goetze, *Complex Dynamics of Glass-Forming Liquids: A Mode-Coupling Theory* (Oxford University Press, Oxford, 2009)
5. R. Zwanzig, R.D. Mountain, J. Chem. Phys. **43**(12), 4464 (1965)
6. M. Baggioli, M. Vasin, V. Brazhkin, K. Trachenko, Phys. Rep. **865**, 1 (2020).
7. J. Boon, S. Yip, *Molecular Hydrodynamics*. Advanced Book Program (McGraw-Hill, New York, 1980)
8. M. Baggioli, M. Landry, A. Zaccone, Phys. Rev. E **105**, 024602 (2022)

---

[9] This will explain the power-law exponents in $G(t)$ larger than $1/2$ in the numerical calculations.

9. R. Zwanzig, J. Statist. Phys. **9**(3), 215 (1973)
10. R. Zwanzig, R.D. Mountain, J. Chem. Phys. **43**(12), 4464 (1965)
11. A. Caldeira, A. Leggett, Ann. Phys. **149**(2), 374 (1983)
12. A. Rognoni, R. Conte, M. Ceotto, J. Chem. Phys. **154**(9), 094106 (2021)
13. T. Damart, A. Tanguy, D. Rodney, Phys. Rev. B **95**, 054203 (2017)
14. H.C. Andersen, J. Chem. Phys. **72**(4), 2384 (1980)
15. J.R. Ray, A. Rahman, J. Chem. Phys. **80**(9), 4423 (1984)
16. A. Lemaître, C. Maloney, J. Statist. Phys. **123**(2), 415 (2006)
17. B. Cui, J. Yang, J. Qiao, M. Jiang, L. Dai, Y.J. Wang, A. Zaccone, Phys. Rev. B **96**, 094203 (2017)
18. A. Zaccone, E. Scossa-Romano, Phys. Rev. B **83**, 184205 (2011)
19. V.V. Palyulin, C. Ness, R. Milkus, R.M. Elder, T.W. Sirk, A. Zaccone, Soft Matt. **14**, 8475 (2018)
20. R. Milkus, A. Zaccone, Phys. Rev. E **95**, 023001 (2017)
21. I. Kriuchevskyi, V.V. Palyulin, R. Milkus, R.M. Elder, T.W. Sirk, A. Zaccone, Phys. Rev. B **102**, 024108 (2020)
22. R.S. Hoy, Phys. Rev. Lett. **118**, 068002 (2017)
23. K. Kremer, G.S. Grest, J. Chem. Phys. **92**(8), 5057 (1990)
24. R. Ranganathan, Y. Shi, P. Keblinski, Phys. Rev. B **95**, 214112 (2017)
25. R.M. Stratt, Acc. Chem. Res. **28**(5), 201 (1995)
26. T. Keyes, J. Phys. Chem. A **101**(16), 2921 (1997)
27. R.M. Elder, A. Zaccone, T.W. Sirk, ACS Macro Lett. **8**(9), 1160 (2019)
28. F. Spaepen, D. Turnbull, Annu. Rev. Phys. Chem. **35**(1), 241 (1984)
29. F.R.N. Nabarro, F. de Villiers, *Physics of Creep and Creep-Resistant Alloys* (Taylor and Francis, London, 1995)
30. C. Zener, *Elasticity and Anelasticity of Metals* (University of Chicago Press, Chicago, 1948)
31. Z. Wang, B.A. Sun, H.Y. Bai, W.H. Wang, Nat. Commun. **5**, 5823 (2014)
32. J. Qiao, Y.J. Wang, J.M. Pelletier, L.M. Keer, M.E. Fine, Y. Yao, Acta Mater. **98**, 43 (2015)
33. L. Sjogren, A. Sjolander, J. Phys. C: Solid State Phys. **12**(21), 4369 (1979)
34. M. Mendelev, M. Kramer, R. Ott, D. Sordelet, Philos. Mag. **89**(2), 109 (2009)
35. M. Parrinello, A. Rahman, J. Appl. Phys **52**(12), 7182 (1981)
36. Z. Li, X. Bian, X. Yang, G.E. Karniadakis, J. Chem. Phys. **145**(4), 044102 (2016)
37. Z. Li, H.S. Lee, E. Darve, G.E. Karniadakis, J. Chem. Phys. **146**(1), 014104 (2017)
38. G. Jung, M. Hanke, F. Schmid, J. Chem. Theory Comput. **13**(6), 2481 (2017)
39. H. Meyer, T. Voigtmann, T. Schilling, J. Chem. Phys. **147**(21), 214110 (2017)
40. S. Izvekov, J. Chem. Phys. **146**(12), 124109 (2017)
41. J. Hansen, I. McDonald, *Theory of Simple Liquids* (Elsevier Science, Amsterdam, 2006)
42. R. Ranganathan, Y. Shi, P. Keblinski, J. Appl. Phys. **122**(14), 145103 (2017)
43. B. Cui, Z. Evenson, B. Fan, M.Z. Li, W.H. Wang, A. Zaccone, Phys. Rev. B **98**, 144201 (2018)
44. L. Cipelletti, K. Martens, L. Ramos, Soft Matt. **16**, 82 (2020)
45. R. Cabriolu, J. Horbach, P. Chaudhuri, K. Martens, Soft Matt. **15**, 415 (2019)
46. D. Bonn, M.M. Denn, L. Berthier, T. Divoux, S. Manneville, Rev. Mod. Phys. **89**, 035005 (2017)
47. O. Adeyemi, S.. Zhu, L.. Xi, Phys. Fluids **34**(5), 053107 (2022)
48. T. Sentjabrskaja, P. Chaudhuri, M. Hermes, W.C.K. Poon, J. Horbach, S.U. Egelhaaf, M. Laurati, Sci. Rep. **5**(1), 11884 (2015)
49. B.P. Tighe, Phys. Rev. Lett. **109**, 168303 (2012)
50. J. Weiss, D. Amitrano, Phys. Rev. Materials **7**, 033601 (2023)
51. R.H. Pritchard, E.M. Terentjev, J. Rheol. **61**(2), 187 (2017)
52. A. Jaishankar, G.H. McKinley, Proc. R. Soc. A Math. Phys. Eng. Sci. **469**(2149), 20120284 (2013)
53. C.S. O'Hern, L.E. Silbert, A.J. Liu, S.R. Nagel, Phys. Rev. E **68**, 011306 (2003)
54. S. Franz, G. Parisi, P. Urbani, F. Zamponi, Proc. Natl. Acad. Sci. **112**(47), 14539 (2015)

Wave Propagation and Damping

4

Abstract

Propagation of acoustic mechanical waves and other more energetic excitations in amorphous solids is strongly dominated by anharmonicity and intricate scattering processes, leading to strong damping and attenuation phenomena. We shall start with a mathematical description based on the hydrodynamic or effective field theory framework and then build a more microscopic first-principles derivation of damping in amorphous solids based on nonaffine atomic motions. This derivation will show that the often-observed Rayleigh-type damping of acoustic transverse waves in amorphous solids, e.g., glasses, is rooted in the nonaffine nature of microscopic motions. We shall also consider important concepts of quasiparticle kinetic theory (e.g., Ioffe-Regel crossover) in connection with the crossover from ballistic propagation to diffusive propagation of vibrational excitations. At variance with previous chapters, in this chapter, Cartesian components will be denoted with Latin indices instead of Greek indices. Furthermore, the microscopic friction coefficient will be denoted with the Greek letter ζ, instead of ν.

4.1   Sound Attenuation

Soon after the finding of "anomalous" behavior in the low-temperature thermal properties of amorphous solids such as a linear in T specific heat at low temperatures and a peak in the Debye-normalized specific heat, much attention was paid to the acoustic and dielectric properties at low temperature. A review of those first experiments can be found in Chap. 6 of [1]. Specifically, many measurements of internal friction and sound velocity in a wide frequency range (typically between 0.1 KHz and 100 MHz) and down to mK temperatures were performed mainly in thin films of metallic and semiconducting glasses, polymers, and oxide glasses.

© The Author(s), under exclusive license to Springer Nature Switzerland AG 2023
A. Zaccone, *Theory of Disordered Solids*, Lecture Notes in Physics 1015,
https://doi.org/10.1007/978-3-031-24706-4_4

Those experiments revealed again a striking *universal* glassy behavior, in turn extremely different from their crystalline counterparts. For instance, the internal friction was orders of magnitude higher than in crystals, with another universal *plateau* in the range 0.1–10 K with a height $\approx 5 \times 10^{-4}$ always within one order of magnitude for all studied cases. The interested reader is referred to the recent reviews [2, 3]. Let us mention, nevertheless, that all those glassy features in the low-temperature ultrasonic properties were later also reported in disordered crystals, including a quasicrystal [2], as well as in polycrystalline metals and superconductors [4], a fact that is often forgotten.

However, we will focus here on the acoustic attenuation at higher frequencies, namely, in the hypersonic range ($\approx$ GHz) and in the vibrational range around the boson peak (≈ 1 THz). As we shall see with much greater detail in the next chapter, a fundamental ingredient for the appearance of the boson peak anomaly (excess of vibrational modes on top of the Debye prediction, typically in the Terahertz regime) is the presence of a strong sound attenuation constant Γ. In particular, the nature of this parameter is essential to distinguish the various theoretical paradigms and to understand the mechanism behind glassy anomalies. Importantly, the fluctuating elasticity theory of Schirmacher and collaborators [5] identifies the low temperature Rayleigh scattering mechanism, producing a characteristic $\Gamma \sim k^4$ scaling of the sound attenuation constant, as the fundamental origin of the boson peak anomaly, and in particular it predicts the location of the boson peak frequency as the crossover point between the low-frequency Rayleigh scattering regime and the high-frequency quadratic one $\Gamma \sim k^2$. Is this case really supported by the simulations and experimental data? Despite some evidence of this feature in vitreous silica [6], a more comprehensive exploration of the existing literature seems to say otherwise.[1]

In other words, it is not obvious why the assumption of a quartic regime in k, confined in a very low-frequency range, is necessary to explain the appearance of a boson peak anomaly at frequencies much higher than that. Similar concerns have been already raised in the literature (see, e.g., the conclusions of [15]). The problem is even more severe, since the Rayleigh regime disappears very fast by increasing the temperature T and it gets completely washed out by the most common quadratic scaling (see [10]). Nevertheless, the boson peak is reported experimentally in a range of temperature, which spans more than two decades, much beyond the low temperature regime. How does Rayleigh scattering explain the persistence of the BP anomaly up to such large temperatures at which its effects are totally negligible and definitely not visible? A new perspective which could be able to put low-frequency [16] and high-frequency [17] sound dissipation under the same umbrella is based

---

[1] Moreover, fluctuating elasticity theory underestimates the amplitude of low-temperature acoustic scattering in glasses by about two orders of magnitude [7], and some of its assumptions have been shown to be untenable by careful analysis of molecular simulations of model glasses in [8]. A long list of results [9–14] unambiguously shows that the boson peak frequency is generally within the quadratic regime and far away from the Rayleigh one (see Fig. 4.1).

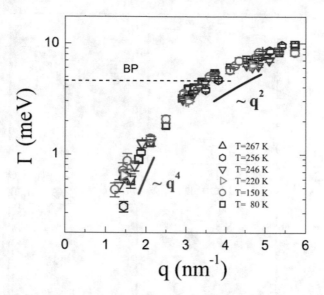

Fig. 4.1 Top: Inelastic X-ray scattering results for the longitudinal acoustic excitations of glassy sorbitol at the indicated temperatures. Adapted from [14] with permission from the American Institute of Physics

on the dominant role of nonaffine microscopic dynamics on the elastic acoustic scattering in amorphous systems.

Not only do nonaffine motions produce the ubiquitous quadratic term in the sound attenuation constant [17], but they even provide a contribution to the Rayleigh term [18], which, accordingly to the numerical analysis of [16], is dominant over that coming from the elastic fluctuations, i.e., heterogeneous elasticity theory [19]. As discussed already in Chap. 2, this nonaffine dynamics is active also in ordered structures because of thermal fluctuations. Older studies [20–22] proved already that sound relaxation in glasses is compatible with a damping only (or at least mainly) produced by the anharmonic coupling of the sound waves with thermally excited modes.

Continuing with our previous discussion, the quadratic-in-k damping has been observed experimentally in a lot of different systems and in different ranges of temperature [23–25].

To be more specific, a compilation of many experiments with X-ray and light scattering shows that the wavenumber dependence of the longitudinal sound attenuation coefficient, $\Gamma_L(k)$, is in general divided into three regimes [6,11,12,26–30]: (i) $\Gamma_L(k) \sim k^2$ at low k; (ii) $\Gamma_L(k) \sim k^4$ at an intermediate k regime; and (iii) $\Gamma_L(k) \sim k^2$ for large k. It has been observed that the k^4 to k^2 transition for sound attenuation in the higher-frequency regime is mainly "harmonic." In contrast, the k^2 dependence in the low-frequency regime, as already pointed out above, has been demonstrated to be related to viscous attenuation caused by anharmonicity. See Fig. 4.2 for experimental evidence of these crossovers.

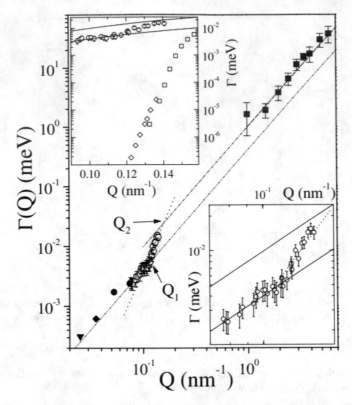

Fig. 4.2 The crossovers from quadratic Akhiezer damping at low wavenumber Q to quartic Rayleigh damping and then again back to a quadratic law at high wavenumbers, measured experimentally in silica glass. Adapted from Ref. [29] with permission from the American Physical Society

In the following, we will present theoretical derivations of the first two regimes, i.e., the anharmonic Akhiezer damping $\Gamma \sim k^2$ occurring at low k (the so-called hydrodynamic regime of large wavelengths) and the Rayleigh-type attenuation $\Gamma \sim k^4$. The crossover between the two regimes can be predicted from first-principles analysis of nonaffine motions within the nonaffine response formalism, as will be shown in this chapter.

4.2   Akhiezer Damping

Dissipative processes at the atomic and molecular level are responsible for the viscosity of solids [31]. At the microscopic level, the long-range (e.g., Coulomb, van der Waals, etc.) interactions between atoms lead to a dynamic coupling between the tagged atom and other atoms. By schematically describing such dynamic coupling in terms of Caldeira-Leggett particle-bath Hamiltonian, as we did in Chap. 3, Sect. 3.2.1, one obtains a Langevin-type equation for the microscopic motion of the tagged atom/molecule [32], cfr. Eq. (3.23). By Fourier transforming this equation and upon applying definitions of stress and strain, a viscoelastic analogue of Eq. (2.3) has been derived in Chap. 3 (cfr. Eq. (3.36)):

$$G(\omega) = G_\infty - 3\rho \int_0^{\omega_D} \frac{g(\omega_p)\Gamma(\omega_p)}{m\omega_p^2 - m\omega^2 + i\omega\zeta} d\omega_p \qquad (4.1)$$

where ω indicates the externally applied oscillation frequency and $G_\infty \equiv G_A$ is the infinite-frequency shear modulus.[2] Equation (4.1) has been shown to quantitatively predict the frequency-dependent viscoelastic moduli of model computer glasses in parameter-free agreement with molecular simulations of the deformation process [33, 34] as discussed in the previous chapter.

From the point of view of elastodynamics, the viscous contribution to the overall stress can be added in a phenomenological way [31, 35]:

$$\rho \ddot{u}_i = \nabla_j \sigma_{ij} + f_i^{ext}(\mathbf{r}) \qquad (4.2)$$

where now Latin indices are used to denote spatial components (e.g., $ij = xy$ for a shear deformation). The above equation expresses the fact that the internal stress force $\nabla_k \sigma_{ik}$ plus the external force density $f_i^{ext}(\mathbf{r})$ is equal to the acceleration of the elastic displacement field u_i times the density ρ of the medium. In all real solids (crystals with or without defects, glasses), there is a dissipative component σ_{ij}' to the stress tensor due to the finite velocity of the motion (i.e., the finite deformation rate), which causes dissipation, with the form $\sigma_{ij}' = \eta_{ijkl} \nabla_k \dot{u}_l$, which can be derived by symmetry arguments and using the Rayleigh dissipation function [31]. Here η_{ijkl} represents the viscosity tensor.

For an isotropic elastic medium, such as a glass or a cubic crystal, the viscosity tensor can be written as [31] $\eta_{ijkl} = \zeta \delta_{ij}\delta_{kl} + \eta(\delta_{ik}\delta_{jl} + \delta_{il}\delta_{jk} - 2\delta_{ij}\delta_{kl}/d)$ with d the number of spatial dimensions. Here ζ denotes the bulk viscosity,[3] while η is the shear viscosity coefficient. The total stress tensor is then the sum of the elastic and of the dissipative components, $\sigma_{ij} \Rightarrow \sigma_{ij}^{el} + \sigma_{ij}'$, where $\sigma_{ij}^{el} = 2Gu_{ij} + \lambda u_{ll}\delta_{ij}$, where u_{ij} is the linearized strain tensor of elasticity theory, G is the shear modulus,

---

[2] In this chapter, we use the Greek letter ζ to denote the microscopic friction instead of ν used in other chapters.

[3] Not to be confused with the Langevin-type microscopic friction coefficient, also denoted as ζ in the chapter.

and λ is the longitudinal Lamé coefficient. Hence we have:

$$\nabla_j \sigma_{ij}^{el} = G \nabla^2 u_i + \frac{(\lambda + 2G)}{2(1 - \nu)} \nabla_i \nabla_l u_l$$

$$\nabla_j \sigma_{ij}' = \nabla_j \eta \left(\delta_{ik} \delta_{jl} + \delta_{il} \delta_{jk} - \frac{2}{d} \delta_{ij} \delta_{kl} \right) \nabla_k \dot{u}_l \qquad (4.3)$$

where ν is the Poisson ratio.

These expressions are replaced in Eq. (4.2), and the displacement field is split into transverse (T) and longitudinal (L) components $u_{T,L}^i = P_{T,L}^{ij} u_j$ using the projectors $P_T^{ij}(k) = \delta^{ij} - k^i k^j / k^2$, $P_L^{ij}(k) = k^i k^j / k^2$, where $\mathbf{k}$ is the wavevector. Next, we first take the divergence of both sides noting $k_i u_T^i = 0$ (or equivalently in vectorial notation $\nabla \cdot \mathbf{u}_T = 0$); we are left, after some algebra and after taking the Fourier transform in space of both sides, with:

$$\ddot{u}_L + v_L^2 \, k^2 \, u_L + \frac{1}{2\rho} \left[\zeta + \frac{2(d-1)}{d} \eta \right] k^2 \, \dot{u}_L = f_L^{ext}$$

$$\ddot{u}_T + v_T^2 \, k^2 \, u_T \frac{\eta}{\rho} k^2 \, \dot{u}_T = f_T^{ext}. \qquad (4.4)$$

These equations are in the familiar mathematical form of a forced damped harmonic oscillator. The Green's function is readily found by replacing the external force with a δ-function source:

$$[\partial_t^2 + (\eta/\rho) \, k^2 \partial_t + v_T^2 \, k^2] \mathcal{G}_T(k, t - t') = \delta(t - t') \qquad (4.5)$$

and upon Fourier-transforming in time, we get:

$$\mathcal{G}_T(k, \omega) = \frac{1}{-\omega^2 + (G/\rho) \, k^2 - i \, \omega \, (\eta/\rho) \, k^2} \qquad (4.6)$$

and, mutatis mutandis, an analogous expression for the longitudinal (L) Green's function.

In general, we thus have the Green's functions for L and T modes in the following generic form:

$$\mathcal{G}_{L,T}(k, \omega) = \frac{1}{\Omega_{L,T}^2(k) - \omega^2 - i \, \omega \, \Gamma_{L,T}(k)}, \qquad (4.7)$$

with the resonances providing the following set of dispersion relations for transverse and longitudinal phonons, respectively:

$$\Omega_{L,T} = v_{L,T} \, k - i \, D_{L,T} \, k^2, \tag{4.8}$$

$$D_T = \frac{\eta}{2\rho}, \quad D_L = \frac{1}{2\rho}\left[\zeta + \frac{2(d-1)}{d}\eta\right].$$

The respective speeds of sound are given by:

$$v_T^2 = \frac{G}{\rho}, \quad v_L^2 = \frac{K + \frac{2(d-1)}{d}G}{\rho} \tag{4.9}$$

where G is the shear modulus and K is the bulk modulus. In general, $v_L > v_T$, and $G < K$, for solids with Poisson ratio in the usual range $0 < \nu < 1/2$. Using Eqs. (4.7) and (4.8), we therefore identify $\Omega_{L,T}(k) = v c_{L,T} \, k$ and

$$\Gamma_{L,T}(k) = 2D_{L,T} \, k^2, \tag{4.10}$$

i.e., a diffusive viscous damping, known as *Akhiezer sound damping* [36], and widely observed in amorphous solids [37]. The root cause of Akhiezer damping is *anharmonicity* (cfr. Chap. 1, Sect. 1.1.4).

The Akhiezer phonon damping coefficient D_L can be directly related to the anharmonicity of the interatomic potential via the Grüneisen parameter and Eq. (1.19) of Chap. 1.

The above derivation follows a hydrodynamic approach [31]; by comparing with the result of a microscopic approach based on the Boltzmann transport equation for phonons, it has been shown that [38]:

$$D_L = \frac{C_v T \tau_U}{2\rho}\left(\frac{4}{3}\langle\gamma_{xy}^2\rangle - \langle\gamma_{xy}\rangle^2\right) \approx \frac{C_v T \tau_U}{2\rho}\langle\gamma_{xy}^2\rangle \tag{4.11}$$

where the last approximation for acoustic modes can be motivated with the typical wild fluctuations of the Grüneisen parameter γ for low-frequency vibrational excitations in both crystals [39] and metal alloys [40]; hence $\langle\gamma_{xy}\rangle \approx 0$.

Here we neglected the contribution from bulk viscosity ζ, since normally $\eta \gg \zeta$. Furthermore, $\langle\ldots\rangle$ indicates averaging with respect to the Bose-Einstein distribution as a weight, while γ_{xy} is the xy component of the tensor of Grüneisen constants. Also, C_v is the specific heat at constant volume, while τ_U is the average time interval between two Umklapp scattering events. Since $\tau_U \sim T^{-1}$ (which is an experimental observation for most solids [38, 41]), the diffusive constant D_L and also the sound damping are weakly dependent on temperature, i.e., a well-known experimental fact [41], in the Akhiezer regime.

A substantially equivalent expression for the damping of longitudinal phonons, in terms of an average Grüneisen constant of the material γ_{av}, was proposed by

Boemmel and Dransfeld [41]

$$D_L \approx \frac{C_v T \tau_U}{2\rho} \gamma_{av}^2 \tag{4.12}$$

and provides a good description of the Akhiezer damping measured experimentally in quartz at $T > 60$ K [41].

Importantly, in this theory, we recover the ubiquitously observed $\Gamma \sim k^2$ scaling of the acoustic phonon linewidth, without any unnecessary assumption about disorder or fluctuating elastic constants. Furthermore, Eq. (4.10) represents the lowest order of an effective field theory (EFT) or hydrodynamic expansion where symmetry (total energy conservation) forbids odd powers of k.

4.3   Microscopic Theory of Sound Attenuation in Amorphous Solids

Since the mid-twentieth century, it is well known that wave attenuation in the acoustic regime of low frequencies and wavelengths is dominated by anharmonic processes at the lowest wavevectors (the so-called hydrodynamic regime). In this regime, the sound attenuation, or damping, scales with the wavevector as $\Gamma \sim k^2$. This is a diffusive law, which can be derived by using conservation equations, i.e., with the methods of hydrodynamic theory and effective field theory [42], as we did in Sect. 4.2.

As already mentioned, in the context of crystals, this is also known as Akhiezer damping [36], whereas in liquids, this is known as the Brillouin linewidth. The same phenomenon is known to dominate sound attenuation in amorphous solids, such as glasses, at low k, where it plays an important role in determining the thermal conductivity, as we shall see later in Chap. 6 [43]. More recently, the diffusive nature of vibrational excitations in glasses has been pointed out in numerical simulations [44–46] and used as the starting point for theoretical models of the boson peak caused by anharmonicity in glasses and crystals [47, 48].

In glasses, upon going to higher wavevectors, a crossover from $\sim k^2$ to a $\sim k^4$ regime is typically observed experimentally [29], where the $\sim k^4$ scaling has been interpreted as Rayleigh-type scattering from random fluctuations of some (usually macroscopic) quantity. In this sense, the fluctuating elasticity theory has provided a derivation of this Rayleigh-type damping based on the assumption of macroscopic or, at best, mesoscopic, Gaussian spatial fluctuations of the shear modulus [49]. This theory, however, is entirely at the continuum level; hence, it does not account for the microscopic structural order/disorder [50] nor for the underlying microscopic nonaffine particle dynamics, as highlighted in recent works [51, 52].

Importantly, it has been recently shown using numerical simulations [16] that the "harmonic-type" random elastic fluctuations described by fluctuating elasticity theory are not the only mechanism behind the ubiquitous Rayleigh term $\sim k^4$ and may be relegated to a subsidiary role in view of their being two orders of magnitude smaller compared to the observed damping. Nonaffine motions, which arise from the dynamics in non-centrosymmetric environments [50] as discussed extensively in Chap. 2, provide a decisive contribution to the Rayleigh-type damping, and they are indeed dominant with respect to the fluctuating-elasticity contribution, even at $T = 0$ [16]. In the following, we will provide such derivation from microscopic particle motions.

While the previous models of wave damping in amorphous solids are invariably at the continuum level (or at best effective medium theories [9]), the *nonaffine response formalism* or nonaffine lattice dynamics [33, 53–57] presented in the previous chapters provides a more microscopic description. As we saw in Chap. 3, this framework provides quantitatively accurate predictions of the viscoelastic moduli of glasses with no fitting parameters in good agreement with simulations [33]. It is by now recognized that nonaffine motions play a central role in determining the dynamics of glasses at the microscopic level.

In the following, we present the theory of acoustic wave attenuation in amorphous solids based on nonaffine motions, and we analytically predict the contributions from anharmonic nonaffinity to the damping $\Gamma(k)$ as a function of k. The theory recovers the hydrodynamic diffusive damping at low k, including the prefactor, which is related to important physical parameters such as the Debye frequency, the nonaffine correction to the shear modulus, and the microscopic friction due to anharmonicity derived in Sect. 3.2.1. The theory provides a useful closed-form analytical expression for the Rayleigh contribution $\Gamma \sim k^4$ arising from nonaffinity and identified in the simulations of [16] as the dominant effect on sound damping in amorphous solids. Finally, the theory also predicts the experimentally observed [29] crossover from diffusive $\sim k^2$ damping to Rayleigh $\sim k^4$ damping at higher k and yields an estimate of the critical wavelength where the crossover occurs, in terms of physical parameters.

4.3.1 Linear Response Theory

In linear response theory, the time-dependent expectation value of the stress tensor $\sigma^{ij}(t)$ is given as a function of the history-dependent strain in terms of a linear convolution:

$$\langle \sigma^{ij}(t) \rangle \simeq \int_{-\infty}^{\infty} \chi_{\sigma\sigma}^{ijkl}(t - t') \, \gamma^{kl}(t') + O(\gamma^2). \qquad (4.13)$$

with the strain tensor γ^{ij}, whose kernel is given by the dynamic response function (sometimes also labelled two-point function, correlator, or Green function), $\chi_{\sigma\sigma}^{ijkl}(t)$, i.e., the stress autocorrelation function [58]. Neglecting dissipation, the zero-frequency limit of the Fourier-transformed response function reduces to the elastic tensor.

From now on, we will focus only on the shear response ($ij = xy$), and therefore Latin indices will be omitted to simplify the notation. Upon Fourier transforming the previous equation, we have:

$$\sigma(\omega) = \chi_{\sigma\sigma}(\omega)\,\gamma(\omega),\tag{4.14}$$

which is valid in the linear regime, to leading order in the external shear strain γ. Importantly, in general, $\chi_{\sigma\sigma}(\omega)$ is a complex-valued function. Its real and imaginary components are related to the reactive and dissipative parts of the response function, respectively, and it coincides with the complex dynamic modulus used in viscoelasticity theory (cfr. Chap. 3). The real and imaginary components are mutually related via the standard Kramers-Kronig relations [59] (reflecting causality). In a simple viscoelastic system, to leading order in ω, they are given by (see e.g. Ref. [60]):

$$\chi_{\sigma\sigma}(\omega) = G + i\,\omega\,\eta + O\left(\omega^2\right)\tag{4.15}$$

where G is the static shear modulus and η the shear viscosity.

4.3.2 Dynamic Response Function: From the Stress to the Displacement Correlator

Let us consider a generic operator ϕ and its conjugate external field δh. We define the dynamic response function $\chi_{\phi\phi}$ associated with such operator as:

$$\delta\langle\phi\rangle(\omega, k) = \chi_{\phi\phi}(\omega, k)\,\delta h(\omega, k)\tag{4.16}$$

where linear response is assumed as well as space-time translational invariance. The response function, as already mentioned, is complex valued, $\chi = \chi' + i\chi''$, and the Kramers-Kronig relations are read as:

$$\chi'(\omega) = \mathcal{P}\int_{-\infty}^{\infty}\frac{\chi''(\omega')}{\omega' - \omega}\frac{d\omega'}{\pi}.\tag{4.17}$$

Let us consider the case of transverse phonons and elasticity (cfr. Section 7.3.1 in Ref. [58]). By denoting the transverse displacement as $\mathbf{u}_T$ and the external force as $\mathbf{f}$, we obtain:

$$\chi_{\mathbf{u}_T\mathbf{u}_T}(k, \omega) \equiv \frac{\mathbf{u}_T}{\mathbf{f}} = \frac{1}{-\rho\,\omega^2 + G\,k^2 - i\omega\,\eta\,k^2}\tag{4.18}$$

which can be generalized to:

$$\chi_{\mathbf{u}_T\mathbf{u}_T}(k, \omega) = \frac{1}{\rho}\frac{1}{-\omega^2 + \Omega(k)^2 - i\omega\,\Gamma(k)}\tag{4.19}$$

where:

$$\Omega(k)^2 = v_T^2 k^2 + \dots, \qquad \Gamma(k) = D k^2 + \dots \tag{4.20}$$

with the transverse speed of sound $v_T^2 = G/\rho$ and Akhiezer diffusive coefficient $D = \eta/\rho$, as expected.

We now go back to Eq. (4.18) and power-expand it at low frequency, to get:

$$\chi_{\mathbf{u}_T \mathbf{u}_T}(k, \omega) = \frac{1}{k^2 G} + i \frac{\eta \omega}{k^2 G^2} + \dots \tag{4.21}$$

At $\omega = 0$, this gives the static susceptibility:

$$\chi_{\mathbf{u}_T \mathbf{u}_T}(k, \omega = 0) = \frac{1}{k^2 G}, \tag{4.22}$$

which coincides with Eq. 6.4.24 in Ref. [58].

Now, comparing Eq. (4.21) with Eq. (4.15), we notice that the two response functions are related via:

$$\chi_{\sigma_T \sigma_T}(k, \omega) = G^2 k^2 \chi_{\mathbf{u}_T \mathbf{u}_T}(k, \omega) \tag{4.23}$$

a result that will be used in the following manipulations on the nonaffine part of the stress and the displacement.

4.3.3 From Nonaffine Motions to the Susceptibility

We should recall from Chap. 2 that, in disordered solids, each particle is displaced to a position dictated by the macroscopic strain tensor $\mathbf{F}$, according to $\mathbf{r}_i = \mathbf{F}(\gamma) \cdot \mathbf{r}_{i,0}$. This position is called the *affine* position. While a tagged particle i is moving under applied strain toward its affine position, also its nearest neighbors are being displaced in the local force field of interaction. Therefore, the net force acting on i in the affine position is not zero, due to the lack of inversion symmetry, as discussed in Chap. 2, Sect. 2.4.1. This force vector is denoted as Ξ_i and is called the affine force field since it represents the forces that trigger the nonaffine displacements, cfr. Eq. (2.18) in Chap. 2.[4]

As discussed in Chap. 2, a useful way of representing the nonaffine displacements is by defining [54]:

$$\mathbf{r}_i = \mathbf{F}(\gamma) \cdot \mathring{\mathbf{r}}_i(\gamma) \tag{4.24}$$

---

[4] One can show that the mean squared Ξ_i is proportional to the mean squared nonaffine displacement, where the nonaffine displacement is defined as $\delta \mathbf{r}_{NA}$ in $\mathbf{r}_i(\gamma) = \mathbf{F}(\gamma) \cdot \mathbf{r}_{i,0} + \delta \mathbf{r}_{NA}$, which gives the final position of the particle i in the deformed frame.

where the new variable $\mathring{\mathbf{r}}_i$ does the book-keeping of the nonaffine displacements as measured in the undeformed frame.

As was derived in Chap. 2, the affine force vector can be written as follows [54]:

$$\Xi_i = -\frac{\partial U}{\partial \mathring{\mathbf{r}}_i \partial \gamma}\bigg|_{\gamma \to 0}. \tag{4.25}$$

Working through the same steps reported in Chap. 3, Sect. 3.2.2, leading to Eq. (3.34), one arrives at (cfr. Eq. (4.14)):

$$\sigma(\omega) = \chi_{\sigma\sigma}(\omega)\,\gamma(\omega)$$

$$= \left(G_A + \sum_p \frac{\hat{\Xi}_p^2}{\omega^2 - \omega_p^2 - i\omega\zeta} \right) \gamma(\omega) \tag{4.26}$$

where ζ is the (Langevin-type) microscopic damping coefficient for particle motion (cfr. Sect. 3.2.1). As before (cfr. Sect. 3.2.2), $\hat{\Xi}_p$ is the projection of the $3N$-dimensional vector Ξ onto the $3N$-dimensional eigenvector of the Hessian matrix $|\mathbf{p}\rangle$. Using the Dirac's bra-ket notation introduced in Chap. 2, $\hat{\Xi}_p = \langle\Xi|\mathbf{p}\rangle$. As usual, G_A is the affine shear elastic modulus, ω_p denotes the p-th eigenfrequency of the solid, and $\omega_p^2 = \lambda_p$, where λ_p is the p-th eigenvalue associated with eigenvector $|\mathbf{p}\rangle$. Prefactors with dimension of volume have been omitted, and we assumed that particles masses are all equal to one. We recall from Chap. 2 that the Hessian matrix is defined as (cfr. Eq. (2.17)):

$$\mathbf{H}_{ij} = \frac{\partial U}{\partial \mathring{\mathbf{r}}_i \partial \mathring{\mathbf{r}}_j}\bigg|_{\gamma \to 0} = \frac{\partial U}{\partial \mathbf{r}_i \partial \mathbf{r}_j}\bigg|_{\mathbf{r} \to \mathbf{r}_0} \tag{4.27}$$

since $\mathring{\mathbf{r}}(\gamma)|_{\gamma \to 0} = \mathbf{r}_0$. We recall that this is a $3N \times 3N$ matrix with $p = 1, \ldots, 3N$ eigenvalues λ_p and associated eigenvectors $|\mathbf{p}\rangle$. From our derivation of viscoelastic moduli in Chap. 3, it is clear that there exists two distinct contributions, one coming from affine displacements and encoded in the affine (infinite-frequency) shear modulus G_A and a second contribution (in bracket) arising from *nonaffine* motions, which is controlled by the quantity $\hat{\Xi}_p^2$, i.e., the square of the affine force field projected in the basis of the eigenvectors of the Hessian. Again consistent with what we derived in Chap. 3, the real part of this second term in bracket represents the (negative, softening) nonaffine contribution to the shear modulus G_{NA} (or, with alternative notation, G_{NA}), defined consistent with:

$$G = G_A - G_{NA} \tag{4.28}$$

which is the total zero-frequency shear modulus containing both affine and nonaffine contributions. As shown in Chap. 3, the nonaffine term may include a dissipative (damping) component. In particular, we shall demonstrate that the imaginary part of the second term in the bracket of Eq. (4.26) is non-zero, and it gives a finite contribution to the sound attenuation constant. This contribution, clearly, arises entirely from nonaffine motions.

Hence, we can identify a susceptibility component arising from nonaffine motions, as follows:

$$\chi_{NA}(\omega) = \sum_p \frac{\hat{\Xi}_p^2}{\omega^2 - \omega_p^2 - i\omega\zeta}. \tag{4.29}$$

where from now, we will drop the σ labels for ease of notation.

As done in Chap. 3, Sect. 3.2.2, we can turn the discrete sum over eigenmodes in Eq. (4.29) into a continuous integral over eigenfrequency. This is done, as usual, by introducing the vibrational density of states (VDOS) of the solid [54]. We obtain [61]:

$$\chi_{NA}(\omega) = 3\rho \int_0^{\omega_D} \frac{g(\omega_p)\,\xi(\omega_p)}{\omega^2 - \omega_p^2 - i\omega\zeta} d\omega_p \tag{4.30}$$

where $\rho = N/V$ is the particle density and ω_D is the Debye frequency, and consistent with what we did in Sect. 3.2.2, we have defined:

$$\xi(\omega_p) = \langle\hat{\Xi}_p^2\rangle_{\omega_p \in [\omega_p + d\omega_p]} \tag{4.31}$$

where the average is performed for all the projections of Ξ on eigenvectors $|\mathbf{p}\rangle$ with eigenfrequency $\omega_p \in [\omega_p + d\omega_p]$, cfr. Eq. (3.33). Note that in the previous literature as well as in Chap. 3, $\xi(\omega_p)$ was denoted as $\Gamma(\omega_p)$, and the present change of notation is done to avoid confusing it with the sound attenuation constant discussed in this chapter, also denoted as Γ.

In Chap. 2 (and originally in Ref. [56]), it was shown that for amorphous solids in d space dimensions, cfr. Eq. (2.46):

$$\langle\Xi|\mathbf{p}\rangle\langle\mathbf{p}|\Xi\rangle = d\,\kappa\,R_0^2\,\lambda_p \sum_\alpha B_{\alpha,xyxy} \tag{4.32}$$

where $\alpha = x, y, z$ and $B_{\alpha,xyxy}$ represents a set of coefficients which originate from an angular average over nearest-neighbor bond orientation vectors. The proportionality to λ_p has been verified in numerical simulations of different glassy systems in [33, 61]. Note that $G_{NA} = \frac{1}{V}\sum_p \frac{\langle\Xi|\mathbf{p}\rangle\langle\mathbf{p}|\Xi\rangle}{\lambda_p}$, which is Eq. (2.48) in Chap. 2.

The above equation therefore implies that $\xi(\omega_p) \propto \lambda_p \propto \omega_p^2$ and consequently:

$$\chi_{NA}(\omega) = c \int_0^{\omega_D} \frac{\omega_p^4}{\omega^2 - \omega_p^2 - i\omega\zeta} d\omega_p \tag{4.33}$$

where we used the Debye law for the VDOS, $g(\omega_p) \sim \omega_p^2$, which we anticipate from Chap. 5 (cfr. Sect. 5.1.1 and Eq. (5.4)), and we lumped all the prefactors into an overall prefactor c with units $[Pa \cdot s^3]$ whose physical interpretation will become clear in the following. The scaling $\xi(\omega_p)g(\omega_p) \sim \omega_p^4$ has been successfully verified in the numerical simulations of [16].

4.3.4 Acoustic Wave Propagation

In this section, we will identify nonaffine corrections to the speed of sound and their contribution to the sound attenuation. After having determined the stress-stress autocorrelation function, we can use it to obtain the transverse displacement autocorrelation function $\chi_{\mathbf{u}_T \mathbf{u}_T}(\omega, k)$, which in the following will be denoted as $C(\omega, k)$. As shown also in Section 4.2, the latter can be written in a damped harmonic oscillator (DHO) form as [16, 27, 58, 62]:

$$C(\omega, k) = \frac{1}{m} \frac{1}{-z^2 + E(k)^2 - \Sigma(\omega, k)}. \tag{4.34}$$

Here $z \equiv \omega + i\epsilon$ is the complex-valued frequency, $E(k)^2 = v^2 k^2$ (with v the Born or high-frequency speed of transverse sound in this case). $\Sigma(\omega, k)$ denotes the self-energy (cfr. Eq. (1.21) in Chap. 1) and m the mass density. By setting the self-energy to zero, solving for the poles of the Green function in Eq. (4.34) yields $\omega = \pm v\,k$, i.e., the dispersion relation of the acoustic phonons in which the effects of nonaffinity are neglected (the corresponding speed of sound v is proportional to the square root of the affine modulus). Again, explicit indices of the branches (L or T) are not shown, and we consider only the transverse phonons correlation function.

Next, we decompose the self-energy into real and imaginary parts:

$$\Sigma(\omega, k) = \Sigma'(\omega, k) + i\,\Sigma''(\omega, k), \tag{4.35}$$

so that the dispersion relation of the sound mode can be retrieved as a solution to the following equation:

$$\omega^2 = v^2 k^2 - \Sigma'(\omega, k) - i\,\Sigma''(\omega, k). \tag{4.36}$$

This should be compared with the general form:

$$\omega^2 = \Omega(k)^2 - i\,\omega\,\Gamma(k). \tag{4.37}$$

where $\Gamma(k)$ is the sound attenuation [58].

A simple comparison shows that:

$$\Omega(k)^2 = v^2 k^2 - \Sigma'(vk, k), \qquad \Gamma(k) = \frac{1}{vk}\Sigma''(vk, k). \tag{4.38}$$

The expressions in Eq. (4.38) are valid only in the region where a linear dispersion relation between phonon frequency and momentum holds. Equation (4.38) is in agreement with the discussion in [16] and will serve as the starting point to derive a microscopic expression for the sound attenuation based on the nonaffine formalism in the next steps.

As a consistency check, we can study the nonaffine correction to the sound speed given by:

$$v'^2 = v^2 - \frac{\Sigma'(vk, k)}{k^2} = \frac{G_A - G_{NA}}{m}, \tag{4.39}$$

and verify that it correctly reaches a constant value in the limit $k \to 0$. For that to be true, it is necessary that:

$$\lim_{k \to 0} \Sigma'(vk, k) \sim k^2. \tag{4.40}$$

Now, using the above results in Eq. (4.34) and sticking to the linear dispersion assumption, we get:

$$\text{Re}\, C(vk, k) = -\frac{1}{m} \frac{\Sigma'(vk, k)}{\Sigma'(vk, k)^2 + \Sigma''(vk, k)^2} \tag{4.41}$$

and then:

$$\text{Re}\, \chi_{NA}(vk, k) \sim -\frac{k^2 \,\Sigma'(vk, k)}{\Sigma'(vk, k)^2 + \Sigma''(vk, k)^2} \tag{4.42}$$

where all the dimensionful parameters have been omitted.

From Eq. (4.33), we see that, upon taking the limit of zero frequency (which corresponds also to zero momentum), the real part of the nonaffine dynamic response function approaches a negative constant, $\text{Re}\, \chi_{NA}(vk, k) \sim -\beta$, with $\beta > 0$. Therefore, we have that:

$$\lim_{k \to 0} \frac{k^2 \,\Sigma'(vk, k)}{\Sigma'(vk, k)^2 + \Sigma''(vk, k)^2} \sim \beta > 0. \tag{4.43}$$

Now, we know that $\Sigma'' \sim k^3$, in the regime of low frequency and momentum, which implies:

$$\lim_{k \to 0} \frac{k^2 \,\Sigma'(vk, k)}{\Sigma'(vk, k)^2 + k^6} \sim \beta > 0. \tag{4.44}$$

By power-expanding at small wavenumber k, we get $\Sigma' \sim k^2/\beta > 0$, which indeed gives a constant and importantly a *negative* nonaffine correction to the speed of sound and crucially linked with the negative nonaffine correction to the static shear modulus (cfr. Eq. (4.39)), in agreement with all previous works [16, 33, 54–56, 61] and with Eqs. (2.30) and (2.50) derived in Chap. 2.

Based on the above analysis, we can now rewrite the dimensionful parameter c in terms of underlying physical quantities. By restoring all the prefactors in the above formulas, one, at low energy, gets:

$$\text{Re}\,\chi_{NA}(vk, k) = -\frac{G_{NA}^2 k^2}{m} \frac{1}{\Sigma'(vk, k)}. \tag{4.45}$$

Furthermore, by using Eq. (4.39), we can rewrite the real part of the self-energy as:

$$\Sigma'(vk, k) = \frac{G_{NA}}{m} k^2 + \ldots \tag{4.46}$$

Equations (4.46) and (4.45) imply:

$$\text{Re}\,\chi_{NA}(vk, k) = -G_{NA} \tag{4.47}$$

which allows us to determine c in terms of physical quantities as:

$$c = \frac{3\,G_{NA}}{\omega_D^3}. \tag{4.48}$$

The result above proves useful to express the sound attenuation constant in terms of fundamental physical quantities, as we shall see later.

4.3.5   Acoustic Wave Attenuation

Sound attenuation is encoded in the imaginary part of the susceptibility. We start by taking the imaginary part of the autocorrelation function in Eq. (4.34) and obtain:

$$\text{Im}\,C(vk, k) = \frac{1}{m} \frac{\Sigma''(vk, k)}{\Sigma''(vk, k)^2 + \Sigma'(vk, k)^2} \tag{4.49}$$

where we also used the low-energy expression $\omega = vk$.

As demonstrated in [58] and above in Eq. (4.23), there is a simple relation between the stress autocorrelation function $\chi_{\sigma\sigma}$ and the displacements autocorrelation function χ_{uu}. Restricting this relation to the nonaffine components (since anyway the affine ones do not contribute to dissipative properties), that yields:

$$\chi_{\sigma\sigma}^{NA}(\omega, k) = G_{NA}^2 k^2 \chi_{uu}^{NA}(\omega, k) = G_{NA}^2 k^2 C_{NA}(\omega, k). \tag{4.50}$$

This result (upon recalling that the affine part of $\chi(\omega)$ is independent of ω [54]) finally implies:

$$\text{Im}\,\chi_{NA}(vk, k) = \frac{G_{NA}^2\,k^2}{m}\,\frac{\Sigma''(vk, k)}{\Sigma''(vk, k)^2 + \Sigma'(vk, k)^2}. \tag{4.51}$$

Now, using Eq. (4.38) together with Eq. (4.28), one can easily verify that the real part of the self-energy at low momentum is given by $\Sigma'(vk, k) = G_{NA}/m\,k^2$. Hence, using $\Sigma'' = vk\Gamma$, from Eq. (4.38), we get:

$$\text{Im}\,\chi_{NA}(vk, k) = \frac{\Gamma(k)\,k\,G_{NA}^2\,m\,v}{k^2 G_{NA}^2 + \Gamma(k)^2 m^2 v^2}. \tag{4.52}$$

Now, let us assume that Γ grows faster than linear in k (as observed in all experimental and simulations results); therefore, at low k we have:

$$\text{Im}\,\chi_{NA}(vk, k) = m\,v\,\frac{\Gamma(k)}{k}. \tag{4.53}$$

This clearly implies that a linear in k term in the imaginary part of the self-energy will produce a diffusive $\sim k^2$ term in the damping, while a term k^3 will produce the Rayleigh damping $\sim k^4$ contribution. We shall now verify this by direct application of the nonaffine response theory outlined above.

We recall Eq. (4.30), and by performing the integral analytically, we obtain:

$$\chi_{NA}(\omega) = c\int_0^{\omega_D}\frac{\omega_p^4}{\omega^2 - \omega_p^2 - i\,\omega\,\zeta}\,d\omega_p = -\frac{c\,\omega_D^3}{3} - c\,\omega_D\,\omega\,(\omega - i\zeta)$$

$$+ c\,\omega^{3/2}(\omega - i\zeta)^{3/2}\tanh^{-1}\left(\frac{\omega_D}{\sqrt{\omega(\omega - i\zeta)}}\right). \tag{4.54}$$

We first focus on the behavior at the lowest order in frequency. In that limit, the above integral in Eq. (4.54) becomes:

$$\chi_{NA}(\omega) = -\frac{c\,\omega_D^3}{3} + i\,\zeta\,c\,\omega_D\,\omega + \dots \tag{4.55}$$

By comparing this with Eq. (4.53), we obtain:

$$\Gamma(k) = c\,\frac{\zeta\,\omega_D}{m}\,k^2, \tag{4.56}$$

which is the Akhiezer damping derived, this time, based on a microscopic theory (cfr. with results of Sect. 4.2 and in particular Eq. (4.10), where it was derived based on a hydrodynamic or continuum approach). This result can be simplified further by means of Eq. (4.48), leading to:

$$\Gamma(k) = \frac{3\zeta\, G_{NA}}{m\,\omega_D^2} k^2 . \tag{4.57}$$

This is an important result: it shows that the coefficient for the diffusive (Akhiezer) sound damping is proportional to the microscopic (Langevin) friction ζ, which in turn is related to long-range anharmonic couplings between atoms as discussed in Chap. 3, Sect. 3.2.1. It is also proportional to G_{NA}, meaning that in a perfect centrosymmetric crystal at zero temperature (where $G_{NA} = 0$), the sound damping would be identically zero. This is consistent with the expectation that sound damping is present only in anharmonic crystals with defects and/or thermal fluctuations[5] and in amorphous solids.

Hence, we found a contribution from nonaffine motions to the ubiquitous *diffusive* Akhiezer damping, which has been observed in countless experimental and simulation studies of glasses, liquids, and supercooled liquids [28, 45, 66]. Importantly, the microscopic theory explains that the nonaffine contribution to the coefficient D in $\Gamma(k) = Dk^2$ is proportional to the microscopic friction coefficient for particle motion, $D \propto \zeta$, which further clarifies the close connection of the Akhiezer damping with anharmonicity of particle motion.

Furthermore, the coefficient D of the Akhiezer damping obeys a diffusive law:

$$D = \frac{\langle l_{NA}^2 \rangle}{\tau_{\text{micro}}}, \qquad \tau_{\text{micro}}^{-1} \equiv \zeta, \qquad \langle l_{NA}^2 \rangle \equiv \frac{3\, v_{NA}^2}{\omega_D^2} \tag{4.58}$$

where τ_{micro} and $\langle l_{NA}^2 \rangle$ are the microscopic relaxation time and the nonaffine mean squared displacement, respectively, and where we have defined $v_{NA}^2 \equiv G_{NA}/m$ in analogy to its affine counterpart.

In general, the full expression in Eq. (4.54) is rather cumbersome and not particularly illuminating. Nevertheless, it becomes quite simple in the limit of small microscopic friction, $\zeta \ll 1$. The imaginary part of the $\tanh^{-1}$ in the regime of $\omega \gg \zeta$ and $\omega \ll \omega_D$ (small friction and low frequency) is approximately constant and equal to $-\pi/2$, leading to:

$$\text{Im}\chi_{NA}(\omega) = c\,\omega_D\,\omega\,\zeta + c\,\frac{\pi}{2}\,\omega^3 + \dots \tag{4.59}$$

This result implies that there exists a smooth crossover between a diffusive damping at low wavevector, $\Gamma(k) \sim k^2$, and a Rayleigh one, $\Gamma(k) \sim k^4$, at larger wavevector [18]:

$$\Gamma(k) = \Gamma_2 k^2 + \Gamma_4 k^4 . \tag{4.60}$$

---

[5] The latter cause nonaffinity also in perfect crystals as discussed in [63–65] and in Chap. 2.

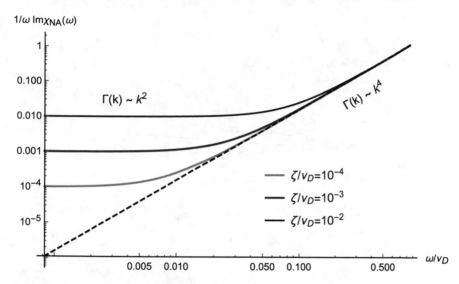

Fig. 4.3 The imaginary part of the nonaffine stress autocorrelation function as a function of the frequency ω. The different colors represent different values of the microscopic damping for particle motion, ζ. The dashed line guides the eye toward the Rayleigh damping scaling. The different regimes of sound damping Γ, Akhiezer followed by Rayleigh, predicted by the theory (via Eq. (4.52)), are indicated. The dimensionful parameter c is set to unity. The Debye frequency in the plots is indicated as v_D instead of ω_D as in the main text. Adapted from Ref. [18] with permission from the Institute of Physics Publishing

Importantly, the weight of the diffusive Akhiezer term at low wavevector is controlled by the magnitude of the microscopic Langevin friction ζ:

$$\frac{\Gamma_2}{\Gamma_4} = \frac{2}{\pi}\frac{\omega_D\,\zeta}{v^2} \tag{4.61}$$

and it vanishes in absence of microscopic friction, $\zeta = 0$.

The behavior of the imaginary part of the nonaffine stress correlation function is shown in Fig. 4.3 for different values of the microscopic friction coefficient ζ. The full function plotted in the figure displays a smooth crossover from a diffusive-Akhiezer regime (Im$\chi \sim \omega$) to a Rayleigh regime (Im$\chi \sim \omega^3$). The location of the crossover is controlled by the value of ζ: the larger the ζ, the more extended (and therefore more important) the diffusive regime. At very low values of the microscopic friction ζ, the function is well approximated by Eq. (4.59), and the diffusive regime is pushed to very low frequencies, whereas the Rayleigh damping regime becomes predominant.

For comparison, the low-frequency behavior of the nonaffine stress autocorrelation function is shown in Fig. 4.4 for different values of the microscopic friction ζ.

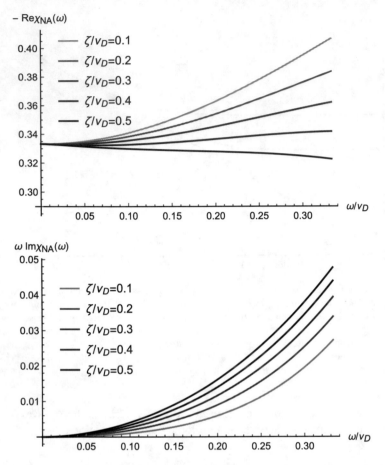

Fig. 4.4 Real and imaginary parts of the nonaffine stress autocorrelation function in the low-frequency regime. Different colors represent different values of the microscopic friction coefficient ζ. The dimensionful coefficient c is set to unity. The Debye frequency in the plots is indicated as v_D instead of ω_D as in the main text. Adapted from Ref. [18] with permission from the Institute of Physics Publishing

In the small friction limit $\zeta \ll 1$, we can also deduce the contribution of nonaffinity to the Rayleigh $\sim k^4$ term in the sound attenuation. In particular, the prefactor of the Rayleigh $\sim k^4$ law is given by:

$$\Gamma_4 = \frac{3\pi}{2} \frac{v^2 G_{NA}}{m \omega_D^3} \tag{4.62}$$

where G_{NA} is the nonaffine part of the shear modulus and m is the mass density. Interestingly, this contribution does not vanish in the limit $\zeta \to 0$, and it is therefore present even at $T = 0$ for athermal amorphous solids, as directly shown in the simulations of [16]. This is a fundamental relationship between the prefactor of the Rayleigh damping of sound waves and the nonaffine correction to the shear modulus.

Finally, in the limit of low microscopic friction ζ, our theory provides an analytic estimate of the crossover point k_* between the quadratic-in-k (Akhiezer) and quartic-in-k (Rayleigh) regimes in Eq. (4.60) given by:

$$k_*^2 = \frac{2\,\omega_D\,\zeta}{\pi\,v^2}\,. \tag{4.63}$$

Since the Debye frequency is proportional to the Debye temperature (cfr. Chap. 5, Sect. 5.1.2), this relation predicts that materials with a higher Debye temperature will have the crossover shifted to higher k values. The same happens for strong anharmonic couplings encoded in the friction parameter ζ.

In summary, we presented an analytical theory of sound attenuation in amorphous solids starting from single particle motion and taking into account the inherently nonaffine dynamics. We derived in closed form the contributions from nonaffinity to both the hydrodynamic diffusive term $\sim\Gamma_2 k^2$ (Eq. (4.57)) and the Rayleigh term $\sim \Gamma_4 k^4$ (Eq. (4.62)), with the prefactors Γ_2, Γ_4 expressed as functions of important physical parameters. In particular, the theory provides a direct link between sound damping and the nonaffine softening corrections to the shear modulus.

The Rayleigh attenuation due to nonaffine motions, Eq. (4.62), survives in the limit of zero temperature, or equivalently zero microscopic friction $\zeta = 0$, and it has been recently shown via simulations to be the dominant process of sound attenuation in amorphous solids in Ref. [16]. Furthermore, the theory analytically predicts the crossover from diffusive to Rayleigh damping at larger values of wavevector k (Eq. (4.63)), with the crossover being proportional to the microscopic friction for particle motion (cfr. Chap. 3, Sect. 3.2.1) and to the Debye temperature of the material (cfr. Eq. (5.8) in Chap. 5) via the Debye frequency.

The analytical prediction of Rayleigh damping from nonaffinity presented above is supported by recent numerical simulations data [16], which also confirm the fundamental importance and dominance of this mechanism, with respect to other effects, in the determination of sound attenuation in amorphous solids. In particular, the traditional Rayleigh scattering from large-scale fluctuations has been shown to provide a contribution orders of magnitude smaller than the one of nonaffine dynamics [16].

Additionally, the main result in Eq. (4.57), namely, the low-k sound attenuation constant being quadratic in the frequency, or wavevector, and linear in the microscopic damping ζ, is confirmed by numerical simulations of Ref. [17].

4.4   Ioffe-Regel Crossover

Historically, the concept of collective excitations as independent quasiparticles in solids has revived concepts of the kinetic theory of gases, such as the concept of mean free path ℓ, i.e., the average distance a quasiparticle travels between collisions. For example, the resistivity of metals, ρ, increases with temperature T, as is well known. Within the quasiparticle concept, this is equivalent to saying that the mean free-path ℓ becomes shorter as the number of scattering events increases with temperature T. In ordinary metals, defects are the dominant scatterers of electrons at low temperature, whereas phonons are the dominant scatterers at high temperatures, and accordingly, $\rho(T)$ is found to increase linearly with T above some typical phonon energy. In considering the problem of low mobilities in certain semiconductors, Ioffe and Regel [67] pointed out that a the decrease in ℓ with increasing temperature cannot continue indefinitely. They argued that ℓ can never become shorter than the interatomic spacing a, since in that situation, the assumption of coherent quasiparticle motion no longer holds. Similar arguments were expressed by Mott shortly thereafter, and the notion of a minimum metallic conductivity compatible with a minimum mean free path $\ell_{min} = a$ became known as the Mott-Ioffe-Regel (MIR) limit in the context of semiconductor and metal physics. In the same context, different definitions of the crossover have been used, including $k_F \ell_{min} \sim 1$ and $k_F \ell_{min} \sim 2\pi$, where k_F is the Fermi wavevector. A striking manifestation of the MIR criterion is represented by the phenomenon of resistivity saturation in metals. In certain metallic elements, alloys and intermetallic compounds characterized by high resistivity, $\rho(T)$, deviates from the linear-in-T behavior at high T, with a temperature dependence, which becomes increasing weaker and eventually "saturates," i.e., it approaches a constant value consistent with $\ell_{min} = a$, a phenomenon known as "resistivity saturation."

Since phonons are quasiparticles, it is natural to define a similar Ioffe-Regel crossover also for phonon propagation. In the case of phonons in amorphous solids, the mean free path ℓ is controlled by a complex interplay of disorder and anharmonicity. This is because both disorder and anharmonicity can promote scattering, and we have seen that there is an intimate connection between disorder and anharmonicity in disordered solids such as metallic glasses (cfr. Chap. 1, Sects. 1.1.4 and 1.6).

Let us consider a collective phonon mode moving in a solid. At any finite temperature, simple arguments from effective field theory and hydrodynamics [58] imply the presence of a finite relaxation rate $\Gamma = \tau^{-1}$ (which appears to be quadratic in the momentum k, $\Gamma \sim k^2$, as discussed in Sect. 4.2, cfr. Eq. (4.10)). This is tantamount to saying that the phonon will stop propagating ballistically at a certain frequency, known as the *Ioffe Regel frequency* ω_{IR} [67], and defined as the energy at which the mean free path of the phonon l becomes comparable to its wavelength λ:

$$\ell(\omega_{IR}) = \lambda(\omega_{IR}). \qquad (4.64)$$

The exact location of the IR frequency is actually a matter of definition. Another definition used in the literature is to set the crossover at the frequency at which the phonon mean free path becomes equal to half the wavelength of the phonon:

$$\ell = \frac{\lambda}{2}. \tag{4.65}$$

By identifying the dispersion relation of the damped phonon as the root of the following simple equation:

$$\omega^2 = v^2 k^2 - i\omega\,\Gamma(k), \tag{4.66}$$

then we have:

$$k = \frac{1}{v}\sqrt{\omega^2 - i\Gamma\omega} = k' - ik'', \tag{4.67}$$

where the wavenumber $k = k' - ik''$ picks up an imaginary part due to damping. By plugging $k = k' - ik''$ into the standard plane wave form $\psi \sim \exp[i(\omega t - kx)]$, the damped phonon wave amplitude is given by:

$$\psi(x,t) \sim e^{-k''x} e^{i(\omega t - k'x)} \tag{4.68}$$

where the spatial decay length:

$$k'' = \frac{\Gamma}{2v} \tag{4.69}$$

can be identified with the mean free path:

$$\ell = \frac{\Gamma}{2v}. \tag{4.70}$$

Using the criterion $\ell = \lambda$, together with Eq. (4.70), $v = \Omega/k$ and $k = 2\pi/\lambda$, we thus obtain the IR crossover as the point at which:

$$\Omega = \pi\,\Gamma \tag{4.71}$$

and it can be defined independently for the transverse and longitudinal branches.

A neat determination of the IR crossover by applying Eq. (4.71) to acoustic dispersion relations and corresponding linewidths (obtained from Lorentzian fittings of the dynamic structure factors of the phonons $S(k,\omega)$) is shown in Fig. 4.5.

The first observation of a strong, and possibly universal, correlation between the IR frequency of the transverse branch and the boson peak frequency (see the next Chap. 5, Sect. 5.2) appeared in [28] using Brillouin scattering of light and x rays in lithium diborate glass.

Fig. 4.5 Longitudinal and transverse dispersion relations in a model simulated glass. The gray band indicates the location of the boson peak frequency (see Chap. 5, Sect. 5.2). The Ioffe-Regel crossover in the transverse phonons appears around $q \approx 0.5$ and is estimated using Eq. (4.71). Figure adapted with permission from [45]

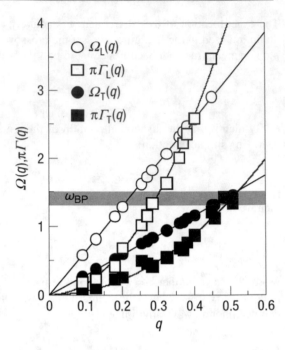

Despite some initial controversy [68], the universal correlation between the IR frequency of the transverse acoustic modes and the boson peak frequency has been confirmed using different experimental and simulation techniques [25, 28, 45, 69], and it is now widely accepted, at least in the ballpark of amorphous systems and glasses with structural disorder.

References

1. W.A. Phillips, *Amorphous Solids: Low-Temperature Properties* (Springer, Berlin, 1981)
2. R.O. Pohl, X. Liu, E. Thompson, Rev. Mod. Phys. **74**, 991 (2002)
3. U. Buchenau, G. D'Angelo, G. Carini, X. Liu, M.A. Ramos (2020). arXiv:2012.10139
4. P. Esquinazi, R. König, *Influence of Tunneling Systems on the Acoustic Properties of Disordered Solids* (Springer, Berlin, 1998), pp. 145–222
5. W. Schirmacher, G. Ruocco, T. Scopigno, Phys. Rev. Lett. **98**, 025501 (2007)
6. G. Baldi, V.M. Giordano, G. Monaco, B. Ruta, Phys. Rev. Lett. **104**, 195501 (2010)
7. C. Caroli, A. Lemaître, Phys. Rev. Lett. **123**, 055501 (2019)
8. Y.C. Hu, H. Tanaka, Nat. Phys. **18**(6), 669 (2022)
9. E. DeGiuli, A. Laversanne-Finot, G. Düring, E. Lerner, M. Wyart, Soft Matt. **10**, 5628 (2014)
10. H. Mizuno, G. Ruocco, S. Mossa, Phys. Rev. B **101**, 174206 (2020)
11. G. Monaco, S. Mossa, Proc. Natl. Acad. Sci. **106**(40), 16907 (2009)
12. G. Monaco, V.M. Giordano, Proc. Natl. Acad. Sci. **106**(10), 3659 (2009)
13. H. Mizuno, A. Ikeda, Phys. Rev. E **98**, 062612 (2018)
14. B. Ruta, G. , V.M. Giordano, L. Orsingher, S. Rols, F. Scarponi, G. Monaco, J. Chem. Phys. **133**(4), 041101 (2010)

15. A. Moriel, G. Kapteijns, C. Rainone, J. Zylberg, E. Lerner, E. Bouchbinder, J. Chem. Phys. **151**(10), 104503 (2019)
16. G. Szamel, E. Flenner, J. Chem. Phys. **156**(14), 144502 (2022)
17. T. Damart, A. Tanguy, D. Rodney, Phys. Rev. B **95**, 054203 (2017)
18. M. Baggioli, A. Zaccone, J. Phys. Condensed Matt. **34**(21), 215401 (2022)
19. W. Schirmacher, G. Ruocco, Heterogeneous elasticity: the tale of the boson peak, in *Low-Temperature Thermal and Vibrational Properties of Disordered Solids: A Half-Century of Universal "Anomalies" of Glasses*, ed. by M.A Ramos (World Scientific, Singapore 2020)
20. E. Rat, M. Foret, G. Massiera, R. Vialla, M. Arai, R. Vacher, E. Courtens, Phys. Rev. B **72**, 214204 (2005)
21. R. Vacher, E. Courtens, M. Foret, Phys. Rev. B **72**, 214205 (2005)
22. R. Vacher, J. Pelous, F. Plicque, A. Zarembowitch, J. Non-Crystall. Solids **45**(3), 397 (1981)
23. D. Fioretto, U. Buchenau, L. Comez, A. Sokolov, C. Masciovecchio, A. Mermet, G. Ruocco, F. Sette, L. Willner, B. Frick, D. Richter, L. Verdini, Phys. Rev. E **59**, 4470 (1999)
24. G. Ruocco, F. Sette, R. Di Leonardo, D. Fioretto, M. Krisch, M. Lorenzen, C. Masciovecchio, G. Monaco, F. Pignon, T. Scopigno, Phys. Rev. Lett. **83**, 5583 (1999)
25. B. Rufflé, M. Foret, B. Hehlen, Low-frequency vibrational spectroscopy of glasses, in *Low-Temperature Thermal and Vibrational Properties of Disordered Solids: A Half-Century of Universal "Anomalies" of Glasses*, ed. by M.A Ramos (World Scientific, Singapore 2020), pp. 227–298
26. G. Carini, M. Federico, A. Fontana, G.A. Saunders, Phys. Rev. B **47**, 3005 (1993)
27. G. Baldi, V.M. Giordano, G. Monaco, Phys. Rev. B **83**, 174203 (2011)
28. B. Rufflé, G. Guimbretière, E. Courtens, R. Vacher, G. Monaco, Phys. Rev. Lett. **96**, 045502 (2006)
29. C. Masciovecchio, G. Baldi, S. Caponi, L. Comez, S. Di Fonzo, D. Fioretto, A. Fontana, A. Gessini, S.C. Santucci, F. Sette, G. Viliani, P. Vilmercati, G. Ruocco, Phys. Rev. Lett. **97**, 035501 (2006)
30. G. Baldi, V.M. Giordano, B. Ruta, G. Monaco, Phys. Rev. B **93**, 144204 (2016)
31. L. Landau, E. Lifshitz, *Theory of Elasticity: Volume 6* (Pergamon Press, Oxford, 1986)
32. R. Zwanzig, J. Statist. Phys. **9**(3), 215 (1973)
33. V.V. Palyulin, C. Ness, R. Milkus, R.M. Elder, T.W. Sirk, A. Zaccone, Soft Matt. **14**, 8475 (2018)
34. I. Kriuchevskyi, V.V. Palyulin, R. Milkus, R.M. Elder, T.W. Sirk, A. Zaccone, Phys. Rev. B **102**, 024108 (2020)
35. P. Chaikin, T. Lubensky, *Principles of Condensed Matter Physics* (Cambridge University Press, Cambridge, 2000)
36. A.I. Akhiezer, J. Phys. **1**, 277 (1939)
37. Y.M. Beltukov, D.A. Parshin, V.M. Giordano, A. Tanguy, Phys. Rev. E **98**, 023005 (2018)
38. H.J. Maris, *Physical Acoustics*, vol. VIII (Academic, London, 1971)
39. D. Cuffari, A. Bongiorno, Phys. Rev. Lett. **124**, 215501 (2020)
40. Z.Y. Yang, Y.J. Wang, A. Zaccone, Phys. Rev. B **105**, 014204 (2022)
41. H.E. Bömmel, K. Dransfeld, Phys. Rev. **117**, 1245 (1960)
42. P.C. Martin, O. Parodi, P.S. Pershan, Phys. Rev. A **6**, 2401 (1972)
43. J.M. Ziman, *Electrons and Phonons: The Theory of Transport Phenomena in Solids.* Oxford Classic Texts in the Physical Sciences (Oxford University Press, Oxford, 2001)
44. P.B. Allen, J.L. Feldman, J. Fabian, F. Wooten, Philos. Mag. B **79**(11–12), 1715–1731 (1999)
45. H. Shintani, H. Tanaka, Nat. Mater. **7**(11), 870 (2008)
46. Y.M. Beltukov, V.I. Kozub, D.A. Parshin, Phys. Rev. B **87**, 134203 (2013)
47. M. Baggioli, A. Zaccone, Phys. Rev. Lett. **122**, 145501 (2019)
48. M. Baggioli, A. Zaccone, Phys. Rev. Res. **2**, 013267 (2020)
49. W. Schirmacher, Europhys. Lett. **73**(6), 892 (2006)
50. R. Milkus, A. Zaccone, Phys. Rev. B **93**, 094204 (2016)
51. C. Caroli, A. Lemaître, Phys. Rev. Lett. **123**, 055501 (2019)
52. C. Caroli, A. Lemaître, J. Chem. Phys. **153**(14), 144502 (2020)

53. B.A. DiDonna, T.C. Lubensky, Phys. Rev. E **72**, 066619 (2005)
54. A. Lemaître, C. Maloney, J. Statist. Phys. **123**(2), 415 (2006)
55. C.E. Maloney, A. Lemaître, Phys. Rev. E **74**, 016118 (2006)
56. A. Zaccone, E. Scossa-Romano, Phys. Rev. B **83**, 184205 (2011)
57. S. Saw, S. Abraham, P. Harrowell, Phys. Rev. E **94**, 022606 (2016)
58. P.M. Chaikin, T.C. Lubensky, *Principles of Condensed Matter Physics* (Cambridge University Press, Cambridge, 1995)
59. J.S. Toll, Phys. Rev. **104**, 1760 (1956)
60. A.C. Pipkin, *Lectures on Viscoelasticity Theory*, vol. 7 (Springer Science & Business Media, Cham, 2012)
61. R. Milkus, A. Zaccone, Phys. Rev. E **95**, 023001 (2017)
62. Y. Xu, J.S. Wang, W. Duan, B.L. Gu, B. Li, Phys. Rev. B **78**, 224303 (2008)
63. T. Das, S. Sengupta, M. Rao, Phys. Rev. E **82**, 041115 (2010)
64. S. Ganguly, S. Sengupta, P. Sollich, M. Rao, Phys. Rev. E **87**, 042801 (2013)
65. D. Squire, A. Holt, W. Hoover, Physica **42**(3), 388 (1969)
66. J. Hansen, I. McDonald, *Theory of Simple Liquids* (Elsevier Science, Amsterdam, 2006)
67. A. Ioffe, A. Regel, Prog. Semicond. **4**, 237 (1960)
68. T. Scopigno, J.B. Suck, R. Angelini, F. Albergamo, G. Ruocco, Phys. Rev. Lett. **96**, 135501 (2006)
69. Y.M. Beltukov, V.I. Kozub, D.A. Parshin, Phys. Rev. B **87**, 134203 (2013)

Phonons and Vibrational Spectrum

5

Abstract

In Chap. 4, we considered how elastic (acoustic) waves propagate through an isotropic amorphous solid, and we paid special attention to the dissipative mechanisms, which are responsible for the damping of the waves. In this chapter, we shall consider how the frequencies of the vibrational waves that can propagate in an amorphous solid are distributed, a problem that can be solved quantitatively by studying and computing the normal modes of vibration of the microscopic entities (again, atoms, molecules, colloids, or grains), which constitute the material.

5.1   The Debye Model of Solids

5.1.1   The Debye Density of States

Debye was the first to realize that in order to correctly describe the vibrational and thermal properties of a solid, one has to identify and count the different modes present in the material. The key assumption is that vibrational modes behave like standing waves in the solid. This is mainly due to the discretization of wavenumber brought about by the hard-wall boundary conditions for wave propagation in the solid. Therefore, the possible wave numbers for a particular crystallographic orientation are given by:

$$k = 0, \pm \frac{2\pi}{L}, \pm \frac{4\pi}{L}, \dots, \pm \frac{N\pi}{L} \tag{5.1}$$

where L is the linear size of the solid and N the number of atoms in that orientation. For an isotropic solid, these are the same in all three Cartesian directions. By sheer counting, it is clear that there is one k value per $2\pi/L$ or, in 3D, one allowed k

© The Author(s), under exclusive license to Springer Nature Switzerland AG 2023   179
A. Zaccone, *Theory of Disordered Solids*, Lecture Notes in Physics 1015,
https://doi.org/10.1007/978-3-031-24706-4_5

Fig. 5.1 Schematic of the allowed points in k-space under the assumption of periodic "hard-wall" boundary conditions used in the Debye model. The dashed horizontal line indicates approximately the limit of linearity of the dispersion relation, after which the flattening toward the Brillouin zone boundaries at $k = \pm\pi/a$ occurs

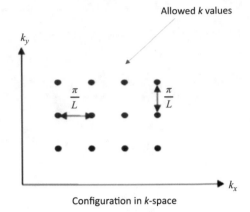

Configuration in k-space

value per $(2\pi/L)^3$. The situation is schematically depicted in Fig. 5.1 below for a 2D section of the material in the $x - y$ plane.

Hence the number density of modes in a 3D isotropic solid will be $(\frac{L}{2\pi})^3$. To get the total number of allowed values of wavevector in the solid, we need to multiply this number density by the volume of reciprocal space occupied within a spherical domain of radius k:

$$N(k) = \left(\frac{L}{2\pi}\right)^3 \left(\frac{4}{3}\pi k^3\right) = \frac{Vk^3}{6\pi^2} \tag{5.2}$$

The density of states is the derivative of N with respect to frequency ω:

$$g(\omega) = \frac{dN}{d\omega} = \frac{Vk^2}{2\pi^2}\frac{dk}{d\omega}. \tag{5.3}$$

Using the linear dispersion relation for acoustic modes, $\omega = vk$, where v is the speed of sound (see Fig. 5.2), inside the above relation, one finally obtains:

$$g(\omega) = \frac{V\omega^2}{2\pi^2 v^2}\frac{d}{d\omega}\frac{\omega}{v} = \frac{V}{2\pi^2 v^3}\omega^2 \tag{5.4}$$

which is the celebrated Debye law for the vibrational density of states (VDOS) of solids. In isotropic solids, we have two transverse acoustic modes and one longitudinal; hence, the Debye law becomes:

$$g_{\text{Debye}}(\omega) = \frac{\omega^2 V}{2\pi}\left(\frac{2}{v_T^3} + \frac{1}{v_L^3}\right) \tag{5.5}$$

where v_L and v_T are the longitudinal and the transverse speed of sound, respectively.

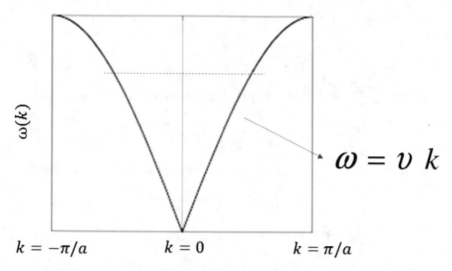

Fig. 5.2 Schematic of the allowed points in k-space under the assumption of periodic "hard-wall" boundary conditions used in the Debye model

5.1.2 Debye Frequency, Momentum, and Temperature

The problem with the Debye model, as it stands, is that it lets the specific heat (which is given as an integral over the VDOS, cfr. Chap. 6) grow indefinitely as $\sim T^3$, whereas from experimental results, it was known that the specific heat drops off to $3k_B N$ at very high temperatures, which is the famous Dulong-Petit law.

In the original Debye model, the $\sim \omega^2$ behavior of the VDOS was thus somewhat artificially capped with a cutoff ω_D, where the VDOS is sharply forced to zero.

Assuming an acoustic isotropic dispersion relation (valid for a cubic lattice or for an isotropic amorphous solid): $\omega = vk$, the value of the Debye frequency is defined as follows.

For a one-dimensional monatomic chain, with lattice constant a, the Debye frequency is given as:

$$\omega_D = vk_D = v\pi/a = v\pi N/L = v\pi\rho_l, \tag{5.6}$$

with N the total number of atoms in the chain and ρ_l the linear number density (recall that $L = Na$ holds). Also, k_D is the Debye momentum or Debye wavenumber.

For a monatomic cubic crystal in $d = 3$, the Debye frequency is given by:

$$\omega_D^3 = \frac{6\pi^2}{a^3}v^3 = \frac{6\pi^2 N}{V}v^3 \equiv 6\pi^2\rho v^3. \qquad (5.7)$$

The Debye temperature θ_D is related to the Debye frequency via:

$$\theta_D = \frac{\hbar}{k_B}\omega_D, \qquad (5.8)$$

and represents, literally, the temperature associated with the highest vibrational eigenmode of the solid.

5.1.3 Van Hove Singularities

At sufficiently low ω, most solids conform to Debye's $\sim \omega^2$ law for the VDOS. In most crystals, at higher ω, edgy peaks appear, which are related to Van Hove singularities, i.e., points where $\frac{d\omega}{dk} = 0$, which happens, e.g., at the Brillouin or pseudo-Brillouin zone boundaries.

We may define the density of states $g(k)dk$ in terms of wavenumber k, as the number dN of standing waves with wavenumber between k and $k + dk$: $dN = g(k)dk = \frac{L}{2\pi}$.

Moving now to generally anisotropic situations in 3D, the density of states for a sample of side length L is given by:

$$d^3N = g(\mathbf{k})d^3k = \frac{L^3}{(2\pi)^3}d^3k \qquad (5.9)$$

where d^3k is the volume element in reciprocal space. By a gradient expansion, the VDOS in frequency space can be written as:

$$d\omega = \nabla_{\mathbf{k}}\omega \cdot d\mathbf{k} \qquad (5.10)$$

where $\nabla_{\mathbf{k}}$ is the gradient in k-space.

Clearly, the mapping between ω values and points in k-space can be recast as $\omega(\mathbf{k}) = const$, which forms a surface in k-space. Then, by standard geometry of surfaces, the gradient of $\omega(\mathbf{k}) = const$ will be a vector orthogonal to this surface at every point.

The vibrational density of states as a function of ω viewed as a surface of points in k-space thus satisfies:

$$g(\omega)d\omega = \iint_{\partial\omega} g(\mathbf{k})\,d^3k = \frac{L^3}{(2\pi)^3}\iint_{\partial\omega} dk_x\,dk_y\,dk_z \qquad (5.11)$$

where $\partial\omega$ denotes the surface $\omega(\mathbf{k}) = const$ over which the integral is taken.

One can choose a new coordinate system k'_x, k'_y, k'_z in such a way that k'_z is orthogonal to the $\omega(\mathbf{k}) = const$ surface and therefore parallel to $\nabla_{\mathbf{k}}\omega$. If this Cartesian coordinate system is obtained simply from a rotation of the original coordinate system, then the volume element in k-prime space is just the same as in the original coordinate system.

Then using the primed system in Eq. (5.10), it is evident that, by construction, dk'_z is parallel to the gradient of ω, while the latter is orthogonal to dk'_x and dk'_y. This, therefore, leaves us with:

$$d\omega = |\nabla_{\mathbf{k}}\omega|\, dk'_z. \tag{5.12}$$

By replacing this in the expression for $g(\omega)$:

$$g(\omega)d\omega = \iint_{\partial\omega} g(\mathbf{k})\, d^3k = \frac{L^3}{(2\pi)^3} \iint_{\partial\omega} dk'_x\, dk'_y\, dk'_z \tag{5.13}$$

we get:

$$g(\omega) = \frac{L^3}{(2\pi)^3} \iint \frac{dk'_x\, dk'_y}{|\nabla_{\mathbf{k}}\omega|} \tag{5.14}$$

where $dk'_x\, dk'_y$ is an area element on the constant-ω surface.

The clear implication of Eq. (5.14) for the VDOS is that at the points in $\mathbf{k}$-space where the dispersion relation $\omega(\mathbf{k})$ has derivative equal to zero, the integrand in the VDOS expression diverges. This is precisely the cause of the edgy peaks visible in the VDOS of real solids, and this happens when the phonon dispersion relation goes through a local maximum or a local minimum or simply becomes flat. The latter case is the most common occurrence, since, e.g., acoustic dispersion relations in crystalline solids become flat at the Brillouin zone boundaries.

5.2   The Boson Peak in the VDOS

In this section, we provide a historical introduction to the much studied anomalies in the low-energy vibrational spectra of glasses. We are going to anticipate a quantitative link between the specific heat $C_p(T)$ and the vibrational density of states $g(\omega)$, which will be derived and presented in Chap. 6 (cfr. Eq. (6.1)).

Deviations from Debye's $\sim \omega^2$ law at low frequency, which cannot be obviously interpreted as Van Hove peaks [1], have attracted much attention, in particular when these deviations show up as a broad peak upon plotting $g(\omega)/\omega^2$, i.e., the VDOS normalized by the Debye law. This broad peak is known as the "boson peak" since in early Raman experiments on glasses at low temperature, its scaling with temperature was shown to follow the Bose-Einstein distribution.

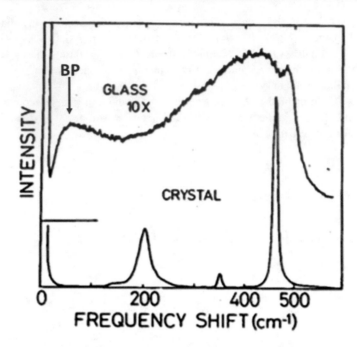

Fig. 5.3 Raman spectra for SiO_2 in glass and crystal states, respectively. The boson peak (BP) feature is marked by an arrow. Adapted with permission from [4]

The abovementioned vibrational feature known as the boson peak originally was referred to a broad band ubiquitously observed in low-frequency Raman scattering of glasses. Already in the 1950s, several light-scattering experiments, especially on fused quartz or vitreous silica (a-SiO_2), showed the appearance of a broad and intense band at ≈ 50 cm^{-1} in the Raman spectrum at room temperature [2,3]. This Raman feature, observed at low frequencies in nearly all glasses studied until now, is completely different from the Raman spectrum of the corresponding crystals in the same frequency range (see Fig. 5.3).

This broad and ubiquitous peak in the low-frequency range of Raman spectra for glasses was soon ascribed to the breakdown of the wavevector selection rules for Raman scattering in crystals (where it is second order), which makes Raman scattering in glasses to be first order, thus exhibiting the whole continuous vibrational spectrum. Specifically, it was established [5] that Raman scattering in non-crystalline solids was directly proportional to the vibrational density of states $g(\omega)$ and to the Bose-Einstein thermal population factor $n(\omega, T) = \left(e^{\hbar\omega/k_B T} - 1\right)^{-1}$, weighted by the light-to-vibration coupling parameter $C(\omega)$ describing the coupling of the vibrational modes of angular frequency ω to the incident light for a particular polarization geometry:

$$I_{Raman} = C(\omega)\, g(\omega)\, \frac{1 + n(\omega, T)}{\omega}. \tag{5.15}$$

As a matter of fact, some early measurements of low-temperature specific heat already indicated a clear excess over the Debye elastic contribution $\propto T^3$, leading some authors to claim that the observed excess heat capacity was due to those *optical modes of very low frequency* observed by Raman scattering [3] or that the thermal properties of glasses at low temperatures in general were dominated by the existence of *localized vibrational modes* of very low frequency [6].

Nevertheless, most of the literature in the field for decades focused only on the boson peak observed by Raman scattering measurements on glasses or non-crystalline solids.

Later on, Martin and Brenig [7] proposed a theory for the unknown photon-phonon coupling coefficient $C(\omega)$ between the scattered light and the vibrational spectrum $g(\omega)$, which was assumed to follow the Debye model, i.e., acoustic phonons with $g(\omega) \propto \omega^2$. In this important theoretical framework, the "boson peak" was the consequence of a maximum in $C(\omega)$, attributed to a correlation length of the various mechanical and electrical fluctuations involved. Many Raman experiments have been wrongly interpreted in terms of this debatable form for two decades.

However, Buchenau and coworkers [8–11] showed by means of exhaustive inelastic neutron scattering (INS) experiments the existence of a clear excess in the low-frequency vibrational density of states for the studied glasses over that predicted by Debye's theory. The assumption of the theory by Martin and Brenig that vibrations in amorphous solids follow Debye's law $g(\omega) \sim \omega^2$ is incorrect [12, 13]. On the contrary, such vibrational excess manifested as a broad peak in the Debye-reduced density of states $g(\omega)/\omega^2$, which accounted very well for the broad maximum in the constant-pressure, reduced specific heat C_p/T^3 measured in glasses at few Kelvins. These experimental findings are closer to the earlier interpretations of the boson peak and the low-temperature thermal properties proposed in the 1950s and 1960s and mentioned above.

Following the abovementioned seminal INS experiments, this and other experimental techniques have been employed to study this ubiquitous and controversial feature of glassy materials over the last 30 years. At least, it was undoubtedly confirmed that the boson peak originates from a broad maximum in the Debye-reduced vibrational density of states $g(\omega)/\omega^2$, beyond the uncertainty in the evaluation of the thermal and frequency factors of Raman scattering.

In order to facilitate the discussion of different approaches and theories about the boson peak and the low-temperature properties of glasses, we provide some relevant experimental facts and typical data, which reflect the often referred to as "universal" character of these glassy "anomalies," since they appear so similar in a broad range of compounds and materials.

Soon after the pioneering Buchenau INS experiments quoted above, other authors confirmed the close relation between the Raman-scattering boson peak and the maximum observed in $g(\omega)/\omega^2$ obtained from INS measurements for insulating, semiconducting, and metallic glasses [12]. Furthermore, by applying a simple rescaling of the maximum position and height, all boson peaks merge into a master curve (Fig. 5.4), whose shape follows rather closely a log-normal distribution [14]. It is to be stressed that the frequency dependence of the coupling coefficient $C(\omega)$ in

Fig. 5.4 Low-energy spectra to the energy scale normalized by E_{max}: **(a)** $g(E)/E^2$ spectra, (1) As$_2$S$_3$ ($E_{max} = 2.65$ meV), (2) SiO$_2$ (5.1 meV), (3) Mg$_{70}$Zn$_{30}$ (5.5 meV); **(b)** Raman spectra (1) As$_2$S$_3$, (2) SiO$_2$, coupling factor $C(E)$ dependence for As$_2$S$_3$ (1') and SiO$_2$, (2'), dashed line is $C(E)$ with the contribution of quasi-elastic scattering taken into account. Adapted from [12] with permission

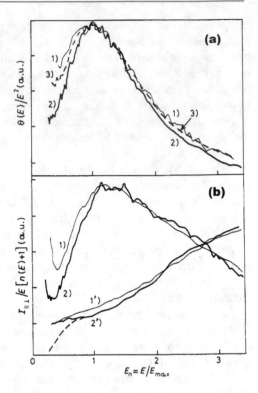

Eq. (5.15) is controversial and not well established. Different theoretical approaches have been proposed, but none has received unanimous recognition. Experimental results tend to indicate a monotonic increase of $C(\omega)$ with frequency in most cases, as can be seen in Fig. 5.4. As a consequence, the position of the boson peak directly observed in the reduced Raman spectra occurs at higher frequencies than the corresponding peak observed in $g(\omega)/\omega^2$ from, e.g., inelastic neutron scattering.

On the other hand, Granato proposed a framework [15], where liquids can be modeled as being "equivalent" to crystals having a small concentration of dumbbell interstitials, which become frozen in the glassy state. Self-interstitial resonance modes were suggested to be the physical realization of the soft modes responsible for the boson peak in this model. Within the simplest approximation, taking a single frequency for the resonance modes, the maximum in C_p/T^3 was predicted to appear at a temperature $T_{max} \approx \Theta_D/35$. Several authors [16–18] have tested this relation for different collections of experimental data, where they found good agreement with the proposed correlation. Moreover, the position T_{max} of the maximum in C_p/T^3 and its counterpart in the Debye-reduced $g(\omega)/\omega^2$ (boson peak at ω_{BP}) have been found to approximately follow the relation $\hbar \, \omega_{BP} \approx 4.5$ to $5 \, k_B T_{max}$ for all studied glasses [19, 20].

To better illustrate this issue, we will assume a vibrational density of states given by the Debye law plus a boson peak modelled emirically by a log-normal distribution [12, 14], with a height that is twice the Debye level, resembling what is observed in paradigmatic glasses as amorphous such as SiO_2 or glycerol:

$$\frac{g(\omega)}{\omega^2} = \frac{g_D(\omega)}{\omega^2} + \frac{g_{BP}(\omega)}{\omega^2}$$

$$= \frac{3}{\omega_D^3}\left[1 + 2\exp\left[-\frac{\ln^2(\omega/\omega_{BP})}{2\sigma^2}\right]\right]. \quad (5.16)$$

We will typically fix $\omega_{BP} = \omega_D/7.5$ and put $\sigma^2 = 0.24$, as found by [14]. The resulting Debye-reduced VDOS is shown in the inset of Fig. 5.5. Then, by performing the integral of Eq. (6.1), the specific heat is obtained, also shown in Fig. 5.5.

As can be seen in Fig. 5.5, the maximum in the Debye-reduced specific heat C_p/T^3 is found at $T_{max} \approx \Theta_D/35$ when $\omega_{BP} = \omega_D/7.5$ for the assumed log-

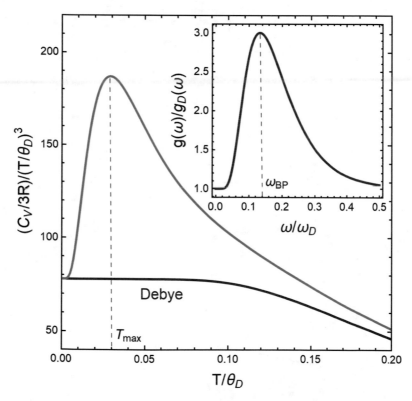

Fig. 5.5 Specific heat calculated from a pure Debye density of states (red) and from the one in the inset for the boson peak (green), using Eq. (6.1). **Inset:** plot of the Debye-reduced low-frequency density of states $g(\omega)/g_D(\omega)$ given by Eq. (5.16). Courtesy of Dr. Matteo Baggioli

normal distribution for the boson peak in $g(\omega)/\omega^2$, that is, $\hbar\omega_{BP} \approx 4.7 k_B T_{max}$, as found in the experiments.

A central issue in current research is to unveil the relevance of the anharmonic character of most of the low-energy excitations dominant in glasses (and even in many more-or-less disordered crystals, as we will see below). In this section, we will focus on a few meaningful examples taken from the literature, before providing a more theoretical quantitative discussion in following sections. From INS experiments by Buchenau et al. [10] in vitreous silica, the harmonic behavior of the contributing vibrational modes was inspected by plotting the scattering intensity versus the Bose factor times the Debye-Waller factor. Vibrational modes at 220 GHz (nothing to do with the assumed two-level states (TLS) of very low energies) exhibited a dramatic anharmonic behavior, as shown in the lower panel of Fig. 5.6. Only when approaching the frequency of the boson-peak maximum, the vibrational

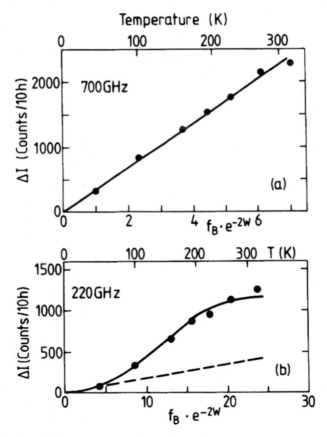

Fig. 5.6 Temperature dependence of inelastic neutron scattering intensities in vitreous silica, plotted versus the Bose factor multiplied by the Debye-Waller factor, at **(a)** 700 GHz and **(b)** 220 GHz. The dashed line in (b) denotes the harmonic contribution. Adapted from [10]

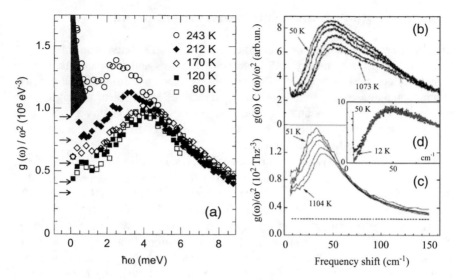

Fig. 5.7 (a) VDOS of glycerol-D3 from measurements at different temperatures below and above $T_g = 190$ K, plotted as $g(\omega)/\omega^2$. The arrows mark the limiting Debye value, as derived from ultrasonic data. The shaded area indicates schematically the onset of quasielastic scattering in the 243 K spectra. Adapted with permission from [21]. (b) Reduced Raman spectra $I(\omega)/(\omega(n(\omega,T)+1))$, Eq. (5.15), in the (0 to 150) cm^{-1} range at different temperatures from top to bottom: (45, 323, 423, 523, 873, and 1073) K. (c) Reduced density of states $g(\omega)/\omega^2$ as obtained by inelastic neutron scattering in the frequency range as in (b) at different temperatures (from top to bottom 51, 318, 523, 873, and 1104 K). The dashed line is the calculated Debye contribution as obtained from Brillouin scattering data at 50 K. (d) Reduced Raman scattering (in arbitrary units) at 12 and 50 K. Adapted with permission from [22]

modes gradually appeared to be more harmonic like. These findings are consistent with the general framework for low-energy excitations in glasses postulated by the soft-potential model, as we discuss below, and with the more recent anharmonic model that we will present in Sect. 5.4.3.

In Fig. 5.7, some typical examples taken from the literature [21, 22] are reproduced, both from Raman and inelastic neutron scattering, both below and above the glass transition temperature T_g, showing clear deviations from a temperature-independent curve in the boson-peak frequency range, providing a signature of *anharmonic* behavior. Only at frequencies at or above the boson peak maximum the curves merge. The effect is obviously more dramatic above T_g in the supercooled liquid regime, where the boson peak tends to disappear being overcome by quasielastic structural relaxation [23, 24].

5.3   Case Study: Simple Lattices with Randomness

The abovementioned experimental studies clearly presented numerous challenges in terms of separating authentically "generic" features and mechanisms, from a formidably complex interplay of external factors that range from complex experimental setups based on radiation-matter interactions to chemical complexity. In this sense, simple numerically solvable systems provide the only way to access truly generic features and prove useful tools to directly test assumptions used in theoretical models and interpretations.

We shall therefore consider the three model nearest neighbor lattices of harmonic springs presented in Chap. 1, the random network, the randomly depleted fcc, and the fcc with random vacancies, presented in Sect. 1.5.2 (cfr. Figs. 1.13 and 1.14). These lattices can be easily constructed using simple Monte-Carlo type codes, and the eigenavalue distribution, from which the vibrational density of states (VDOS) is directly obtained, is computed from direct diagonalization of the Hessian matrix. In the calculations, $m = 1$, the lattice constant is $a = 1$, implying a nearest neighbor distance in the fcc lattive given by $r_0 = 1/\sqrt{2}$, and $N = 4000$ lattice nodes in a box of volume $V = L^3 = 10^3$ implies a density $\rho = 4$.

It is instructive to begin with the bond-depleted fcc lattice, which provides a basic understanding of how the VDOS changes from a perfect fcc lattice (in the limit of no bonds being removed, with $z = 12$) down to the isostatic point where $z = 6$. The VDOS spectrum is shown in Fig. 5.8 and was first studied in Ref. [25].

The dashed line in the figure shows the VDOS of a perfect fcc lattice, with an extended $\sim \omega^2$ Debye behavior at low energy, followed by prominent, edgy Van Hove peaks corresponding to the flattening of the acoustic phonon dispersion

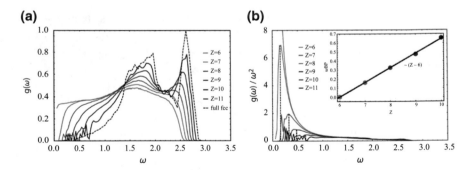

Fig. 5.8 VDOS of the randomly depleted fcc lattice for different average coordination numbers z of the lattice. By comparing the spectra of the depleted lattices with the VDOS of a perfect fcc lattice (dashed line), we can identify the edgy peaks as remnants of the two characteristic van Hove singularities of the ordered fcc lattice. Panel (b) shows the Debye-reduced VDOS plots of the same systems, which clearly show the boson peak (BP) anomaly. The dashed lines indicate the position of the BP, which is plotted against the average coordination number in the inset. The numerical results for these simple lattices reproduce the well-known $\omega_{BP} \sim (z - 6)$ scaling first reported for random jammed packings of soft spheres in Ref. [26]. Plots courtesy of Dr. Rico Milkus

relations at the Brillouin-zone boundaries, according to the mechanism discussed in Sect. 5.1.3.[1]

It is observed that the VDOS undergoes quite dramatic changes upon going from the perfect fcc lattice with $z = 12$ to the most depleted lattice with vanishing rigidity at the isostatic point $z = 6$. The static shear modulus for these athermal lattices changes all the way from a purely affine $G = G_A = \frac{1}{30} \frac{N}{V} \kappa R_0^2 z$ with $z = 12$ for the perfect fcc lattice to $G = G_A - G_{NA} = \frac{1}{30} \frac{N}{V} \kappa R_0^2 (z - 6) = 0$ at the marginal rigidity (isostatic) point with $z = 6$ (cfr. Chap. 2 for the quoted formulae).

Furthermore, the Van Hove peaks get strongly smeared out by the added disorder, until they almost (but not entirely) disappear as z approaches 6. A substantial amount of softer modes, much above the Debye law, is seen to develop in the region $\omega = 0.0 - 1.5$, which gives rise to the boson peak (shown in Fig. 5.8b), which increases spectacularly as $z = 6$ is reached. The boson peak position ω_{BP} drifts toward lower frequencies as z decreases according to the same law by which G vanishes,[2] i.e., $\omega_{BP} \sim (z - 6)$, and virtually goes to zero frequency in the $z = 6$ isostatic limit. In this limit, the non-normalized VDOS develops a plateau reaching down to $\omega = 0$, which is visible in Fig. 5.8a. This behavior is rather universal and not specific of random-depleted lattices and was reported initially for random jammed packings of soft spheres upon approaching $z = 6$ in the pioneering work of Ref. [26].

This universality for simple nearest neighbor systems becomes striking upon comparing the random network system with the randomly depleted fcc. The comparison is shown in Fig. 5.9.

This comparison is surprising if one thinks of the obvious structural differences between the random network (RN), Fig. 1.13, and the bond-depleted fcc lattice, Fig. 1.14, which feature very different values of bond-orientational order parameter (cfr. Chap. 1, and Fig. 1.16), with $F_6 = 1$ for the depleted fcc lattice and $F_6 \approx 0.3$ for the RN system. In spite of these large differences in the lattice structure, the two systems display exactly the same spectrum at low frequency and exactly the same boson peak (cfr. Ref. [25]). A reason for this striking universality, first discovered in Ref. [25], may be identified in the fact that the two lattices have exactly the same values of inversion-symmetry breaking order parameter F_{IS} at all z values considered, cfr. Fig. 1.16. In turn, the local breaking of inversion symmetry controls the shear modulus, which varies as $G \sim (z - 6)$ for both systems and possibly also the phonon scattering and damping in the system (see later on the theoretical models based on phonon inelastic scattering or anharmonicity).

---

[1] Recall that the fcc lattice has only acoustic phonons and no optical phonons because a given lattice has $3s$ acoustic phonon branches and $3s - 3$ optical phonon branches, where s is the number of atoms per primitive cell. Since $s = 1$ for the fcc lattice, clearly there are no optical modes.

[2] This is a manifestation of the fact that the boson peak is closely linked to the transverse shear modes and to their strong nonaffinity, as will be discussed later on.

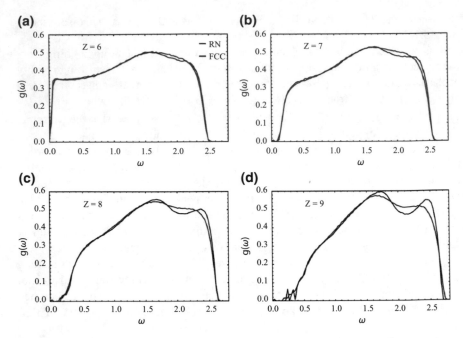

Fig. 5.9 Comparison of VDOS between depleted random network (red) and depleted fcc lattice (black) at different values of mean coordination number z from (**a**) to (**d**) as indicated in the plots. We can easily see the striking similarities, especially at low frequencies. The random-depleted fcc has a noisier Debye regime, since, due to its more regular structure, some eigenfrequencies tend to cluster at low frequencies. Some differences arise in correspondence of the Van Hove peak relics, again due to the structural differences. Plots courtesy of Dr. Rico Milkus, for the original see [25]

In the next sections, we shall explore theoretical models in analytical form that have been proposed to describe the low-energy part of the VDOS of amorphous solids.

5.4   Theoretical Models

We have seen, in the previous section, how nearest neighbor lattices with different structure, also reflected in very different values of bond-orientational order, possess exactly the same low-frequency VDOS spectrum. In this section, we shall present theoretical models, which attempt to capture this universality in the vibrational spectrum of disordered solids. We will present essentially two theoretical frameworks. One is based on random matrix theory and its phenomenological application to disordered solids, where the intricate phonon scattering events due to the disorder are taken care of by the random-matrix behavior of the eigenvalue spectrum of the Hessian matrix. The other framework, instead, is even more agnostic with respect to the microscopic structure and disorder and in fact is applicable even to crystalline solids where the boson peak has often been observed. In this latter framework, the intricacies of the scattering events are described using an effective field theory approach, by modelling the anharmonic damping or linewidth of the phonons, through the Akhiezer model presented in Chap. 4.

5.4.1 The Random Matrix Theory (RMT) Model

Random matrix theory (RMT) is a subfield of mathematics which studies the properties of matrices whose elements are random variables. RMT has widespread applications in physics, ranging from high-energy physics to solid-state physics, quantum optics, and chaos, besides, obviously, in statistical physics and disordered systems. The theory of random matrices was initiated by Eugene Wigner, who proposed that the lines in the spectra of heavy nuclei can be modelled in terms of the spacings between eigenvalues of random matrices. For real symmetric matrices, the assumption that each entry is an independent identically distributed (i.i.d.) random variable leads to defining the so-called Gaussian orthogonal ensemble (GOE), which is one of the most studied ensembles in RMT. Similarly, for unitary random matrices, one has the Gaussian unitary ensemble (GUE). Many random symmetric matrices with i.i.d. entries have an asymptotic (in limit of large N) eigenvalue distribution given by the Wigner semicircle distribution.

In our problem, the Hessian matrix of the disordered solid under study can be regarded as a random matrix. The main difference with the most studied random matrices in math and physics is that these matrices are highly sparse and, importantly, they are rather random *block* matrices, since in $\mathbf{H}_{ij}$ each ij entry is actually a $d \times d$ block, where d is the space dimension.

In all models or random spring networks of disordered solids, the elastic energy is a quadratic function of the displacements of the particles from "frozen" positions in the amorphous lattice (where the reference state can be, and often is, a "stressed" one [27, 28]). The stiffness matrix or Hessian matrix $\mathbf{H}_{ij}$ is a Laplacian random symmetric matrix where each row is comprised of a small and random number of nonzero coefficients. The off-diagonal entries $\mathbf{H}_{ij}$, with $i < j$, are identical independent random variables, whereas the diagonal entries are given by $\mathbf{H}_{ii} = -\sum_{j \neq i} \mathbf{H}_{ij}$. The latter requirement is dictated by enforcing mechanical equilibrium on every atom i in the lattice. A typical exercise in RMT is the study of the spectrum of the adjacency matrix or the Laplacian matrix of an Erdos-Renyi graph with N vertices in the limit of large order of the matrices (the large N limit). The only parameter in the model is the probability p/N of a link in the random graph to be present, whereas the dimension d of the embedding space $\mathbb{R}^d$ of the amorphous material or the random spring model is not considered.

For a RMT model to be fully accurate and realistic, the block dimensionality should therefore be taken in consideration, which poses formidable challenges to analytical theory. Let us recall, from Chap. 2, the form of the Hessian, in Cartesian components α, β, for a random spring network:

$$H_{ij}^{\alpha\beta} = \delta_{ij} \sum_s \kappa c_{is} n_{is}^{\alpha} n_{is}^{\beta} - (1 - \delta_{ij}) \kappa c_{ij} n_{ij}^{\alpha} n_{ij}^{\beta} \tag{5.17}$$

where c_{ij} is the adjacency matrix (with entries equal to only 1 or 0 depending on whether two particles i and j are bonded, or not). It is clear that this is a real symmetric matrix of dimension $dN \times dN$, where each row or column has N random block entries, each being a $d \times d$ matrix given by $\mathbf{n}_{ij} \mathbf{n}_{ij}^T$. For a z-coordinated lattice (on average, of course, fluctuations in z can be present), every $d \times d$ off-diagonal block has probability $1 - z/N$ of being a null matrix and a probability z/N of being a rank one matrix.

Most results are either directly based on or related to analytical results obtained in the Gaussian or Wishart ensembles, for which spectral distributions converge to the form of the well-known Wigner semi-circle and Marchenko-Pastur distribution, respectively. Recently, however, it has been rigorously demonstrated [29] that the Marchenko-Pastur spectral distribution corresponds to random Laplacian block matrices with $d \times d$ blocks, where d is the space dimension, and with connectivity z, only in the limit $d \rightarrow \infty$ with $z/d \rightarrow \infty$ fixed. The Wigner semi-circle is recovered for adjacency matrices in the limit $z/d \rightarrow \infty$ with d fixed, while for the same matrices in the limit $d \rightarrow \infty$ with $z/d \rightarrow \infty$ fixed, one recovers the effective medium approximation of Ref. [30].

Clearly, the Hessian matrix of an amorphous solid can be most realistically represented by a Laplacian random block matrix with 3×3 blocks (in $d = 3$). Nevertheless, as shown in [29], the Marchenko-Pastur distribution, which is exact only for $z/d \rightarrow \infty$, still captures the salient qualitative features also of the spectral distribution at finite d.

It has been shown in Ref. [29] that the effect of block dimensionality d is important for the overall shape of the eigenvalue distribution, hence of the VDOS. However, it has been also shown that the low-frequency part of the spectrum is more weakly dependent on d. This may be an explanation for the fact that, as we shall see next, high-dimensional RMT models are able to capture the low-energy part of the eigenvalue spectrum fairly well. This finding explains why high-dimensional theories, which become exact in $d \rightarrow \infty$, such as the replica theory [31], seem able to capture certain features of the VDOS of random jammed packings. These theories were shown to reduce to the well-known Marchenko-Pastur eigenvalue distribution, which is the asymptotic limit of the eigenvalue distribution of random matrices in the Wishart ensemble.

We shall now demonstrate that the RMT model does provide, indeed, a good approximation to the numerical data presented in the previous section for the random network nearest neighbor lattice, with some modifications. The latter that are needed since the RMT model can be regarded, at best, as a mean-field approximation that becomes exact only in the limit of infinitely large blocks with $d \rightarrow \infty$.

We work in the Wishart ensemble, which is defined as the ensemble of square $n \times n$ matrices $M = AA^T$, where A is a rectangular matrix $n \times m$, with $m \geq n$ and A^T is its transpose. The entries of A are all i.i.d. random variables. We will take M as the starting point to provide an approximation of the Hessian $\mathbf{H}_{ij}$, and we shall check and refine this approximation by direct comparison with numerical data. Then, the asymptotic (large n) limit of the eigenvalue distribution of M is given by the Marchenko-Pastur distribution:

$$p(\lambda) = \frac{\sqrt{((1 + \sqrt{\rho})^2 - \lambda)(\lambda - (1 - \sqrt{\rho})^2)}}{2\pi\rho\lambda} \tag{5.18}$$

where we defined the parameter $\rho = m/n$. Since we are interested in the vibrational density of states $g(\omega)$ of the eigenfrequencies $\omega = \sqrt{\lambda}$, we transform $p(\lambda)$ to the frequency space: $p(\lambda)d\lambda = g(\omega)d\omega$:

$$g(\omega) = \frac{\sqrt{((1 + \sqrt{\rho})^2 - \omega^2)(\omega^2 - (1 - \sqrt{\rho})^2)}}{\pi\rho\,\omega}. \tag{5.19}$$

As was originally shown by Parshin and coworkers [32, 33], this "bare" random matrix spectrum does not possess mechanical stability, since it does contain acoustic phonons and cannot describe elasticity correctly. This can be improved by adding a positive-definite matrix to M with a multiplicative prefactor, which correlates positively with the shear modulus [32, 33]. Following a similar procedure, we shift the distribution in the frequency space by an amount δ and introduce the width of the spectrum b with the transformation: $\omega \rightarrow \frac{2}{b} (\omega - \delta)$, which gives:

$$g(\omega) = \frac{\sqrt{((1 + \sqrt{\rho})^2 - \frac{4}{b^2}(\omega - \delta)^2)(\frac{4}{b^2}(\omega - \delta)^2 - (1 - \sqrt{\rho})^2)}}{\pi\rho\frac{2}{b}|\omega - \delta|}. \tag{5.20}$$

This is the shifted spectrum of a matrix M' that can be deduced from the original Wishart matrix M in the following way [32, 33]: $(M')^{1/2} = \frac{b}{2}M^{1/2} + \delta\mathbf{1}$, where δ and b both depend on the minimal and maximal eigenfrequencies of the system (without accounting for phonons), $\omega_- = \frac{b}{2}(1 - \sqrt{\rho}) + \delta$ and $\omega_+ = \frac{b}{2}(1 + \sqrt{\rho}) + \delta$, respectively, which define the support of the random matrix spectrum.

The value of δ controls the shift of the lower limit of the support of the random matrix spectrum, and thus its value controls the frequency $\omega^* \approx \omega_-$, which is directly linked with the boson peak.

The shift of the support of the random matrix spectrum toward higher frequency is a reflection of the underlying existence of Goldstone modes, i.e., the phonons. The random-matrix part of the VDOS, described by Eq. (5.20), must have a gap, if the system is fully rigid, i.e., above the isostatic point, $z > 6$. This is because the gap is populated with the Goldstone excitations (the acoustic phonons). These arise, by Goldstone's theorem, from symmetry breaking of translations due to the short-range

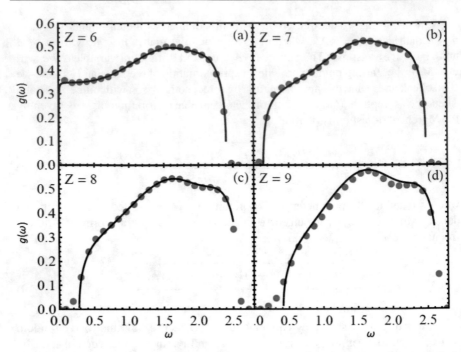

Fig. 5.10 Numerical VDOS (symbols) of the random network (RN) system and the analytical fitting based on the RMT model provided by Eq. (4) (solid lines) for different values of the nearest neighbor spring network connectivity, z. Adapted from Ref. [34] with permission of the American Physical Society

order due to the coordination of the lattice of equal length bonds, and follow the ω^2 Debye law starting from $\omega = 0$ up to the point where the random-matrix part of the spectrum sets in. Obviously, the acoustic phonons cannot be present in the random-matrix part of the spectrum (i.e., Eq. (5.20)) of the VDOS, which is dominated by intense scattering due to the intrinsic randomness and where phonons can no longer propagate ballistically over long distances. The fact that δ is an increasing function of $(z - 6)$ is certainly consistent with previous work, e.g., simulations of Ref. [26], where the low-frequency phononic part of the spectrum described by the Debye law ω^2 extends up to larger frequencies as z is increased. The trend of the width b is also consistent with those numerical data.

By choosing $\rho = 1.6$, from the fitting to the numerical VDOS spectra (shown in Fig. 5.10), one gets: $\delta = (2.72 + 0.074(z - 6))$ and $b = (2.4 - 0.056(z - 6))^2$.

In order to fit the numerical VDOS data accurately, two additional modifications are required: first we need to correct the lower limit of the frequency by a factor that behaves like $\sim \omega^{-1/2}$ for $\omega \to 0$ and like ~ 1 for $\omega \gg 0$. Then, we need to add peak functions to model the relics of the Van Hove singularity peaks, which become more prominent for systems with high values of z because the structure of the random network becomes influenced by the regular fcc lattice limit, to which any lattice will converge when $z = 12$. This further correction is achieved by modelling the two relics of the Van Hove peaks with two Gaussian functions. The final result for the fitting formulae of the VDOS reads as:

$$g(\omega) = \frac{\sqrt{[(1 + \sqrt{\rho})^2 - \frac{4}{b^2}(\omega - \delta)^2][\frac{4}{b^2}(\omega - \delta)^2 - (1 - \sqrt{\rho})^2]}}{\pi \rho \frac{2}{b} |\omega - \delta|}$$

$$\times \left(\left(\frac{0.65}{\omega} \right)^2 + 0.25 \right)^{1/4} + G_1(z, \omega) + G_2(z, \omega) \qquad (5.21)$$

where $G_1 = (0.011(z - 6)^2 + 0.175)\sqrt{\frac{2}{\pi}} \exp(-2(\omega - 1.6)^2)$ and $G_2 = (0.011(z - 6) + 0.045)\sqrt{\frac{8}{\pi}} \exp(-8(\omega - 2.3 - 0.07(z - 6))^2)$ are the two Gaussian functions used to model the Van Hove peaks. This equation was first proposed in Ref. [34].

The comparison between Eq. (5.21) and the numerical data is shown in Fig. 5.10.[3] It is seen that the model parametrization given by Eq. (5.21) is fairly good and provides an accurate description of the data for all z values considered, in a broad range of connectivity z from $z = 9$ down to the isostatic point (unjamming transition) at $z = 6$. In particular, the expressions of all the parameters δ, b, ρ, and those inside G_1, G_2, either remain fixed upon changing z or evolve with z. Hence, Eq. (5.21) captures the variation of the VDOS spectrum upon varying the coordination number z of the network.

In the comparison, the solid line describes the entire vibrational spectrum with the exception of the Goldstone modes, i.e., the acoustic phonons, which obviously cannot be captured by RMT. At $z = 9$ it is very clear that the symbols follow the Debye law $\sim \omega^2$ that describes the Goldstone phonons as explained in the first part of this chapter. The acoustic phonons are governed by the Debye law regardless of whether the solid is crystalline or amorphous, because translations are broken in both cases.

The shrinking of the phononic spectrum as $z \to 6$ goes hand in hand with the lower limit of the random matrix support going to zero frequency, with the law $\sim (z - 6)$. This is evidently related to the vanishing of the shear modulus G, with the same law, controlled by nonaffinity (cfr. Chap. 2, Eq. (2.50)), since the shear modulus controls the speed of sound of the transverse acoustic phonons, which, therefore, disappear altogether as $G \to 0$. This consideration explains the shrinking of the phononic $\sim \omega^2$ regime to zero at $z = 6$, because the longitudinal phonon has a sound speed controlled by the bulk modulus, which is much larger than the shear modulus (cfr. Chap. 2, Sect. 2.7). In turn, this implies a much smaller contribution

---

[3] The comparison between RMT model and numerical simulations in terms of the corresponding eigenvalue distribution $p(\lambda)$ can be found in the Appendix of [34]. In that comparison, it is seen that at the isostatic point $z = 6$, the random matrix behavior $p(\lambda) \sim \lambda^{-1/2}$ extends all the way down to $\lambda = 0$. This fact has been used in Chap. 3 to derive the creep modulus power-law Eq. (3.56).

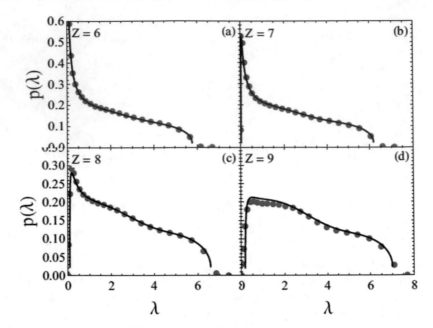

Fig. 5.11 Numerical eigenvalue spectrum (symbols) for random networks with varying values of coordination number z and RMT model fitting (solid lines) using Eq. (5.21) and the transformation $p(\lambda) = \frac{1}{2} \frac{g(\sqrt{\lambda})}{\sqrt{\lambda}}$ to convert the VDOS $g(\omega)$ into the eigenvalue spectrum $p(\lambda)$. Recall that $\omega = \lambda^2$, since mass $m = 1$ is used in the numerics. Adapted from Ref. [34] with permission of the American Physical Society

to the VDOS (practically negligible compared to that of the transverse phonons), in view of the fact that $g(\omega) \sim v^{-3}$, with v the speed of sound of the sound mode being considered, according to Eq. (5.4).[4]

The eigenvalue spectra $p(\lambda)$ corresponding to the VDOS spectra of Fig. 5.10 are shown in Fig. 5.11 together with the RMT model fitting (same parameters are used in both fittings).

In the plots for $z = 6, 7, 8$, the cusp behavior $p(\lambda) \sim \lambda^{-1/2}$, which is typical of the Marchenko-Pastur spectral distribution, is clearly visible. This is the signature of the random-matrix behavior of the eigenvalue spectrum of a disordered solid, which, upon changing variable from eigenvalue λ to eigenfrequency ω, transforms to the behavior $g(\omega) = A\omega + B$ in the VDOS. This scaling becomes dominant at the marginal stability or isostatic point, i.e., upon approaching the unjamming limit $z = 6$, and we shall discuss this singular behavior in more detail in the next section. For fully rigid systems with $z > 6$, this behavior is still present, although less visible and relegated to frequencies above the Goldstone phonons, as discussed above.

---

[4] Clearly, v is much larger for the longitudinal phonons since $K \gg G$, which implies a comparably negligible contribution to $g(\omega)$.

We shall also note that, in Fig. 5.10, the crossover from the phononic regime to the random-matrix regime is roughly coincident with the Ioffe-Regel crossover from ballistic phonons to highly scattered quasilocalized excitations discussed in Chap. 4. As discussed in Chap. 1, Sect. 1.5.2, the network connectivity z controls the local degree of inversion symmetry breaking, which, in turn, controls the probability that a certain node or particle in the system acts as a scattering center (the lower the inversion symmetry around a particle, the more likely it is that it may act as a scattering center for an incoming phonon plane wave). In future work, microscopic theories of wave scattering in disordered environments could be developed to study and reproduce this crossover from a more microscopic point of view.

5.4.2 Singular Behavior Near Marginal Stability

We have seen, in the previous section, that random-matrix scaling of the eigenvalue spectrum, as prescribed by the Marchenko-Pastur distribution, becomes clearly visible as the random lattices of springs approach the isostatic limit with coordination $z = 6$. The associated characteristic square-root cusp scaling $p(\lambda) \sim \lambda^{-1/2}$ then bridges down to $\lambda = 0$ as Goldstone phonons basically disappear from the spectrum. Correspondingly, the VDOS exhibits a plateau $g(\omega) \sim const$ down to the limit $\omega = 0$. This behavior was first detected at the jamming point of soft repulsive sphere packings in the pioneering work of Ref. [26].

This form of the spectrum at marginal stability can be given an effective description in terms of a soft mode instability. Similar to what happens at structural phase transitions (e.g., ferroelectric phase transitions), there are phonons (in this case the acoustic ones) that become unstable as their eigenfrequency $\omega = vk \sim \sqrt{(z-6)} \to 0$, as a consequence of $v = \sqrt{G/\rho} \sim \sqrt{(z-6)} \to 0$ at $z = 6$ (recall Eq. (2.50) in Chap. 2 for the vanishing of the shear modulus due to nonaffinity) and thus pick up purely imaginary frequencies upon crossing $z = 6$ from above.[5]

Based on these considerations, we model the softening acoustic transverse modes as:

$$\omega = vk - i\,\Gamma(k) = vk - i\,\Gamma_0 - i\,Dk^2 + O(k^4) \qquad (5.22)$$

in terms of a generic damping coefficient Γ, which depends on the momentum k of the mode.[6]

---

[5] Something analogous happens in ferroelectric crystals at the soft mode phase transition from paraelectric to ferroelectric, upon decreasing the temperature across the transition temperature T_c, where the phonon frequency of the softening mode behaves as $\omega^2 \sim (T - T_c)$. Cfr. Sect. 1.1.4 and footnote 3 in Chap. 1.

[6] This relation can be derived from the equation of elastodynamics of a solid augmented with a viscous contribution [35] or from macroscopic balance equations that allow for energy dissipation [36].

We consider the limit where the imaginary (unstable) part of the above dispersion relation effectively dominates over the real part (representing the acoustic, elastic, propagative component), which is exactly what happens upon approaching the isostatic or unjamming point $z = 6$. We shall see that by taking this limit, the characteristic properties of marginally stable (jammed) solids [26] can be reproduced.

We take a hydrodynamic approach to express the low-momentum limit of the damping coefficient in terms of a momentum-independent term Γ_0 and a diffusion constant D. Higher order in k terms can be ignored since they do not contribute to the low energy properties of the system.

Hence, in the limit $z \to 6$, where diffusion dominates, the above dispersion relation becomes purely imaginary:

$$\omega_{\text{diff}} = -i\,\Gamma(k) = -i\,\Gamma_0 - i\,D\,k^2 + O(k^4). \tag{5.23}$$

In the opposite limit, where the imaginary part is negligible and the shear modulus is finite, the VDOS of the Debye modes can be retrieved using standard methods such as Eq. (5.5).

Going back to the imaginary modes Eq. (5.23), we need to anticipate Eq. (5.38) that will be derived in the next section, so that their contribution to the VDOS can be evaluated as:

$$g_{\text{diff}}(\omega) = \frac{2\,\omega}{\pi\,k_D^3}\,\text{Im} \int_0^{k_D} \frac{k^2}{\omega^2 - i\,\omega\,\Gamma_0 - i\,D\,k^2}\,dk \tag{5.24}$$

where k_D is the Debye wavenumber.

The integral can be performed analytically, from which we get the following expression (omitting uninteresting constant prefactors):

$$g_{\text{diff}}(\omega) = \text{Re}\left[\frac{\sqrt{D}\,k_D - \sqrt{\Gamma_0 + i\,\omega}\,\tan^{-1}\left(\frac{\sqrt{D}\,k_D}{\sqrt{\Gamma_0 + i\,\omega}}\right)}{D^{3/2}}\right]. \tag{5.25}$$

Importantly, the contribution of the unstable modes to the VDOS displays the low-frequency behavior:

$$g_{\text{diff}}(\omega) = A + B\,\omega^2 + O(\omega^4) \tag{5.26}$$

with:

$$A = \frac{k_D}{D} - \frac{\sqrt{\Gamma_0}\,\tan^{-1}\left(\frac{\sqrt{D}\,k_D}{\sqrt{\Gamma_0}}\right)}{D^{3/2}}. \tag{5.27}$$

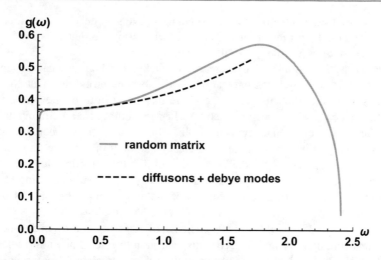

Fig. 5.12 Comparison between the best-fitting RMT model of $z = 6$ numerical data previously shown in Fig. 5.10 (solid line) and the predicted contribution (dashed line) of unstable soft modes (diffusons) originating from the vanishing of shear modulus at $z = 6$, $\omega \sim \sqrt{(z - 6)}$. There is some discrepancy at $\omega < 0.1$ due to the fact that the numerical network has z not exactly 6 and slightly larger, whereas the theory assumes a fully marginally stable system for which there is no going down to zero, in agreement with previous findings [26]. Adapted from Ref. [37] with permission from the American Physical Society

These results were first presented in [37]. The main result here is that the unstable modes provide a constant contribution to the VDOS, which remains finite even at zero frequency.[7]

In Fig. 5.12, the contribution of Eqs. (5.26)–(5.27) to the VDOS at $z = 6$ is shown in comparison with the RMT model best fitting of the numerical data of random spring networks presented in the previous section (Fig. 5.10).

This clearly shows that transverse acoustic phonons that become unstable at the unjamming isostatic point $z = 6$ are directly responsible for the flattening of the VDOS to a constant finite value at $\omega \to 0$. The description presented in this section provides an alternative (but substantially equivalent to the random-matrix description), hydrodynamic description of the singular behavior of the vibrational spectrum of marginally stable systems where rigidity vanishes.

5.4.3 The Damped Phonon Model of the VDOS

In this chapter so far, we have discussed glassy features found in a wide variety of materials that have in common the structural disorder. More specifically, the boson

---

[7] One should note that the dispersion relation Eq. (5.23) and consequently the result Eq. (5.27) are not reliable in the limit of very large values $D \to \infty$ where higher-order terms in Eq. (5.23) have to be considered.

peak (extra-low-frequency modes with respect to Debye's model) has been reported for glasses lacking long-range translational and orientational order regardless of the nature of the bonding (covalent, metallic, ionic).

Nevertheless, the boson-peak excess over the Debye level has been also found in crystalline solids, leading to glassy features in the thermal properties. Due to these similarities, it is important to determine if the physical origin of the boson peak vibrational modes in glasses can be distinguished from that observed in crystals or if a more universal common origin is possible – this is currently an open and active topic of research. In particular, a genuine glassy anomaly in a perfect crystal will be such only if it can be established that it is not due to crystalline features, e.g., lowered and smeared Van Hove singularities in the VDOS resulting from extrema or saddle points in $\omega(k)$ (dispersion curves) of the underlying lattice, as discussed above in Sect. 5.1.3.

A seminal work [38] pointed out the existence of glassy anomalies in the specific heat, thermal conductivity, and dielectric dispersion in the well-known ferroelectric material PMN (Pb$_3$(MgNb$_2$)O$_9$). Subsequently, other ferroelectric materials were studied [39]. Interestingly, two groups of ferroelectrics were identified. Ferroelectric compounds with a "diffuse" phase transition clearly exhibited glassy behavior in their thermal properties, whereas those with a sharp phase transition used to show the expected crystalline behavior. Recent results have enlarged the number of ferroelectrics displaying glassy anomalies, where glassy features such as the specific heat excess as well as the thermal conductivity plateau are evidenced for the ferroelectric mixed crystals Ba$_{1-x}$Sr$_x$Al$_2$O$_4$, when $x > 0$ [40, 41].

As proposed by Ackerman et al. [38], these anomalies were ascribed to the coupling between (acoustic) phonons and localized excitations. Recent simulations on CaF$_2$, PbF$_2$ [42, 43], or superheated crystalline Ni [44] also evidenced the existence of a boson peak in the VDOS.

Crystals of organic molecules are also typical representatives of this kind of materials with glassy anomalies in their vibrational spectra [20, 45, 46].

Other well-ordered systems featuring a boson peak have been uncovered recently. They include quite different types of materials such as ferroelastic metallic alloys, organic-inorganic thermoelectrics, quasicrystals, etc. Shape memory allows for partially frozen ferroelastic (parent phase)/martensitic domains (the frozen order parameter concerns the lattice strain) displaying glassy anomalies including a boson peak in the specific heat around 10 K associated with the phonon softening of the frozen ferroelastic domain surrounded by the martensitic low-temperature structure [47]. Softening of the transverse acoustic branch was found to be not related to the Van Hove singularity nor with disorder but associated with the anharmonicity of the system. Interestingly, the authors find that the ω_{BP} is uncorrelated with the Ioffe-Regel frequency ($\omega_{IR} \ll \omega_{BP}$) observed along the [110] crystallographic direction, where strong phonon damping occurs.

Thermoelectric materials are also representative of systems displaying glassy features. Recent results show unambiguously the existence of boson peak glassy anomalies in thermal properties for a large class of materials [48–55]. Clathrates are cage-like structures of hosts with large voids in the crystalline lattice, which allow one to encapsulate guest atoms/molecules. The rattling motions of these guest atoms/molecules display localized highly anharmonic modes, which strongly interact with the host lattice acoustic phonons, giving rise to the avoided crossing mechanism [56]. Such strong coupling between those modes produces strong phonon scattering and thus low thermal conductivity despite the underlying well-ordered lattice [54]. For inorganic systems, such as Cu_6Te_3S (similarly for Bi_5CuS_8 and Cr_5Te_8 compounds), the anharmonic and anisotropic displacement of some atoms with partial occupancy (such as Cu in Cu_6Te_3S) give rise to the main contribution of low-energy optical modes, which interact with the acoustic modes of the lattice yielding both a boson peak and a linear-in-T specific heat at low temperature.

Similar to the case of clathrates, interaction between low-energy rattling modes and acoustic phonons in crystalline thermoelectric materials $Ba_8Ga_{16}X_{30}$ (X = Ge,Sn,...) originated by the anharmonic potential experienced by the guest atoms is the origin of the avoided crossing mechanism as the main cause of the low thermal conductivity of the material and, consequently, the appearance of a boson peak [55]. The leading thermoelectric material PbTe, a rock salt with a perfect ordered structure, owes its low thermal conductivity to a strong anharmonic coupling between a low-energy transversal optical (TO) and the longitudinal acoustic (LA) phonons that produce the avoided crossing between LA and TO dispersion phonons [52].

A large variety of thermoelectric materials displaying glassy features phenomena like those described here can be found in recent works [49, 50, 52, 54, 55, 57–62].

Incommensurate crystals with broken translational periodicity have also revealed low-temperature glassy features due to the gapped phase and amplitude modes of the incommensurate structure [63–66].

Based on the above experimental evidence, it is quite tempting to imagine a common origin for the boson peak and the low-energy behavior of the VDOS of glasses and crystalline solids. This common origin, also based on the final considerations of the previous section, could be identified with the role of phonon scattering at the microscopic and mesoscopic level. In perfect crystals, anharmonicity rooted in the interatomic or intermolecular potential leads to inelastic phonon-phonon processes, which can be described using standard many-body theory [67]. In glasses, the anharmonicity is also ubiquitous, not only because of the (inevitable) anharmonicity of the interatomic potential but also, and to a large extent, is an emergent property amplified by the structural disorder. The availability of multiple available "soft channels" for atomic motion in the disordered environment leads to a potential of mean force that is strongly anharmonic and associated with large effective Grüneisen coefficients [68].

In MD simulations, it is possible to measure the values of Grüneisen parameter for single atoms or even single vibrational eigenmodes. The analysis showed that $\gamma_G \approx 2 - 4$, which is indicative of large anharmonicity, for modes contributing to the boson peak around Terahertz frequency. Essentially same values for boson peak modes were found for quite different glasses such as amorphous silicon in [69] and the metallic glass $Cu_{50}Zr_{50}$ in [68]. Furthermore, in [68] it was demonstrated, for $Cu_{50}Zr_{50}$, that long atomic trajectories across strongly anharmonic shallow parts (e.g., saddles) of the energy landscape make up the vibrations that constitute the boson peak. At higher frequency, instead, eigenmodes become much more harmonic-like as reflected in lower values of γ_G in the range between 0 and 1, also in line with experimental evidence [70].

All in all, the direct evidence from simulations and several experimental observations [70–72] provide strong evidence supporting a mathematical description of the boson peak based on anharmonic vibrations.

Furthermore, theories based on purely *elastic* or harmonic-like phonon scattering due to density fluctuations or spatial fluctuations of elastic moduli (such as fluctuating elasticity theory [73]) have been recently shown, by means of quantitative comparison to numerical simulations, to be unable to describe the boson peak in glasses [74, 75].[8]

A mathematical model of the boson peak in glasses and crystals can be developed based on the anharmonic Akhiezer damping mechanism described in Chap. 4, Sect. 4.2. This model was formulated in Refs. [76, 77] and is sometimes referred to as the Baggioli-Zaccone or BZ model [46, 78].

Recalling the derivation of Akhiezer phonon damping in Chap. 4, let us consider a simple toy model in which the dispersion relation of the acoustic phonons is well described by the roots of the following equation:

$$\omega_{T,L}^2 = v_{T,L}^2 k^2 - a_{T,L} k^4 + i \omega_{T,L} D_{T,L} k^2 + \ldots \tag{5.28}$$

Here, $v_{T,L}$ is the sound speed (L = longitudinal, T = transverse), $a_{T,L}$ an effective parameter which takes into account the bending or flattening of the acoustic branches towards larger k, and the appearance of a related Brillouin zone[9] ($d \operatorname{Re}(\omega)/dk = 0$). Finally $D_{T,L}$ is the already mentioned Akhiezer damping coefficient (cfr. Chap. 4), which is an increasing function of the Grüneisen parameter γ_G. Notice how Eq. (5.28) could be blindly derived using the standard effective field theory/hydrodynamics gradient expansion in the wavevector k. The Green functions for the damped phonon modes corresponding to Eq. (5.28) can be written (cfr. [67, 79]) as:

$$G_{T,L}(\omega, k) = \frac{1}{\omega^2 - v_{T,L}^2 k^2 + a_{T,L} k^4 + i \omega D_{T,L} k^2}. \tag{5.29}$$

---

[8] In [74] it was shown that predictions of fluctuating elasticity theory for the Rayleigh scattering damping coefficient are systematically lower by two orders of magnitude compared to simulations, which is sort of expected since spatial large-scale fluctuations alone cannot provide an efficient damping mechanism. In [75], it was demonstrated that several physical assumptions of the fluctuating elasticity theory of, e.g., [73] do not withstand scrutiny when carefully checked in numerical simulations.

[9] Pseudo-Brillouin zone in the case of amorphous solids.

The vibrational density of states (VDOS) is defined as:

$$g(\omega) = \frac{1}{3N} \sum_{\lambda} \delta(\omega - \Omega_{\lambda}) \tag{5.30}$$

where we use the compound index λ representing the pair of indices $\mathbf{k}j$ with j the branch ($j = L, T$) and $\mathbf{k}$ the wavevector. Here N denotes the total number of atoms in the solid. It is convenient to work with the distribution function of the eigenfrequencies squared since the eigenvalue distribution of the equation of motion of lattice dynamics is expressed in terms of the eigenfrequencies squared. In the harmonic case, the diagonalization of the dynamical matrix provides the bare (non-renormalized) eigenfrequencies squared as the eigenvalues of the dynamical matrix. In anharmonic solids, as is well known (cfr. Sect. 1.1.4), the frequencies get renormalized and acquire a small correction with respect to the bare values, which is already accounted for in our "dressed" eigenfrequency $\Omega(\mathbf{k})$. Hence, we have:

$$G(\omega^2) = \frac{1}{3N} \sum_{\lambda} \delta(\omega^2 - \Omega_{\lambda}^2) \tag{5.31}$$

and therefore:

$$g(\omega) = 2\omega G(\omega^2). \tag{5.32}$$

From this we have:

$$g(\omega) = \frac{2\omega}{3N} \sum_{\lambda} \delta(\omega^2 - \Omega_{\lambda}^2). \tag{5.33}$$

For an isotropic cubic crystal with one atom in the primitive cell, or equivalently, for an isotropic amorphous solid, the above expression for the VDOS becomes:

$$g(\omega) = \frac{2\omega}{3N} \sum_{|\mathbf{k}|} \delta(\omega^2 - \Omega(k)^2). \tag{5.34}$$

We can now recall Eq. (5.29), and we also recall the well-known Plemelj identity:

$$\frac{1}{x - x' \pm i\epsilon} = \mp i\pi \delta(x - x') + \frac{P}{x - x'}, \tag{5.35}$$

where P denotes the principal part.

Upon combining Eqs. (5.29), (5.30), and (5.35), we thus obtain:

$$\frac{1}{\omega^2 - \Omega_{L,T}^2(k) + i\,\omega\,\Gamma_{L,T}(k)} = -i\pi \delta(\omega^2 - \Omega_{L,T}^2(k)) \tag{5.36}$$

from which we finally get:

$$g(\omega) = -\frac{2\,\omega}{3\,\pi\,N} \sum_{|\mathbf{k}|<k_D} \mathrm{Im}\,\{2\,\mathcal{G}_T(k,\omega) + \mathcal{G}_L(k,\omega)\}\,, \tag{5.37}$$

where k_D denotes the maximum (Debye) wavenumber in the system, $k_D = (6\pi^2 N/V)^{1/3}$ and $\mathcal{G}_{L,T}$ is given by Eq. (5.29).

This discrete sum over $|\mathbf{k}|$ can be converted into an integral over $|\mathbf{k}| = k$ by means of the standard transformation: $\frac{3}{k_D^3}\int_0^{k_D} k^2 dk = \frac{1}{N}\sum_{|\mathbf{k}|<k_D}$.

Finally, the VDOS can be computed as:

$$g(\omega) = -\frac{2\,\omega}{\pi\,k_D^3} \int_0^{k_D} \mathrm{Im}\,[2\mathcal{G}_T(\omega,k) + \mathcal{G}_L(\omega,k)]\,k^2\,dk \tag{5.38}$$

The results of the computation are shown in Fig. 5.13.

The model, first formulated in [76,77], is clearly able to predict the occurrence of the boson peak in the VDOS, without any assumptions about the microscopic structure and interactions in the solid. The Akhiezer phonon damping, which is an effective description of underlying intricate scattering processes, thus represents a possible common origin for the boson peak observed in glasses and crystals. This appears justified also in light of the giant emergent anharmonicity in amorphous solids as quantified by the Grüneisen coefficient [68,69].

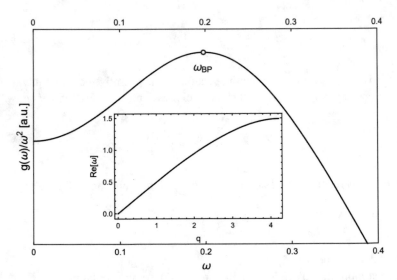

Fig. 5.13 The normalized density of states (VDOS) and the appearance of a BP anomaly well detached from the pseudo-Brillouin zone. Here we have considered a single transverse mode with $v_T = 0.5$, $a_T = 0.007$, $D_T = 0.012$, $k_D = 5.95$. Adapted from Ref. [77] with permission from the American Physical Society

In this model, the attenuated sound wave then has the following properties (we drop the indices L, T for ease of notation):

$$\text{Re}\,\Omega(k) = \frac{1}{2}\sqrt{-(D^2 + 4\,a_4)\,k^4 + 4\,v^2\,k^2}, \tag{5.39}$$

$$\text{Im}\,\Omega(k) = -\frac{1}{2}D\,k^2. \tag{5.40}$$

It is easy to see that Eq. (5.40) implies the presence of a Van Hove-like singularity $(d\omega/dk = 0)$ located at k_{VH}:

$$k_{VH} = \frac{\sqrt{2}v}{\sqrt{4a_4 + D^2}}. \tag{5.41}$$

At larger wavevector values, the group velocity $v_g \equiv d\omega/dk$ becomes negative, and eventually the real part of the dispersion relation vanishes at:

$$k_0 = \frac{2\,v}{\sqrt{4\,a_4 + D^2}} = \sqrt{2}\,k_{VH}. \tag{5.42}$$

For values of k larger than k_0, the imaginary part of the sound wave becomes positive and the model ill-defined. In this regime, the model should be used with caution, and in particular the problem can be avoided by taking the Debye wavevector k_D to be larger than the Van Hove-like peak position k_{VH} but smaller than k_0, which is a physically meaningful assumption.

The model can also be applied to systems where the phonon dispersion relations display some degree of softening such as in strain glass alloys [47] or, in general, if a maximum is present in the acoustic dispersion relation $\omega(k)$, often followed by a roton-like minimum. The presence of a maximum at sufficiently large k was found, e.g., in early numerical studies of Lennard-Jones glasses by Rahman and coworkers [80]. It has been observed ubiquitously in metallic glasses, where it is usually followed (upon further increasing k) by a softening roton-like minimum (see Refs. [81, 82]). These features in amorphous solids are connected with the onset of the characteristic oscillations in the atomic density distribution function (e.g., in the radial distribution function) due to excluded volume, at values of wavenumber k, which become comparable to the inverse of a few multiples of atomic diameter.

Finally, we show how the above model can be applied to describing the entire VDOS of paradigmatic amorphous materials. Thanks to terms like $a_{L,T}\,k^4$ in the phonon propagator Eq. (5.29), which model the flattening of dispersion relations upon approaching pseudo-Brillouin zone boundaries, also the relics of Van Hove-type peaks can be captured.

Fittings of experimental data using the above model are shown in Fig. 5.14 for amorphous silicon (top) and silica glass (bottom).

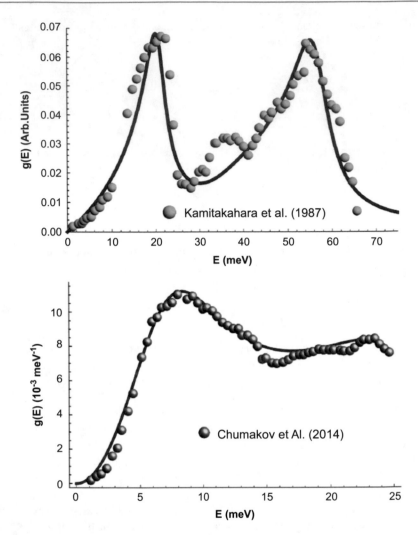

Fig. 5.14 Top panel: Comparison between the VDOS spectrum calculated using the theory presented in this section (Eqs. 5.29 and 5.38) and the one obtained experimentally for amorphous silicon by means of inelastic neutron scattering in [83]. The ratio between the longitudinal and transverse speeds is taken as $v_L/v_T \approx 1.7$, which is compatible with the experimental values for the bulk and shear moduli in amorphous silicon. The values of the parameters are $v_T = 0.31$, $v_L = 0.527$, $k_D = 5.95$, $D_T = 0.017$, $D_L = 0.01$, $a_T = 0.006$, $a_L = 0.007$. Bottom panel: Comparison between the VDOS spectrum calculated using Eqs. (5.29) and (5.38), with $v_L/v_T \approx 1.8$, and experimental data for silica glass at ambient conditions from Ref. [84] (see legend). In this latter plot only the lower 20% of the spectrum is shown. The values of the fitting parameters are $v_T = 0.4$, $v_L = 0.72$, $k_D = 3.3$, $D_T = 0.133$, $D_L = 0.03$, $a_T = 0.009$, $a_L = 0.007$. Adapted from Ref. [85]

The only nontrivial fitting parameter in the model is the diffusion coefficient of the Akhiezer damping $(D_{L,T})$, while the other parameters are constrained to reproduce meaningful shapes of the acoustic dispersion relations. In the case of amorphous silicon, the theoretical model provides a good fit to the entire spectrum, including the two smeared Van Hove peaks. The low-frequency Van Hove peak is due to transverse modes and the upper one to the longitudinal mode. The latter contribution, in a more detailed description, would result from hybridization of the longitudinal mode (which lies higher in energy) with a transverse optical mode, which is not reproduced here in order to keep the model as simple as possible. Also the small peak at ≈ 35 meV is possibly caused by hybridized optical modes that are not taken into account by the model.

References

1. S.N. Taraskin, Y.L. Loh, G. Natarajan, S.R. Elliott, Phys. Rev. Lett. **86**, 1255 (2001)
2. R.S. Krishnan, Proc. Indian Acad. Sci. Sect. A **37**(3), 377 (1953)
3. P. Flubacher, A. Leadbetter, J. Morrison, B. Stoicheff, J. Phys. Chem. Solids **12**(1), 53 (1959)
4. J. Jäckle, *Low Frequency Raman Scattering in Glasses* (Springer, Berlin, Heidelberg, 1981), pp. 135–160
5. R. Shuker, R.W. Gammon, Phys. Rev. Lett. **25**, 222 (1970)
6. A. Leadbetter, Phys. Chem. Glasses **9**, 1 (1998)
7. A.J. Martin, W. Brenig, Phys. Status Solidi (B) **64**(1), 163 (1974)
8. U. Buchenau, N. Nücker, A.J. Dianoux, Phys. Rev. Lett. **53**, 2316 (1984)
9. U. Buchenau, M. Prager, N. Nücker, A.J. Dianoux, N. Ahmad, W.A. Phillips, Phys. Rev. B **34**, 5665 (1986)
10. U. Buchenau, H.M. Zhou, N. Nucker, K.S. Gilroy, W.A. Phillips, Phys. Rev. Lett. **60**, 1318 (1988)
11. W.A. Phillips, U. Buchenau, N. Nücker, A.J. Dianoux, W. Petry, Phys. Rev. Lett. **63**, 2381 (1989)
12. V.K. Malinovsky, V.N. Novikov, P.P. Parshin, A.P. Sokolov, M.G. Zemlyanov, Europhys. Lett. (EPL) **11**(1), 43 (1990)
13. M.A. Ramos, Phys. Rev. B **49**, 702 (1994)
14. V. Malinovsky, V. Novikov, A. Sokolov, Phys. Lett. A **153**(1), 63 (1991)
15. A.V. Granato, *An Interstitialcy Model for the Boson Peak* (World Scientific, Singapore, 1996)
16. G. Carini, G. D'angelo, G. Tripodo, G.A. Saunders, Philos. Mag. B **71**(4), 539 (1995)
17. D.M. Zhu, H. Chen, J. Non-Cryst. Solids **224**(1), 97 (1998)
18. M.A. Ramos, C. Talón, S. Vieira, J. Non-Cryst. Solids **307–310**, 80 (2002)
19. G. Carini, G. Carini, D. Cosio, G. D'Angelo, F. Rossi, Philos. Mag. **96**(7–9), 761 (2016)
20. M. Moratalla, J.F. Gebbia, M.A. Ramos, L.C. Pardo, S. Mukhopadhyay, S. Rudić, F. Fernandez-Alonso, F.J. Bermejo, J.L. Tamarit, Phys. Rev. B **99**, 024301 (2019)
21. J. Wuttke, W. Petry, G. Coddens, F. Fujara, Phys. Rev. E **52**, 4026 (1995)
22. A. Fontana, R. Dell'Anna, M. Montagna, F. Rossi, G. Viliani, G. Ruocco, M. Sampoli, U. Buchenau, A. Wischnewski, Europhys. Lett. (EPL) **47**(1), 56 (1999)
23. A.P. Sokolov, E. Rössler, A. Kisliuk, D. Quitmann, Phys. Rev. Lett. **71**, 2062 (1993)
24. A.I. Chumakov, I. Sergueev, U. van Bürck, W. Schirmacher, T. Asthalter, R. Rüffer, O. Leupold, W. Petry, Phys. Rev. Lett. **92**, 245508 (2004)
25. R. Milkus, A. Zaccone, Phys. Rev. B **93**, 094204 (2016)
26. C.S. O'Hern, L.E. Silbert, A.J. Liu, S.R. Nagel, Phys. Rev. E **68**, 011306 (2003)
27. S. Alexander, Phys. Rep. **296**(2), 65 (1998)
28. B. Cui, G. Ruocco, A. Zaccone, Granul. Matter **21**(3), 69 (2019)

29. G.M. Cicuta, J. Krausser, R. Milkus, A. Zaccone, Phys. Rev. E **97**, 032113 (2018)
30. G. Semerjian, L.F. Cugliandolo, J. Phys. A Math. Gen. **35**(23), 4837 (2002)
31. S. Franz, G. Parisi, P. Urbani, F. Zamponi, Proc. Natl. Acad. Sci. **112**(47), 14539 (2015)
32. Y.M. Beltukov, V.I. Kozub, D.A. Parshin, Phys. Rev. B **87**, 134203 (2013)
33. Y.M. Beltukov, C. Fusco, D.A. Parshin, A. Tanguy, Phys. Rev. E **93**, 023006 (2016)
34. M. Baggioli, R. Milkus, A. Zaccone, Phys. Rev. E **100**, 062131 (2019)
35. P.M. Chaikin, T.C. Lubensky, *Principles of Condensed Matter Physics* (Cambridge University Press, Cambridge, 1995)
36. P.C. Martin, O. Parodi, P.S. Pershan, Phys. Rev. A **6**, 2401 (1972)
37. M. Baggioli, A. Zaccone, Phys. Rev. Res. **1**, 012010 (2019)
38. D.A. Ackerman, D. Moy, R.C. Potter, A.C. Anderson, W.N. Lawless, Phys. Rev. B **23**, 3886 (1981)
39. E. Hegenbarth, Ferroelectrics **168**(1), 25 (1995)
40. M. Simenas, S. Balciunas, S. Svirskas, M. Kinka, M. Ptak, V. Kalendra, A. Gagor, D. Szewczyk, A. Sieradzki, R. Grigalaitis, A. Walsh, M. Maczka, J. Banys, Chem. Mater. **33**(15), 5926 (2021)
41. Y. Ishii, A. Yamamoto, N. Sato, Y. Nambu, S. Ohira-Kawamura, N. Murai, T. Mori, S. Mori, Glasslike phonon excitation caused by ferroelectric structural instability (2021). https://doi.org/10.48550/arXiv.2104.01969
42. A. Gray-Weale, P.A. Madden, J. Phys. Chem. B **108**(21), 6624 (2004)
43. H. Zhang, X. Wang, A. Chremos, J.F. Douglas, J. Chem. Phys. **150**(17), 174506 (2019)
44. H. Zhang, M. Khalkhali, Q. Liu, J.F. Douglas, J. Chem. Phys. **138**(12), 12A538 (2013)
45. A. Jeżowski, M.A. Strzhemechny, A.I. Krivchikov, N.A. Davydova, D. Szewczyk, S.G. Stepanian, L.M. Buravtseva, O.O. Romantsova, Phys. Rev. B **97**, 201201 (2018)
46. Y. Miyazaki, M. Nakano, A.I. Krivchikov, O.A. Koroyuk, J.F. Gebbia, C. Cazorla, J.L. Tamarit, J. Phys. Chem. Lett. **12**(8), 2112 (2021)
47. S. Ren, H.X. Zong, X.F. Tao, Y.H. Sun, B.A. Sun, D.Z. Xue, X.D. Ding, W.H. Wang, Nat. Commun. **12**(1), 5755 (2021)
48. M. Christensen, A.B. Abrahamsen, N.B. Christensen, F. Juranyi, N.H. Andersen, K. Lefmann, J. Andreasson, C.R.H. Bahl, B.B. Iversen, Nat. Mater. **7**(10), 811 (2008)
49. A. Bhattacharya, J. Mater. Chem. C **7**, 13986 (2019)
50. C.W. Li, J. Hong, A.F. May, D. Bansal, S. Chi, T. Hong, G. Ehlers, O. Delaire, Nat. Phys. **11**(12), 1063 (2015)
51. T. Lanigan-Atkins, S. Yang, J.L. Niedziela, D. Bansal, A.F. May, A.A. Puretzky, J.Y.Y. Lin, D.M. Pajerowski, T. Hong, S. Chi, G. Ehlers, O. Delaire, Nat. Commun. **11**(1), 4430 (2020)
52. O. Delaire, J. Ma, K. Marty, A.F. May, M.A. McGuire, M.H. Du, D.J. Singh, A. Podlesnyak, G. Ehlers, M.D. Lumsden, B.C. Sales, Nat. Mater. **10**(8), 614 (2011)
53. Y. Takasu, T. Hasegawa, N. Ogita, M. Udagawa, M.A. Avila, K. Suekuni, T. Takabatake, Phys. Rev. Lett. **100**, 165503 (2008)
54. J.S. Tse, D.D. Klug, J.Y. Zhao, W. Sturhahn, E.E. Alp, J. Baumert, C. Gutt, M.R. Johnson, W. Press, Nat. Mater. **4**(12), 917 (2005)
55. M.A. Avila, K. Suekuni, K. Umeo, H. Fukuoka, S. Yamanaka, T. Takabatake, Appl. Phys. Lett. **92**(4), 041901 (2008)
56. M. Baggioli, B. Cui, A. Zaccone, Phys. Rev. B **100**, 220201 (2019)
57. B. Chazallon, H. Itoh, M. Koza, W.F. Kuhs, H. Schober, Phys. Chem. Chem. Phys. **4**, 4809 (2002)
58. J. Ma, O. Delaire, A.F. May, C.E. Carlton, M.A. McGuire, L.H. VanBebber, D.L. Abernathy, G. Ehlers, T. Hong, A. Huq, W. Tian, V.M. Keppens, Y. Shao-Horn, B.C. Sales, Nat. Nanotechnol. **8**(6), 445 (2013)
59. T. Takabatake, K. Suekuni, T. Nakayama, E. Kaneshita, Rev. Mod. Phys. **86**, 669 (2014)
60. M.A. Avila, K. Suekuni, K. Umeo, H. Fukuoka, S. Yamanaka, T. Takabatake, Phys. Rev. B **74**, 125109 (2006)
61. M. Avila, K. Suekuni, K. Umeo, T. Takabatake, Phys. B Condens. Matter **383**(1), 124 (2006)

62. C.W. Li, O. Hellman, J. Ma, A.F. May, H.B. Cao, X. Chen, A.D. Christianson, G. Ehlers, D.J. Singh, B.C. Sales, O. Delaire, Phys. Rev. Lett. **112**, 175501 (2014)
63. G. Reményi, S. Sahling, K. Biljaković, D. Starešinić, J.C. Lasjaunias, J.E. Lorenzo, P. Monceau, A. Cano, Phys. Rev. Lett. **114**, 195502 (2015)
64. J. Etrillard, J.C. Lasjaunias, K. Biljakovic, B. Toudic, G. Coddens, Phys. Rev. Lett. **76**, 2334 (1996)
65. A. Cano, A.P. Levanyuk, Phys. Rev. B **70**, 212301 (2004)
66. A. Cano, A.P. Levanyuk, Phys. Rev. Lett. **93**, 245902 (2004)
67. T. Tadano, S. Tsuneyuki, Phys. Rev. Lett. **120**, 105901 (2018)
68. Z.Y. Yang, Y.J. Wang, A. Zaccone, Phys. Rev. B **105**, 014204 (2022)
69. J. Fabian, P.B. Allen, Phys. Rev. Lett. **79**, 1885 (1997)
70. G. Baldi, V.M. Giordano, B. Ruta, R. Dal Maschio, A. Fontana, G. Monaco, Phys. Rev. Lett. **112**, 125502 (2014)
71. G. Baldi, V.M. Giordano, G. Monaco, B. Ruta, Phys. Rev. Lett. **104**, 195501 (2010)
72. B. Rufflé, G. Guimbretière, E. Courtens, R. Vacher, G. Monaco, Phys. Rev. Lett. **96**, 045502 (2006)
73. W. Schirmacher, G. Ruocco, T. Scopigno, Phys. Rev. Lett. **98**, 025501 (2007)
74. G. Szamel, E. Flenner, J. Chem. Phys. **156**(14), 144502 (2022)
75. Y.C. Hu, H. Tanaka, Nat. Phys. **18**(6), 669 (2022)
76. M. Baggioli, A. Zaccone, Phys. Rev. Lett. **122**, 145501 (2019)
77. M. Baggioli, A. Zaccone, Phys. Rev. Lett. **127**, 179602 (2021)
78. Q. Guo, H.P. Zhang, Z. Lu, H.Y. Bai, P. Wen, W.H. Wang, Appl. Phys. Lett. **121**(14), 142204 (2022)
79. Y. Xu, J.S. Wang, W. Duan, B.L. Gu, B. Li, Phys. Rev. B **78**, 224303 (2008)
80. G.S. Grest, S.R. Nagel, A. Rahman, Phys. Rev. Lett. **49**, 1271 (1982)
81. N. Kovalenko, Y. Krasny, Phys. B Condens. Matter **162**(2), 115 (1990)
82. N. Kovalenko, Y. Krasny, U. Krey, *Physics of Amorphous Metals* (Wiley-VCH, Berlin, 2001)
83. W.A. Kamitakahara, C.M. Soukoulis, H.R. Shanks, U. Buchenau, G.S. Grest, Phys. Rev. B **36**, 6539 (1987)
84. A.I. Chumakov, G. Monaco, A. Fontana, A. Bosak, R.P. Hermann, D. Bessas, B. Wehinger, W.A. Crichton, M. Krisch, R. Rüffer, G. Baldi, G. Carini Jr., G. Carini, G. D'Angelo, E. Gilioli, G. Tripodo, M. Zanatta, B. Winkler, V. Milman, K. Refson, M.T. Dove, N. Dubrovinskaia, L. Dubrovinsky, R. Keding, Y.Z. Yue, Phys. Rev. Lett. **112**, 025502 (2014)
85. M. Baggioli, A. Zaccone, Phys. Rev. Res. **2**, 013267 (2020)

Thermal Properties

6

Abstract

In Chap. 3, we encountered a first example of a macroscopic property that can be expressed through an integral over the vibrational density of states (VDOS), i.e., the viscoelastic moduli. The thermal properties of solids, the most important of which are the *specific heat* and the *thermal conductivity*, can also be written in terms of integrals over the VDOS, as we shall see in this chapter.

6.1 Specific Heat

The specific heat is defined as the derivative of the internal energy of the solid with respect to temperature. In practice, one starts by considering a set of harmonic oscillators representing atomic vibrations (anharmonic corrections can be included by considering the shift or renormalization of frequencies within self-consistent phonon theory, as discussed in Chap. 1, Sect. 1.1.4). The energy is thus is a sum over modes labeled by the wavevector $\mathbf{k}$. Upon moving from discrete sum to a continuous integral, and from wavevector to frequencies with $d^3\mathbf{k}/(2\pi)^3 = g(\omega)d\omega$ in the integral, one obtains [1]:

$$C_V(T) = 3N \int_0^{\omega_D} \left(\frac{\hbar\omega}{2k_B T}\right)^2 \sinh\left(\frac{\hbar\omega}{2k_B T}\right)^{-2} g(\omega)\, d\omega, \qquad (6.1)$$

where ω denotes, here, the vibrational eigenfrequency. This formula provides a direct link between the T dependence of the specific heat and the VDOS. Using the Debye model for the VDOS, $g(\omega) \sim \omega^2$ in (6.1), one readily obtains the dependence $C_V \sim T^3$ originally derived by Debye. In glasses, at very low T (few Kelvins), one instead observes $C_V \sim T$, for which a derivation has been provided hypothesizing the existence of tunneling two-level states (TLS) [2]. The same result can be

© The Author(s), under exclusive license to Springer Nature Switzerland AG 2023
A. Zaccone, *Theory of Disordered Solids*, Lecture Notes in Physics 1015,
https://doi.org/10.1007/978-3-031-24706-4_6

derived by directly using (6.1) with a VDOS that contains a boson peak excess of low-frequency modes due to the presence of diffusive-like excitations [3, 4]. Furthermore, the Debye-normalized specific heat, $C(T)/T^3$, exhibits a peak which is related to the boson peak in the VDOS.

Importantly, this boson peak in the specific heat has been observed not only in glasses (including ionic, polymer, and metallic glasses [5]) but also in many examples of fully ordered crystals, including (i) molecular, noble gas, and cryocrystals [6] and (ii) thermoelectric "host-guest" clathrates [7]. Obviously, this "boson peak" in perfectly ordered crystals with strong quartic anharmonicity (associated especially with the in-cage rattlers motion of the "host" atoms) cannot be explained with disorder-based models such as heterogeneous elasticity theory [8], since there is no disorder in the material. It can be explained and described, instead, purely based on the phonon damping, which is responsible for the boson peak also in perfectly ordered crystals [9].

Some examples of boson peak in the Debye-normalized specific heat of metallic glasses, from Ref. [5], are shown in Fig. 6.1.

The boson peak in the specific heat occurs at a specific temperature, T_{BP}. Experimentally, in Ref. [5], it has been shown that T_{BP} correlates linearly with the transverse speed of sound, v_T, whereas no correlation is found with the longitudinal speed of sound. The authors of Ref. [5] have demonstrated that this feature is

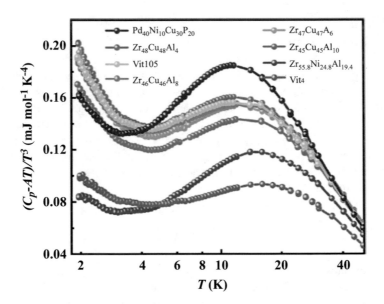

Fig. 6.1 Experimentally measured Debye-normalized boson peak of various metallic glasses. The peak is caused by the boson peak in the VDOS (see the previous Chap. 5) via Eq. (6.1). For metals, it is necessary to subtract the electronic contribution to the specific heat, which, according to Sommerfeld's model, is linear in the temperature T. Adapted from Ref. [5] with permission of the American Institute of Physics

well reproduced by the Baggioli-Zaccone (BZ) model or damped phonon model presented in Chap. 5, Sect. 5.4.3.

6.2 Thermal Conductivity

For the thermal conductivity, one typically starts from kinetic theory and from a heat balance between two regions of the material. By sheer conservation of energy, the heat flux turns out to be proportional to minus the temperature gradient, via a proportionality coefficient, which defines the thermal conductivity as $\kappa = \frac{1}{3}Cv_s\ell$, where C is the specific heat, v_s the speed of sound, and ℓ the mean free path of the phonons. This simple formula applies to a single phonon mode. By considering the various vibrational modes and again using the VDOS to sum up the various contributions, one obtains:

$$\kappa(T) = \frac{1}{V}\int_0^{\omega_D} \frac{1}{3}C(\omega)v_s\ell(\omega)g(\omega)d\omega \tag{6.2}$$

where the single-mode specific heat is given by $C(\omega) = \left(\frac{\hbar\omega}{2k_BT}\right)^2 \sinh\left(\frac{\hbar\omega}{2k_BT}\right)^{-2}$. The mean free path can also be expressed in terms of the phonon relaxation time τ as $\ell(\omega) = v_s\tau(\omega)$. In turn, τ is the inverse of the phonon linewidth, $\tau \sim \Gamma^{-1}$. For the Akhiezer mechanism induced by anharmonicity, one has the diffusive dependence derived in Eq. (4.10), leading to $\ell \sim k^{-2} \sim \omega^{-2}$ [10]. For three-phonon Umklapp processes, one has also a dependence $\ell \sim \omega^{-2}$, since this is also a phenomenon controlled by (cubic) anharmonicity [10]. Equation (6.2) can also be derived from the Peierls-Boltzmann phonon gas model [11].

For perfect crystals, as famously demonstrated first by Peierls [12], the dominant modes are those of (anharmonic) Umklapp processes (compared to which normal scattering processes are insignificant [12]). At low T ($T < \Theta$, where Θ is the Debye temperature), the above integral is dominated by $C(\omega)$, and its dependence controls the one of κ, such that $\kappa \sim T^3$ in that regime. At higher temperatures, $T > \Theta$, the specific heat saturates and becomes independent of T, whereas the mean free path begins to decrease with increasing T, $\ell \sim T^{-n}$, due to scattering processes becoming more important as T rises. The exponent n depends on many factors, including the symmetry and elastic anisotropy of the lattice as discussed by Herring [13]. For the phonon gas, simple kinetic theory gives the high-T limit $n = 1$, which also applies for weakly anharmonic solids dominated by three-phonon scattering. For diffusive type anharmonicity, and also for Umklapp processes, one has $n > 2$. Hence, one has $\kappa \sim T^{-n}$, with the appearance of a peak in κ vs T, which has been one of the most celebrated successes in the quantum theory of solids [14]. For acoustic phonons, the diffusivity D in the Akhiezer damping Eq. (4.10) can be estimated with a simple approximate argument due to Ziman [14], p. 307. The argument goes as follows. The phonon distribution at a given T tends to be distributed around a particular frequency. Assuming a Debye spectrum, the number of quanta in the frequency window $d\omega$ is proportional to $\frac{\omega^2}{1-\exp\hbar\omega/k_BT}$, which has a maximum when $\hbar\omega = \hbar\omega_{dom} = 1.6k_BT$. At a given T, the value of $\ell \sim k^{-2}$ should then be given by $\ell \sim k_{dom}^{-2}$, which then readily gives $\ell \sim T^{-2}$. This scaling coincides with what was found experimentally for quartz in the range $T = 40$–150 K [15].

In glasses, early numerical studies have indicated that localization effects due to the disorder generate a diffusive-like dynamics of phonons very similar, if not indistinguishable, to that of anharmonic processes, as shown e.g. in [16]. On this basis, one would expect similar behaviour as the one controlled by anharmonicity in crystals. The Akhiezer-type diffusive scaling $\ell \sim k^{-2} \sim \omega^{-2}$ has been experimentally measured in glasses by Berman [17] for silica glass and has been confirmed also by numerical studies in amorphous silicon [18]. At $T \gg 10$ K where the specific heat of glasses follows the Debye law $C \sim T^3$, according to (6.2), this leads to the approximately linear relation $\kappa \sim T$, which is observed experimentally in many different glasses, including silica and polymer glasses [17, 19]. It is now accepted that the *diffusons*, characterized by a dissipative dispersion of the Akhiezer type $\Gamma \sim Dk^2$, are the main carriers of heat in glasses for $T > 10$ K and up to room temperature. At $T \sim 10$ K, a shallow crossover (sometimes referred to as the "plateau") occurs from the diffusons regime to a low-T regime with $\kappa \sim T^2$, where quantum tunnelling in the anharmonic potential landscape becomes important and where the TLS theory has provided striking quantitative predictions [2].

It is important to emphasize that the diffusive form $\ell \sim k^{-2}$ emerges also in models of disorder in random lattices [20]. In this case, diffusons are generated by scattering processes controlled by the disorder, which again points toward a universal diffusive anharmonic-type nature of excitations in glasses [16].

Approaches based on "harmonic disorder" such as HET [21], which yield Rayleigh-type scattering dependencies $\ell \sim k^{-4}$ from random fluctuations of elastic moduli, cannot explain the above phenomenology observed experimentally in both glasses and crystals, apart perhaps from substitutional disorder (e.g. isotopic scattering), for which Rayleigh scattering is a more appropriate description. Also, whenever $\ell \sim k^{-n}$ with $n > 2$, the integral in (6.2) is known to diverge, as discussed extensively in [14], which requires extra care and additional assumptions.

In thermoelectric crystals, strong anharmonicity caused by host-guest rattling leads to small values of ℓ, which in turn are responsible for ultralow κ values. This is beneficial to obtaining high thermoelectric energy conversion efficiency as quantified by $z = \alpha^2\sigma/\kappa$, where α is the Seebeck coefficient and σ the electric conductivity [22]. Due to the anharmonicity-induced *glasslike* vibrational and thermal properties of thermoelectrics, also the thermal conductivity is expected to deviate from the $\kappa \sim 1/T$ behaviour typical of crystals. In particular, a much weaker decay with T [23] and sometimes even flat or quasi-linear linear trends (as for glasses, see above) are observed. In general, in spite of anharmonicity playing a dominat role in thermoelectrics, a simple Akhiezer-type argument $\ell \sim k^{-2} \sim \omega^{-2}$ cannot capture the complexity of the scattering processes induced by rattlers, which are spread out over a broad range of frequencies and momenta, and for which *ab initio* self-consistent phonon calculations are required [24].

Even though a simple theory of the ultralow thermal conductivity observed in thermoelectric clathrates is not within reach, the physical origin of this *giant anharmonicity* can be traced back to orbitally driven soft vibrational modes along certain "soft" directions in the crystal lattice, as shown by Delaire and coworkers for tin selenide [25]. In general, these soft modes are similar to lattice instabilities of the ferroelectric type [26].

This "directional" nature of giant anharmonicity in thermoelectrics, which exhibit glasslike thermal properties, lends itself naturally to a connection with structurally disordered systems, based on the degree of centrosymmetry in the force field, introduced in Ref. [27] (cfr. Chap. 1, Sect. 1.5.2), as the "common denominator" between glasses and thermoelectrics. In glasses, this "directional anharmonicity" can be naturally linked with the bonding force with a nearest neighbor not being balanced by a mirror-symmetric bond (evidently missing due to the disorder), and with the concept of string motion and stringlets [28]. In thermoelectrics, the same directional anharmonicity is instead associated with the rattling motion which occurs along certain soft lattice/orbital directions, or with ferroelectric-type soft modes, which also in this case break the mirror symmetry of forces around a tagged atom.

6.3 The Tunneling Model

In 1971, Zeller and Pohl [29] clearly established that thermal properties at low temperatures in amorphous solids were unexpectedly very different from those found in crystals, despite low-temperature properties were supposed to be dominated by long-wavelength acoustic phonons, thus being insensitive to microscopic atomic disorder. As already mentioned above, the specific heat of glasses was found to be much higher than the expected Debye level characteristic of crystals, exhibiting a quasilinear temperature dependence at the lowest temperatures and a broad maximum in $C_p(T)/T^3$, currently associated with the boson peak. The thermal conductivity $\kappa(T)$ presented a strikingly universal behavior, with a ubiquitous plateau in the range $\approx$ 2 K to 20 K, preceded by a $\kappa \sim T^2$ region, and followed by a further increase with temperature above the plateau, opposite to the typical behavior of crystalline solids. Further, acoustic and dielectric properties of amorphous solids also exhibited a rather universal behavior at low temperatures, very different from those in their crystalline counterparts [30].

Only 1 year after the paper by Zeller and Pohl, two independent works proposed what is currently known as the tunneling model [2, 31]. This simple model, essentially having only two fitting parameters, was able to account for most of the abovementioned anomalies of glasses, at least those below 1 K to 2 K. Its fundamental hypothesis is the ubiquitous existence of small groups of atoms in amorphous solids, which can oscillate between two adjacent configurational states of similar energy by quantum-mechanical tunneling through a potential-energy barrier V_B in a double-well potential (DWP). At low enough temperatures, only those two lowest energy levels (two-level states, TLS) will contribute significantly to thermal properties, its energy excitation E being $E = \sqrt{\Delta_0^2 + \Delta^2}$, where Δ is the asymmetry energy of the DWP and Δ_0 is the energy splitting arising from quantum-mechanical tunneling for the case of a symmetric DWP. In the Wentzel-Kramers-Brillouin approximation for high-enough potential barriers V_B, $\Delta_0 = \hbar\,\Omega \exp(-\lambda)$, where Ω can be taken as the angular frequency of oscillation within one single well and the argument of the exponential λ is the tunneling parameter proportional to $V_B^{-1/2}$.

Assuming random disorder, it was natural to postulate a constant distribution function in terms of Δ and λ as independent variables, that is, $P(\Delta, \lambda) = P_0$. Hence the density of TLS per unit energy and unit volume P_0 is the main free parameter of the tunneling model. If we assimilate a random, constant P_0 to a constant density of states n_0, then $C_V(T) = \frac{\pi^2}{6} n_0 k_B^2 T$ is readily obtained. Nonetheless, in order to address the behavior of the thermal conductivity and of the acoustic properties, the interaction of these TLS tunneling states with the sound waves of the atomic lattice (acoustic "phonons") must be taken into account. For a thorough development of the equations of the model, the reader is referred to [32]. In brief, the strain field induced by the elastic sound wave in the TLS system changes the asymmetry Δ of the DWP (neglecting small variations in Δ_0) by an amount characterized by a coupling energy γ_j, where one may distinguish longitudinal and transverse acoustic waves, $j = L, T$. The essential hypothesis of the standard tunneling model is to assume an averaged constant coefficient γ_j for a given substance, this being the second fitting parameter. Then, the relaxation rate of a tunneling state in the thermal bath of phonons can be determined, and hence a more precise expression for the specific heat is given by:

$$C_V(T) = k_B \int_0^\infty \left(\frac{E}{2k_BT}\right)^2 \cosh^{-2}\left(\frac{E}{2k_BT}\right) n(E)\, dE$$

$$\approx \frac{\pi^2}{12} P_0\, k_B^2\, T \ln\left(\frac{4\, t_{\exp}}{\tau_{\min}}\right) \qquad (6.3)$$

with $\tau_{\min}$ being the TLS relaxation time for the symmetric DWP at a given energy E and $t_{\exp}$ the experimental measuring time. Therefore, the tunneling model predicts a quasilinear temperature dependence with a small logarithmic factor that slightly increases with measuring time, what was confirmed, at least qualitatively, in experiments [33]. As in crystalline solids, the heat in insulating amorphous solids is carried by means of propagating phonons, the great difference in their thermal conductivity arising from new, strong scattering mechanisms in glasses, which reduce significantly the mean free path of phonons compared to crystals (see Eq. (6.2)). At very low temperature, the dominant process would be resonant scattering by tunneling states, and the corresponding inverse mean free path of phonons can be shown to lead to:

$$\kappa(T) = \frac{\rho\,k_B^3}{6\pi\,\hbar^2}\left(\sum_j \frac{v_j}{P_0\gamma_j^2}\right)T^2.$$ (6.4)

In conclusion, the tunneling model is able to account for the quadratic temperature dependence of the thermal conductivity below the universal "plateau," though not for the plateau itself, as well as for most acoustic and dielectric measurements at low temperatures [32]. Nevertheless, several theoretical and experimental works have put into question the apparent success of the tunneling model and the general validity of its strikingly simple assumptions (see, for instance, Ref. [34] and references therein).

References

1. D. Khomskii, *Basic Aspects of the Quantum Theory of Solids: Order and Elementary Excitations* (Cambridge University Press, Cambridge, 2010)
2. P.W. Anderson, B.I. Halperin, C.M. Varma, Philos. Mag. **25**(1), 1 (1972)
3. M. Baggioli, A. Zaccone, Phys. Rev. Res. **1**, 012010 (2019)
4. M. Baggioli, R. Milkus, A. Zaccone, Phys. Rev. E **100**, 062131 (2019)
5. Q. Guo, H.P. Zhang, Z. Lu, H.Y. Bai, P. Wen, W.H. Wang, Appl. Phys. Lett. **121**(14), 142204 (2022)
6. M.A. Strzhemechny, A.I. Krivchikov, A. Jeżowski, Low Temp. Phys. **45**(12), 1290 (2019)
7. M. Moratalla, J.F. Gebbia, M.A. Ramos, L.C. Pardo, S. Mukhopadhyay, S. Rudić, F. Fernandez-Alonso, F.J. Bermejo, J.L. Tamarit, Phys. Rev. B **99**, 024301 (2019)
8. A. Marruzzo, W. Schirmacher, A. Fratalocchi, G. Ruocco, Sci. Rep. **3**, 1407 EP (2013). Article
9. M. Baggioli, A. Zaccone, J. Phys. Mater. **3**(1), 015004 (2019)
10. P. Klemens, *Thermal Conductivity and Lattice Vibrational Modes* (vol. 7). Solid State Physics (Academic Press, Cambridge, 1958), pp. 1–98
11. V.L. Gurevich, *Transport in Phonon Systems* (North-Holland, Netherlands, 1986)
12. R. Peierls, Ann. Phys. **395**(8), 1055 (1929)
13. C. Herring, Phys. Rev. **95**, 954 (1954)
14. J.M. Ziman, *Electrons and Phonons: The Theory of Transport Phenomena in Solids*. Oxford Classic Texts in the Physical Sciences (Oxford University Press, Oxford, 2001)
15. H.E. Bömmel, K. Dransfeld, Phys. Rev. **117**, 1245 (1960)
16. S.R. Nagel, G.S. Grest, A. Rahman, Phys. Rev. Lett. **53**, 368 (1984)
17. R. Berman, Phys. Rev. **76**, 315 (1949)
18. P.B. Allen, J.L. Feldman, J. Fabian, F. Wooten, Philos. Mag. B **79**(11–12), 1715 (1999)
19. D.G. Cahill, R.O. Pohl, Phys. Rev. B **35**, 4067 (1987)
20. Y.M. Beltukov, V.I. Kozub, D.A. Parshin, Phys. Rev. B **87**, 134203 (2013)
21. W. Schirmacher, Europhys. Lett. (EPL) **73**(6), 892 (2006)
22. G.J. Snyder, M. Christensen, E. Nishibori, T. Caillat, B.B. Iversen, Nat. Mater. **3**(7), 458 (2004)
23. J. Klarbring, O. Hellman, I.A. Abrikosov, S.I. Simak, Phys. Rev. Lett. **125**, 045701 (2020)
24. T. Tadano, S. Tsuneyuki, Phys. Rev. Lett. **120**, 105901 (2018)
25. C.W. Li, J. Hong, A.F. May, D. Bansal, S. Chi, T. Hong, G. Ehlers, O. Delaire, Nat. Phys. **11**(12), 1063 (2015)
26. O. Delaire, J. Ma, K. Marty, A.F. May, M.A. McGuire, M.H. Du, D.J. Singh, A. Podlesnyak, G. Ehlers, M.D. Lumsden, B.C. Sales, Nat. Mater. **10**(8), 614 (2011)

27. R. Milkus, A. Zaccone, Phys. Rev. B **93**, 094204 (2016)
28. H. Zhang, X. Wang, H.B. Yu, J.F. Douglas, J. Chem. Phys. **154**(8), 084505 (2021)
29. R.C. Zeller, R.O. Pohl, Phys. Rev. B **4**, 2029 (1971)
30. W.A. Phillips, *Amorphous Solids:Low-Temperature Properties* (Springer, Berlin; New York, 1981)
31. W.A. Phillips, J. Low Temp. Phys. **7**(3), 351 (1972)
32. W.A. Phillips, Rep. Prog. Phys. **50**(12), 1657 (1987)
33. M. Meissner, K. Spitzmann, Phys. Rev. Lett. **46**, 265 (1981)
34. M.A. Ramos, Low Temp. Phys. **46**(2), 104 (2020)

Viscosity of Supercooled Liquids

7

Abstract

The viscosity of supercooled liquids exhibits a spectacular increase by up to 13 orders of magnitude upon approaching the glass transition. The temperature dependence of viscosity in supercooled liquids is thus a central object for our understanding of the glass transition, besides being of greatest technological importance. At present, no microscopic quantitative theory of the viscosity of liquids exists, and only phenomenological models are available. We will present one of these phenomenological model in greater detail, the shoving model, since it provides a direct link between viscosity and microscopic bonding and structural parameters. This model also allows for rationalizing the so-called fragility, a parameter that relates the temperature dependence of viscosity to the chemical features of the liquid.

7.1 The Viscosity of Liquids

Viscosity is a basic transport quantity of matter which quantifies the internal frictional forces between adjacent layers of liquid that move relative to each other. These frictional forces are due to the fact that atoms or molecules belonging to a fluid layer are constantly scattered in collisions with atoms or molecules belonging to an adjacent layer.

Newtonian fluids are defined as those fluids where the viscosity is not dependent on the shear rate or flow rate. This is the case of all simple liquids. Complex fluids (e.g., colloidal suspensions, starch, slurries etc) instead exhibit a viscosity which depends on the applied shear rate $\dot{\gamma}$ and can either decrease with $\dot{\gamma}$ (shear thinning) or increase with $\dot{\gamma}$ (shear thickening).

In a standard setup where a liquid rests on a solid surface (bottom) and is driven by an external force F acting on a plate located at the free surface (top), the flow

© The Author(s), under exclusive license to Springer Nature Switzerland AG 2023
A. Zaccone, *Theory of Disordered Solids*, Lecture Notes in Physics 1015,
https://doi.org/10.1007/978-3-031-24706-4_7

velocity is observed to vary linearly from zero at the bottom, i.e., at $y = 0$ to v at the top, say $y = L$. The magnitude of the force F acting on the top plate is then proportional to the speed v and the area A of the plate, and inversely proportional to the separation L:

$$F = \eta A \frac{v}{L}. \tag{7.1}$$

The ratio v/L has dimensions of inverse time and is equal to the shear rate $\dot{\gamma}$, while the proportionality coefficient η defines the liquid's viscosity. From dimensional analysis, since F is a force, it is then clear that the viscosity has dimensions: $\frac{kg}{m \cdot s} = \frac{N}{m^2} \cdot s = Pa \cdot s$, where Pa denotes the SI unit of pressure (Pascal).

If the plate is dragged along the Cartesian x direction, then the velocity profile is linear in y, i.e., $v = \dot{\gamma} y$ is the velocity along x, with $\dot{\gamma}$ the (constant) shear rate. This is a situation encountered in many physical systems, such as Couette laminar flows implemented in the gap between two coaxial cylinders in commercial rheometers employed to measure the viscosity of substances. Linear shear flow is also a good approximation for the local flow profile in channel (Poiseuille) flow. In this case, we can identify the shear stress as $\sigma'_{xy} = F/A$ and noting that $v/L = \dot{\gamma} = \partial v/\partial y$, we thus obtain the Newton law of viscous shear flow:

$$\sigma'_{xy} = \eta \frac{\partial v}{\partial y} \tag{7.2}$$

which further simplifies for linear flow fields to:

$$\sigma'_{xy} = \eta \dot{\gamma}. \tag{7.3}$$

The prime on the shear stress indicates that this is a viscous dissipative contribution to the stress tensor (to distinguish it from the elastic one that may be present, cfr. Sect. 4.2 in Chap. 4).

The viscosity η is directly related to the dissipation of kinetic energy in viscous flow, and the above example of linear shear flow represents perhaps the most fundamental and paradigmatic instance of dissipative systems in physics.

Viscous frictional forces (unlike "solid friction") are proportional to velocity, just think of the Stokes viscous drag on a sphere which is dragged inside a liquid: the force that one has to exert is indeed proportional to the dragging velocity v and to the sphere size.

The frictional forces can be written as derivatives of a quadratic function of the velocities, known as the Rayleigh dissipation function, Ψ. The frictional force corresponding to a certain generalized coordinate q_a of the system is then written as $f_a = -\partial \Psi/\partial \dot{q}_a$. The dissipation function can only depend on the velocity gradient in the system (in our case above, $\dot{\gamma}$), since if the velocities were all constant, then there would be no dissipation at all. Due to symmetry under spatial inversion,[1] the dissipation function must be a quadratic function of the velocity gradients.

For a generic anisotropic system (solid or liquid), one has the following form for the dissipation function:

$$\Psi = \frac{1}{2}\eta_{iklm}\left(\frac{\partial v_i}{\partial x_k} + \frac{\partial v_k}{\partial x_i}\right)\left(\frac{\partial v_l}{\partial x_m} + \frac{\partial v_m}{\partial x_l}\right) \tag{7.4}$$

where η_{iklm} is the fourth-rank viscosity tensor. For isotropic systems, η_{iklm} has only two independent components, the shear viscosity η and the bulk viscosity ζ:

$$\eta_{iklm} = \zeta\delta_{ik}\delta_{lm} + \eta(\delta_{il}\delta_{km} + \delta_{im}\delta_{kl} - 2\delta_{ik}\delta_{lm}/d) \tag{7.5}$$

a result that we already encountered in Chap. 4, Sect. 4.2, when we derived the Akhiezer damping using continuum mechanics.

For shear flow with no volume change ($\delta V = 0$), the above Eq. (7.4) becomes:

$$\Psi = \eta\left(\frac{\partial v_i}{\partial x_k} + \frac{\partial v_k}{\partial x_i}\right)^2 \tag{7.6}$$

This is obviously related to the kinetic energy dissipated in the viscous shear flow. Indeed, the dissipated kinetic energy per unit volume can be shown, from the Navier-Stokes equation, to be given by:

$$\dot{E}_{kin} = -\frac{1}{2}\eta\left(\frac{\partial v_i}{\partial x_k} + \frac{\partial v_k}{\partial x_i}\right)^2, \tag{7.7}$$

and thus:

$$\dot{E}_{kin} \propto -\eta\dot{\gamma}^2, \tag{7.8}$$

which implies that the kinetic energy (e.g., injected into the system by driving the top plate) gets dissipated into heat at a rate which is proportional to the viscosity and to the shear rate squared.

Finally, the viscosity plays an important role in the equation of motion of a viscous laminar fluid, i.e., the linearized Navier-Stokes equation, which reads as:

$$\rho\frac{d\mathbf{v}}{dt} = \eta\nabla^2\mathbf{v} - \nabla p \tag{7.9}$$

where ρ and p are the density and the pressure of the liquid, respectively.

---

[1] Shearing in the $+x$ direction must cause the same dissipation as shearing in the $-x$ direction.

7.2   The Vogel-Fulcher-Tammann Empirical Expression

The Vogel-Fulcher-Tammann (VFT) equation for the temperature dependence of liquid viscosity was proposed in the 1920s and reads as follows:

$$\eta = \eta_0 \exp\left(\frac{B}{T - T_{VF}}\right) \tag{7.10}$$

where η_0, B, and T_{VF} are empirical fitting parameters. For example, for water, one finds $\eta_0 = 0.02939$ mPa $\cdot$ s, $B = 507.88$ K, and $T_{VF} = 149.3$ K. The latter temperature T_{VF} is not too far from the temperature $T_g \approx 130$ K below which water is thought to behave as a glassy amorphous solid.[2]

7.3   Fragility

The concept of fragility was introduced by Angell [1] in an attempt to rationalize the glass-forming behavior of liquids based on the temperature dependence of their viscosity upon approaching the glass transition.

The fragility is defined quantitatively as follows:

$$m \equiv \left(\frac{\partial \log_{10} \eta}{\partial \left(T_g/T\right)}\right)_{T=T_g} \tag{7.11}$$

this is the most widespread definition of fragility, also known as "kinetic fragility" or fragility index. Clearly, it is a measure of the viscosity variation with temperature as the system approaches the glass transition temperature from the high-temperature liquid.

In order to appreciate the above definition of fragility, liquids with widely different chemical composition, bonding, and structure are plotted in Fig. 7.1.

According to the above definition, the most fragile liquids are those with the most pronounced slope and curvature in the plot of viscosity η vs T_g/T. At the opposite end, the least fragile liquids, referred to as "strong" liquids, are those which basically display an Arrhenius dependence on temperature, i.e., $\eta \sim \exp[const/T]$. These are typically network glasses (i.e., systems well described by Zachariasen's continuous random network model presented in Chap. 1) such as oxide glasses. As we shall see in the following, in order to better rationalize the disparate trends of

---

[2] The vitrification temperature of water is still the object of much debate and one of the most unsettled issues in liquid physics.

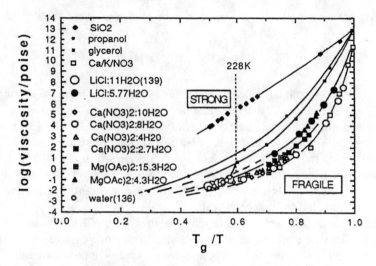

Fig. 7.1   The so-called Angell plot [2] showing the viscosity of various liquid substances plotted as function of temperature in such a way that the fragility index m can be readily extracted according to Eq. (7.11). Strong liquids are seen to follow the Arrhenius dependence $\eta \sim \exp[const/T]$, whereas fragile liquids display a much steeper dependence on temperature upon approaching T_g. Reproduced from Ref. [2] with permission from the American Chemical Society

various liquids, it will be necessary to build on models of viscosity that take the microscopic bonding and structure into account.

7.4   The Adam-Gibbs Model and Growing Length Scales

Adam and Gibbs in 1965 introduced the concept of cooperatively rearranging regions (CRR), defined as "the smallest regions that can undergo a transition to a new configuration without a requisite simultaneous configurational change on and outside its boundary." They also argued that the average size of the CRRs diverges at the Kauzmann temperature T_K,[3] which they argued to coincide with the temperature at which the configurational entropy of the systems becomes equal to zero. Assuming further that the size of the CRR is inversely proportional to the configuration entropy S_c and that the energy barrier W to overcome for flow to occur

---

[3] The Kauzmann temperature was defined by Kauzmann as the temperature at which, by extrapolating liquid data below the glass transition, the entropy of the liquid becomes equal to that of the equilibrium crystal. It is not accessible on experimental time scales because of the intervening glass transition.

is proportional to the CRR size, they proposed the following (Adam-Gibbs) relation:

$$\eta = \eta_0 \exp\left(-\frac{C}{S_c T}\right) \tag{7.12}$$

where C is some constant. By Taylor-expanding the denominator in the exponential about T_K to first order in T, the VFT relation Eq. (7.10) is readily recovered.

The Adam-Gibbs theory has received scarce experimental support [3]. Furthermore, recent stringent numerical simulations tests [4] demonstrate that the relation is somewhat violated (e.g., the T_{VF} significantly deviates from T_K) although these discrepancies may be reduced by resorting to more advanced versions of the theory such as the random first order theory (RFOT) of the glass transition developed by Wolynes, Thirumalai, and coworkers [5].

Nonetheless, the concept of a diverging length scale of CRRs has been one of the most influential concepts in glassy physics in recent years. By resorting to higher-order four-point correlation functions for the spatiotemporal dynamics, it has been possible to show that a growing dynamic length scale accompanying the glass transition becomes visible in experimental data of glass-forming liquids [6].

While these concepts about a growing length scale and growing dynamic heterogeneity at the glass transition have proved very fertile for addressing and characterizing the glass transition problem, their relevance is rather limited if one's goal is to develop more microscopic links between structure, bonding, and material properties.

In the following, we shall explore a different approach, which is not based on entropy but rather on relating the energy barrier for flow events to the elastic shear modulus, which, as we saw in Chaps. 2 and 3, can be then related to microscopic bonding and structural parameters.

7.5 Microscopic Theory of Viscosity

7.5.1 Frenkel's Theory

Y. Frenkel historically had the merit of injecting new deep ideas into the field of liquid state physics, which around the mid-twentieth century had been somewhat left behind compared to the years around the turn of the century when it was one of the most dynamic fields of physics.

Frenkel's seminal idea was that the viscosity of a liquid must be defined as the inverse of the mobility α of the individual particles that constitute the liquid. The mobility in a liquid is directly proportional to the molecular diffusion coefficient D (through the Einstein relation). This then implies that the viscosity of the liquid is inversely proportional to the molecular diffusion coefficient D. Since the latter is directly proportional to temperature (again according to the Einstein relation $D = \alpha k_B T$), this explains why the viscosity of liquids is observed to decrease with temperature.

For liquids, the particle mobility is actually determined by the thermally activated jumping rate over the "cage" formed by its nearest neighbors, characterized by an energy barrier W. Hence according to the Arrhenius law: $\alpha \propto e^{-W/k_B T}$. This then leads to:

$$\eta = A e^{W/k_B T} \tag{7.13}$$

with a proportionality prefactor A.

In order to determine A, Frenkel considered the characteristic hopping time of a particle out of its cage, again a thermally activated process: $\tau = \tau_0 e^{W/k_B T}$, where τ_0 is the molecular "attempt frequency," which is familiar from the transition state theory of chemical reactions. Then, he defined an average translation velocity of the particle as $w = \delta/\tau$, where δ is a characteristic length in the order of the cage size. Using (from dimensional analysis and other considerations) $D = \frac{\delta^2}{6\tau}$ inside the Stokes-Einstein relation for the diffusion coefficient, $D = \frac{k_B T}{6\pi \eta a}$, where a is the molecular size, he then solved for η and arrived at:

$$\eta = \frac{k_B T \tau_0}{\pi a \delta^2} e^{W/k_B T}, \tag{7.14}$$

which identifies $A = \frac{k_B T \tau_0}{\pi a \delta^2}$. Despite its simplicity, the above formula provides a good description of the temperature dependence of viscosity of simple liquids at high temperature (i.e., well above the melting point and the glass transition).

While the energy barrier W set by the cage of nearest neighbors can hardly be evaluated from first principles, the following argument, also due to Frenkel, provides a reasonable estimate in terms of a key physical quantity such as the (affine, high-frequency) shear modulus.

Frenkel borrowed ideas from his work on defects migration in solids and proposed that the barrier energy W for the particle escape from the cage must be equal to the energy required to insert the escaping particle into an "interstitial" position, i.e., a small cavity of free volume just outside the cage. This situation is schematically depicted in Fig. 7.2.

Following Frenkel [7], we estimate the barrier energy W as the elastic energy to insert the tagged particle, escaping from its cage, into a small cavity (interstitial position) just outside the cage. Assuming the cavity to be spherical and the surrounding medium to be an isotropic elastic medium (for liquids, this is true on the short time scales involved in the jumping process, τ_0 and τ considered above), this reduces to an exercise in elasticity theory. One therefore starts from the equation of equilibrium for an isotropic elastic medium, $\partial \sigma_{ik}/\partial x_k = 0$, in the absence of body-volume (e.g., gravitational) forces, where x_k is the k-th Cartesian component of the spatial coordinate and σ_{ik} is the stress tensor. Using the Hooke's law for stress versus strain valid for isotropic media and expressing the strain tensor in terms of

interstitial
position

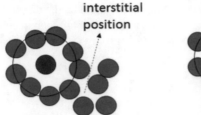

Fig. 7.2 Schematic illustration of an event by which a particle abandons its original quasi-equilibrium position in the cage formed by its nearest neighbors and jumps under the influence of thermal fluctuations to a new quasi-equilibrium position just outside the cage. The new position, in the left frame, is represented by an interstitial position or vacancy or cavity. The energy barrier W can be estimated as the elastic energy needed to accommodate the particle in the cavity, which for simplicity is taken to be spherical. This leads to a quantitative estimate of W

the displacement vector field $\mathbf{u}$ (cfr. its definition in Appendix A), one obtains [8]:

$$(\lambda + \mu)\nabla(\nabla \cdot \mathbf{u}) + \mu\nabla^2\mathbf{u} = 0, \tag{7.15}$$

where λ and μ are the two Lamè coefficients, related to the shear modulus $G \equiv \mu$ and the bulk modulus $K = \lambda + \frac{2}{3}\mu$ (cfr. Appendix A). Since we are in a liquid and we are specifically considering processes occurring on very short time scales, it is clear that $G_A = G_\infty = \mu$, i.e., the infinite frequency affine shear modulus is the relevant one.

Given the spherical nature of the problem, the vector field $\mathbf{u}$ can be written as the gradient of a radial function $\phi(r)$ where r is the radial distance from the center of the spherical cavity (the interstitial position where the particle is jumping into). Hence, $\mathbf{u} = -\nabla\phi$. Replacing this in Eq. (7.15) and making use of standard relations, we get $\nabla\nabla^2\phi = 0$, since $\nabla^2\phi = 0$ must vanish at infinite distance from the cavity (since also $\mathbf{u} \to 0$ as $r \to \infty$). This, in turn, implies $\phi = -\mathcal{A}/r$, where $\mathcal{A}$ is a constant, meaning that the displacement vector field $\mathbf{u}$ around the cavity filled by the particle jumping out of the cage behaves like the electric field of a point charge, and

$$u_x = \frac{\mathcal{A}x}{r^3}, \quad u_y = \frac{\mathcal{A}y}{r^3}, \quad u_z = \frac{\mathcal{A}z}{r^3}. \tag{7.16}$$

This implies that, in spherical coordinates, and upon differentiating, one has:

$$u_{rr} = -\frac{2\mathcal{B}x}{r^3}, \quad u_{\theta\theta} = \frac{\mathcal{B}y}{r^3}, \quad u_{\phi\phi} = \frac{\mathcal{B}z}{r^3}. \tag{7.17}$$

This result indicates that, for a system of constant uniform density, the net isotropic volume change is identically zero, i.e., $\nabla \cdot \mathbf{u} = 0$. In other words, somewhat contrary to one's intuition, the bulk hydrostatic deformation plays no role in this process, which is entirely controlled by the shear deformation energy (as discussed also in

Ref. [9]). The components of the linearized symmetric strain tensor $\mathbf{e} = \frac{1}{2}\left(\mathbf{u}^T + \mathbf{u}\right)$ (cfr. Appendix A) are then computed as:

$$u_{xx} = \frac{\partial u_x}{\partial x} = \frac{\mathcal{A}}{r^3}\left(1 - \frac{3x^2}{r^2}\right)$$

$$u_{xy} = \frac{1}{2}\left(\frac{\partial u_x}{\partial y} + \frac{\partial u_y}{\partial x}\right) = -\frac{3\mathcal{A}xy}{r^5} \tag{7.18}$$

and so on for the other components. We can thus proceed to computing the elastic energy per unit volume (energy density) again according to linear elasticity theory of isotropic bodies:

$$\mathcal{E} = \frac{1}{2}\lambda(\nabla \cdot \mathbf{u})^2 + \mu(u_{xx}^2 + ... + 2u_{xy}^2 + ...). \tag{7.19}$$

Finally, the energy density can be integrated from the cavity of radius r_0 up to infinity, leading to an estimate of the energy cost W of bringing the particle from the initial position in the cage to the interstitial position outside as:

$$W = 8\pi G_A \frac{\mathcal{A}^2}{r_0^3} \tag{7.20}$$

The constant $\mathcal{A}$ can determined by specifying that the net displacement u equal to the mismatch between the spherical cavity radius r_0 and the migrating particle radius r_1 is given by:

$$u = \frac{\mathcal{A}}{r^2} = r_1 - r_0. \tag{7.21}$$

When $r = r_0$, this gives $\mathcal{A} = r_0^2(r_1 - r_0)$, and, finally:

$$W = 8\pi G_A r_0(r_1 - r_0)^2. \tag{7.22}$$

Within the above picture, taking $r_0 = 1$ Å, $r_0 = 2r_1$, and $G_A = 10^{10}$ Pa, Frenkel estimated an energy barrier $W \sim 30$ kcal/mol, which is in the same order of magnitude as the evaporation energy of liquid metals.

7.5.2   Dyre's Shoving Model

J. Dyre, in the 1990s, provided a re-interpretation of the above calculation from a somewhat different angle, although the final result is formally identical or very similar to that of Frenkel's theory.

In Dyre's picture, the elastic energy is not the one required to insert the escaping particle into the interstitial cavity. Instead, it is defined as the elastic energy needed to expand the cage in order for the tagged particle at the center of the cage to escape [10]. In other words, it is the energy cost for the tagged particle to shove aside the surrounding liquid so that it can escape from the cage, hence the name "shoving" model. This is equivalent to the energy cost of elastic deformation that the tagged particle would incur in if its size was growing by an amount ΔR. Denoting the region's size before the expansion by R, one then has the volume of the rearranging region $V = \frac{4}{3}\pi R^3$ and its volume change $\Delta V = 4\pi R^2 \Delta R$. By identifying $(r_1 - r_0)$ in Eq. (7.22) with ΔR and r_0 with R and by comparing with Eq. (7.22), Dyre defines the characteristic volume V_c as:

$$V_c = \frac{2}{3}\frac{(\Delta V)^2}{V} \tag{7.23}$$

such that the energy barrier acquires the appealingly simple form:

$$W(T) = G_A(T)V_c \tag{7.24}$$

and therefore:

$$\eta = \eta_0 \exp\left[\frac{G_A(T)V_c}{k_B T}\right]. \tag{7.25}$$

It is now clear that it is the temperature dependence of the high-frequency shear modulus, G_A, that controls the temperature dependence of the viscosity and hence also the fragility index defined in Eq. (7.11).

As we saw in Chap. 2 and in Chap. 3 when we discussed microscopic theories of elasticity in amorphous solids, the shear modulus G depends on temperature in a nontrivial way. In the present case, the process of jumping out of the cage, as already mentioned, occurs on very short time scales; hence, the relevant form of the shear modulus is the high-frequency, affine modulus G_A. In Chap. 2, we have seen that G_A is related to the average coordination number z and to the bonding interaction via the spring constant κ (Born-Huang formulae). In the next section, we shall develop a quantitative model to determine the temperature-dependent $G_A(T)$ in terms of microscopic bonding and structure as encoded in the interatomic potential (or better, potential of mean force) and the radial distribution function $g(r)$. This will lead to closed-form expressions for the T-dependent viscosity and for the fragility.

7.5.3 Linking Viscosity to Bonding and Structure: The KSZ Model

We start by recalling Eq. (2.30) from Chap. 2:

$$G = \frac{1}{V} \left(\frac{\partial^2 U}{\partial \gamma^2} \bigg|_{\gamma \to 0} - \Xi_i \mathbf{H}_{ij}^{-1} \Xi_j \right). \tag{7.26}$$

from which, in Chap. 2, we derived the closed-form expression for a disordered assembly of spherical particles with z nearest neighbors interacting with a spring-like potential with spring constant κ:

$$G = G_A - G_{NA} = \frac{1}{30} \frac{N}{V} \kappa R_0^2 (z - 6). \tag{7.27}$$

Furthermore, in Chap. 2 and also in Chap. 3, thanks to the nonaffine viscoelastic theory, we established that while $G(T)$ goes to zero at the glass transition temperature, $G(T_g) = 0$, the high-frequency affine shear modulus G_A remains non-zero and only decreases weakly and monotonically as T increases, cfr. Fig. 2.11 in Chap. 2 and Fig. 3.8 in Chap. 3.

We thus focus on G_A given by:

$$G_A = \frac{N}{30V} \kappa R_0^2 z \tag{7.28}$$

and study its temperature dependence. As discussed in Chap. 2, Sect. 2.5.1, the main contribution to the temperature dependence of the affine high-frequency modulus comes from thermal expansion. The first direct effect is to decrease the prefactor N/V, i.e., the density, in the above formula. The other equally important effect is that as T increases, the volume increases, and thus the average nearest neighbor distance also increases. This implies that, statistically, a lower number of nearest neighbor will entertain bonding interactions with the tagged atom at the center of the cage or, in other words, z decreases with temperature.

Recalling the definition of thermal expansion coefficient from Chap. 1, Eq. (1.14), we have that $\alpha_T = \frac{1}{V}(\partial V/\partial T) = -\frac{1}{\phi}(\partial \phi/\partial T)$, with ϕ the packing fraction. As we already did in Chap. 2, we can integrate this simple differential equation to get $\ln(1/\phi) = \alpha_T T + c$.

The change δz in bonded nearest neighbors z can be related to the change $\delta \phi$ in packing fraction by recalling the definition of radial distribution function (rdf) given in Chap. 1, Sect. 1.2, Eq. (1.28). Using as the reference state the point of vanishing rigidity, defined by $\phi = \phi_c$ and $z = z_c$, we get:

$$\delta z = z - z_c \sim \int_1^{1+\delta\phi} r^2 g(r) dr, \tag{7.29}$$

where r is the radial coordinate normalized by the particle diameter σ.

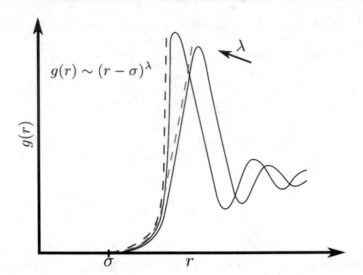

Fig. 7.3 Schematic representation of the power-law approximation, Eq. (7.30) in the text, of the "repulsive" part of the $g(r)$. Higher values of λ are linked with steeper ascending part of the first peak of $g(r)$, which corresponds to steeper repulsion in the potential of mean force for the interatomic interaction via Eq. (1.77)

Since we are interested only in nearest neighbors that are close enough to the tagged particle to be bonded in the interaction minimum, we do not need to have the whole $g(r)$ in analytical form (which would be a formidable task) but only the very short-range part that describes the closest particles. We thus approximate the ascending part of the first peak of $g(r)$, cfr. Fig. 1.6 by means of a simple power law:

$$g(r) \sim (r - \sigma)^{\lambda} \tag{7.30}$$

where λ is a material and composition-dependent exponent that tells the steepness of the "repulsive" side of $g(r)$.[4] The spirit of this approximation is schematically illustrated in Fig. 7.3.

By recalling the definition of potential of mean force, Eq. (1.77) in Chap. 1, we get a relation between the potential of mean force and λ, as $w(r)/k_B T = -\ln g(r) \sim -\ln(r - \sigma)^{\lambda}$. Here the hard-core diameter σ signals the separation at which the repulsion energy between two atoms becomes virtually infinite.

Therefore, λ is a measure of the steepness of the interatomic repulsion and possibly contains further many-body effects encoded in the potential of mean force. It is also apparent that λ is the inverse of the "softness" of the interparticle repulsion. Using Eq. (7.30) in Eq. (7.29) and neglecting the curvature of the spheres (atoms are

---

[4] It has nothing to do with the Lamè coefficient for which the same symbol is used.

taken to be spherical) at short distance (also known as the Derjaguin approximation), we then obtain:

$$\delta z \sim \delta\phi^{1+\lambda}. \tag{7.31}$$

We now make the dependence on ϕ explicit in Eq. (7.28) by recalling that $\phi = \upsilon N/V$ where $\upsilon = \frac{4}{3}\pi R_0^3$ the hard-core volume,

$$G_A = \frac{1}{5\pi}\frac{\kappa}{R_0}\phi z. \tag{7.32}$$

Recalling further that (cfr. also Chap. 2, Sect. 2.5) $\phi(T) \sim e^{-\alpha_T T}$, we therefore get $z(T) \sim e^{-(1+\lambda)\alpha_T T}$, which upon replacing in the above equation yields:

$$G_A(T) \sim \frac{1}{5\pi}\frac{\kappa}{R_0}\phi \exp\left[-(2+\lambda)\alpha_T T\right]. \tag{7.33}$$

We finally have found a connection between the high-frequency shear modulus G_A, the steepness of the interatomic interaction, λ, and the thermal expansion coefficient α_T (the latter in turn related to anharmonicity, cfr. Sect. 1.1.4 in Chap. 1). This result can be rewritten as:

$$G(T) = C_G \exp\left[\alpha_T T_g(2+\lambda)\left(1 - \frac{T}{T_g}\right)\right], \tag{7.34}$$

where $C_G = \frac{\varepsilon}{5\pi}\frac{\kappa}{R_0}e^{-\alpha_T T_g(2+\lambda)}$ is a temperature-independent prefactor, while the constant ε arises from the integration of α_T and from the dimensionful prefactor in the power-law approximation for $g(r)$.

The above expression for $G_A(T)$ can be tested against experimental data of high-frequency shear modulus in metallic glasses obtained from ultrasonic measurements from [11]. The comparison is shown in Fig. 7.4.

From the fitting of Eq. (7.34) to experimental data of different metallic glasses, we therefore obtain an estimate of λ for the different systems. Alternatively, λ can be estimated directly by fitting Eq. (7.30) to data of $g(r)$ from MD simulations, when available [13], or if the static structure factor $S(q)$ is available, through Eq. (1.5).

By using Eq. (7.34) in Eq. (7.25), we are now, finally, in a position to derive an analytical closed-form expression for the viscosity as a function of temperature,

$$\eta(T) = \eta_0 \exp\left\{\frac{V_c C_G}{k_B T}\exp\left[(2+\lambda)\alpha_T T_g\left(1 - \frac{T}{T_g}\right)\right]\right\}, \tag{7.35}$$

where η_0 is a normalization constant. This equation was derived for the first time in Ref. [12] and is sometimes referred to as the KSZ equation or the KSZ model [14, 15]. It is important to note its double-exponential dependence of viscosity

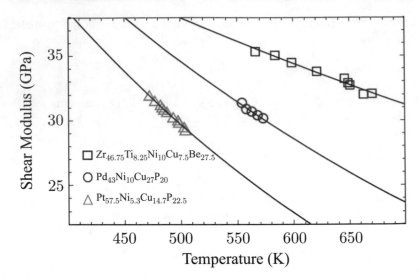

Fig. 7.4 One-parameter comparison between Eq. (7.34) and experimental ultrasonic data for three different metallic glasses from Ref. [11]. Adapted from Ref. [12] with permission from the National Academy of Sciences of the USA

on temperature (a similar double-exponential dependence was found also in other thermodynamically motivated models such as Ref. [16]).

Thanks to its closed form, Eq. (7.35) can be used to obtain a compact closed-form expression also for the fragility index m:

$$m(\lambda) = \frac{1}{\ln 10} \frac{V_c C_G}{k_B T_g} \left[1 + (2 + \lambda)\alpha_T T_g \right]. \tag{7.36}$$

This formula provides a direct link between fragility and quantities that can be, directly or indirectly, related to interatomic interactions and atomic-scale structure, such as α_T, related to anharmonicity, and λ related to the repulsive part of the potential of mean force.

The theoretical fittings of experimental metallic glass data are shown in Fig. 7.5.

Besides λ, there are two nontrivial fitting parameters, which, however, need to take physically meaningful values. One is V_c, which, according to its definition in Eq. (7.23), has to be well within the order of magnitude of the atomic volume, and, indeed, the fitted values are all in the range 0.007–0.015 nm^3, which are compatible with a fraction of the cage volume, as it ought to be. The other one, C_G, has to be in the same order of magnitude as the high-frequency modulus. Finally η_0 is defined as the limiting value of viscosity at high temperature, $\eta_0 = \lim_{T \to \infty} \eta(T)$. The other parameters, α_T and T_g, can be found tabulated for a given substance. When the value of λ is obtained directly from simulations data of $g(r)$, some extra care needs to be put in meaningfully estimating the prefactors of Eq. (7.30) as discussed extensively in Ref. [13].

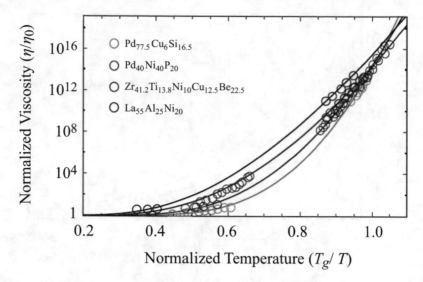

Fig. 7.5 The viscosity as a function of temperature of a set of metallic glasses measured experimentally (symbols) fitted using Eq. (7.35) (solind lines). Adapted from Ref. [12] with permission from the National Academy of Sciences of the USA

Using the fitted λ for the metallic glasses in Fig. 7.5, it is possible to extract an estimate of the repulsive part of the interatomic pseudopotential, plotted in Fig. 7.6.

This is done by approximating the potential of mean force $w(r) = -\lambda \ln(r - 2a_0)$ where a_0 is the Bohr radius, with a repulsive short-range part of the pseudopotential comprising the Born-Mayer short-range repulsion as well as the Ashcroft-Thomas-Fermi screened repulsion (cfr. Chap. 1, Sect. 1.1.3):

$$-\lambda \ln(r - 2a_0) \approx \frac{A\, e^{-q_{TF}(r - 2a_0)}}{r - 2a_0} + B\, e^{-C(r - \bar{\sigma})} \tag{7.37}$$

where we used Eq. (1.4) from Chap. 1. Using λ from viscosity vs T fitting using the KSZ formula Eq. (7.35), one can then graphically[5] determine the values of the parameters entering the Ashcroft-Born-Mayer pseudopotential and thus reverse-engineer the repulsive part of the interatomic potential. This is shown in Fig. 7.6 for the metallic glasses considered in Fig. 7.5.

The theory, exemplified above for the case of metallic glasses, explains the widely observed phenomenon known as "soft particles make hard glasses" [17, 18]. This is the observation that colloidal particles with softer repulsive interactions due, e.g., to lower stiffness or lower bulk modulus of the colloid result in viscosity vs temperature following the Arrhenius law, i.e., in strong glasses according to Fig. 7.1.

---

[5] This was shown in Chap. 1, Fig. 1.4.

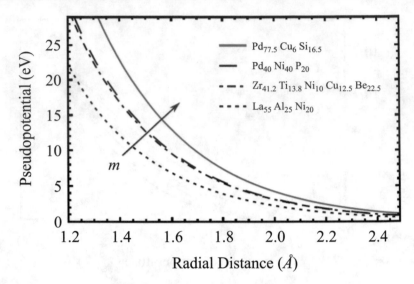

Fig. 7.6 Pseudopotentials (repulsive part only) reverse-engineered from the viscosity vs T fittings reported in Fig. 7.5. Adapted from Ref. [12] with permission from the National Academy of Sciences of the USA

The theory derived here, and in particular the direct proportionality between fragility index m and the steepness λ of repulsion between two particles, Eq. (7.36), provides a theoretical foundation to this widespread observation.

In view of these considerations, the applicability of the above theory is therefore not limited to atomic liquids such as metallic systems. It has been applied to supercooled liquids of charge-stabilized colloidal particles where the relevant controlling variable is not the temperature but the colloid packing fraction ϕ. With colloids, many of the parameters entering the model can be evaluated experimentally by means of confocal microscopy (these include the effective spring constant κ, the potential of mean force $w(r)$, the coordination number in terms of long-lived neighbors z) as shown in Ref. [19]. A similar successful comparison has been presented for simulated systems of charge-stabilized colloids in Ref. [15], where the KSZ model (adapted for colloids) provides a better fitting to simulations data compared to mode-coupling theory (the latter predicts a power-law divergence of viscosity at the glass transition).

Finally, the above theoretical model can be applied equally well to virtually all glass-forming liquids, including organic molecules, polymers, and oxide network-forming glasses. Systematic application of the KSZ model fitting to experimental data for a broad range of materials was reported in Ref. [14]. This kind of systematic analysis of literature data is very useful to confirm theoretical predictions, to unveil general trends across different types of materials, and hence to extract information about chemical-design guidelines for material engineering.

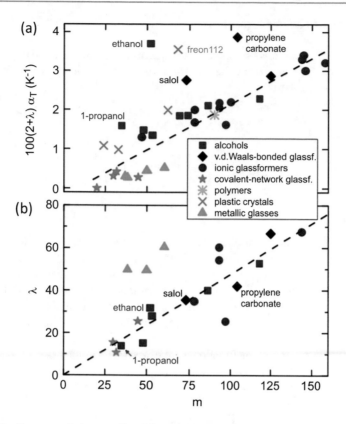

Fig. 7.7 The linear correlations predicted by the KSZ model between fragility m and physical parameters are tested for a wide range of different materials. In (**a**) the linear correlation (dashed line) between fragility index m and the factor $(2+\lambda)\alpha_T$ from Eq. (7.36) is plotted together with data for many glass formers. In (**b**) the direct proportionality between fragility index m and repulsion steepness λ is shown (dashed line, $\lambda = sm$, with $s = 0.48$) together with data from the different systems. Reproduced from Ref. [14] with permission from the American Institute of Physics

The linear trends predicted by the KSZ model for the fragility index m as a function of physical parameters λ and α_T are shown for a variety of different materials in Fig. 7.7.

7.6 First-Principles Frameworks

Frameworks to compute the viscosity of liquids and glassy systems include formalisms deduced from first principles such as the Green-Kubo formalisms and the nonaffine response theory. These frameworks do not lead to compact closed-form expressions and typically are used for numerical calculations based on underlying molecular dynamics simulations.

7.6.1 Green-Kubo Formalism

The Green-Kubo relations provide a direct quantitative link between some macroscopic transport property L and the time-integral of the time-correlation function of some conserved property A:

$$L = \int_0^\infty \langle \dot{A}(t)\dot{A}(0) \rangle \, dt. \tag{7.38}$$

For the viscosity, the Green-Kubo relation reads as [20]:

$$\eta = \beta V \int_0^\infty \langle \sigma_{xy}(t)\sigma_{xy}(0) \rangle dt. \tag{7.39}$$

For a derivation of this formula, see, e.g., [21]. While elegant and invaluable to compute viscosity based on MD simulations of fluids, the above formula does not lead to microscopic closed-form expressions of the kind that we discussed in the previous sections. This is because time-correlation functions such as the stress autocorrelation function are easy to get from MD simulations but are difficult to be evaluated from a theoretical perspective, with, perhaps, the exception of mode-coupling theory.

The stress autocorrelation function in the above formula can be evaluated in MD simulations based on equilibrium static snapshots, i.e., without the need for deforming the simulation box. This is done by taking advantage of the virial stress formula for the stress tensor σ_{xy} [22].

The accuracy of the Green-Kubo estimate suffers in MD simulations because a correlation function has to be integrated, and the long-time tail of the correlation is usually not sampled very well. Fitting tails on correlation functions is, indeed, a kind of an art form, and it can change the predicted transport properties quite significantly.

7.6.2 Nonaffine Response Theory

We recall from Chap. 3, Sect. 3.1, that the viscosity can be obtained from the loss viscoelastic modulus G'' using nonaffine response theory as (cfr. Eq. (3.10)):

$$\eta = \frac{G''}{\omega}. \tag{7.40}$$

The nonaffine response theory developed from first principles in Sect. 3.1 provides the following form for the loss modulus G'' (cfr. Eq. (3.37)):

$$G''(\omega) = 3\rho \int_0^{\omega_D} \frac{g(\omega_p)\,\Gamma(\omega_p)\,\tilde{\nu}(\omega)\,\omega}{m^2(\omega_p^2 - \omega^2)^2 + \tilde{\nu}(\omega)^2\omega^2} d\omega_p. \tag{7.41}$$

For shear deformation ($\kappa\chi = xy$), Eq. (3.34) becomes:

$$\sigma_{xy}(\omega) = G^*(\omega)\gamma(\omega). \tag{7.42}$$

Since $G^* = G' + iG''$ and there is factor ω in the above expression for G'' in Eq. (7.41), we recover, for the dissipative part of the stress, σ'_{xy}, the Newton law of viscous flow (cfr. Eq. (7.3)):

$$\sigma'_{xy} = \eta i\omega\gamma = \eta\dot{\gamma} \tag{7.43}$$

with, indeed, a zero-frequency shear viscosity given by:

$$\eta = 3\rho\tilde{v}(0) \int_0^{\omega_D} \frac{g(\omega_p)\,\Gamma(\omega_p)}{m^2\omega_p^4}d\omega_p. \tag{7.44}$$

For an idealized solid, which follows the Debye law Eq. (5.5):

$$g(\omega_p) = \frac{\omega_p^2\,V}{2\pi}\left(\frac{2}{v_T^3} + \frac{1}{v_L^3}\right)$$

and for which $\Gamma(\omega_p) \sim \omega_p^2$ as derived in [23], we obtain the simple and compact relation (valid up to some undetermined numerical prefactor):

$$\eta \approx \rho\tilde{v}(0)\frac{V}{2\pi}\left(\frac{2}{v_T^3} + \frac{1}{v_L^3}\right)\frac{\omega_D}{m^2}. \tag{7.45}$$

Although highly idealized, this simple relation provides a direct connection between the viscosity η and important physical quantities: the zero-frequency limit of the Fourier-transformed friction kernel, $\tilde{v}(0)$,[6] the Debye frequency, and the longitudinal and transverse speeds of sound.

More generally, realistic predictions can be made using the full Eq. (7.44) with the VDOS $g(\omega_p)$ computed for a realistic system by means of MD simulations or from experimental data. Also $\Gamma(\omega_p)$ has to be evaluated numerically based on the eigenvectors of the Hessian matrix (cfr. Chap. 3, Sects. 3.2 and 3.3), although for disordered systems, the law $\Gamma(\omega_p) \sim \omega_p^2$ might be a reasonable approximation to deduce analytical correlations.

For liquids, the VDOS $g(\omega_p)$ can be computed via MD simulations, and only recently, it has been measured experimentally with inelastic neutron scattering in Ref. [24]. The main complication is that the VDOS of liquids contains a large population of instantaneous normal modes (INMs), introduced in Chap. 3, Sect. 3.3,

---

[6] Which in turn arises from anharmonic couplings between each particle and the other particles via the Caldeira-Leggett or ZCL Hamiltonian (cfr. Sect. 3.2.1. in Chap. 3).

cfr. Fig. 3.5 (see also Refs. [25, 26] for more background information on INMs). Recently, an analytical theory of the VDOS of liquids accounting for INMs has been developed in Ref. [27].

Finally, $\tilde{\nu}(0)$ represents the zero-frequency limit of the one-sided Fourier transform, $\tilde{\nu}(\omega)$ (also known as the "spectral density") of the friction memory kernel $\nu(t)$. There are various ways of estimating $\nu(t)$. Starting from the particle-bath Hamiltonian presented in Chap. 3, cfr. Eqs. (3.22), upon integrating the Euler-Lagrange equations for the coupled dynamics of the tagged particle and the heat-bath oscillators, Zwanzig obtained the identification [28]:

$$\nu(t) = \sum_m \frac{\gamma_m^2}{m\omega_m^2} \cos(\omega_m t) \tag{7.46}$$

where the γ_m is the coupling coefficient in Eq. (3.22) between the tagged particle and the m-th bath oscillator. As usual, one can then move from the discrete set of oscillators frequencies to a continuous integral over the VDOS, leading to Eq. (3.41), previously presented in Chap. 3, Sect. 3.4.4, and reproduced here for convenience:

$$\nu(t) = \int_0^\infty g(\omega_p) \frac{\gamma(\omega_p)^2}{\omega_p^2} \cos(\omega_p t) d\omega_p$$

where $\gamma(\omega_p)$ represents the continuous limit (spectrum) of the discrete set of dynamic coupling constants $\{\gamma_p\}$ in the ZCL Hamiltonian Eq. (3.22). The derivation of Eqs. (7.46) and (3.41) is reported in Appendix C. According to this equation, then, we would have a double dependence of the viscosity η on the VDOS through Eq. (7.44).

The memory function or friction kernel $\nu(t)$ can be evaluated on the basis of MD simulations for which different methods are available. The most immediate way is to apply the fluctuation dissipation theorem (FDT) associated with the governing equation of motion, i.e., the generalized Langevin equation Eq. (3.30). The FDT is derived in Appendix C.3 and reads as follows:

$$\langle F_P(t) F_P(t') \rangle = k_B T \nu(t - t') \tag{7.47}$$

where $\langle F_P(t) F_P(t') \rangle$ is the time autocorrelation function of the as the stochastic force $F_p(t)$.[7] Therefore, by measuring $\langle F_P(t) F_P(t') \rangle$, the above Eq. (7.47) allows one to determine $\nu(t)$ from MD simulations.

Since, in practice, it is easier to measure velocity autocorrelation functions, $\langle v(t)v(0) \rangle$, the following identity (obtained through integration by parts) becomes

---

[7] The averaged time autocorrelation function of a physical variable A is defined as $\langle A(t)A(0) \rangle = \frac{1}{\tau} \int_0^\tau A(t+t')A(t')dt'$. For equilibrium ergodic systems, the time averaging can be replaced with an ensemble average.

very useful [29]:

$$\langle F_P(t)F_P(t')\rangle = -M^2\frac{\partial^2}{\partial t^2}\langle v(t)v(t')\rangle \tag{7.48}$$

where, obviously, F_p and v are, in real situations, 3D vectors. Numerical examples of this kind of reconstruction technique for the memory kernel are discussed in Ref. [29].

Alternatively, one can deduce an equation involving the momentum autocorrelation function (MAF) $C_{pp}(t) = \langle p(t)p(0)\rangle$ and the system force-momentum correlation function (MFC) $C_{pF}(t) = \langle F(t)v(0)\rangle$, where $F(t)$ denotes the conservative force acting on the tagged particle, i.e., $F \equiv -U'(s(t))$, with reference to Eq. (3.30) or $\mathbf{H}_{ij}(t)\mathbf{s}_j(t)$ with reference to Eq. (3.30), both in Chap. 3. The equation reads as [30]:

$$\dot{C}_{pp}(t) = C_{pF}(t) - \int_0^t v(t-t')C_{pp}(t)dt', \tag{7.49}$$

from which, upon Fourier transformation, one gets:

$$\tilde{v}(\omega) = \frac{1 + \tilde{C}_{pF}(\omega)}{\tilde{C}_{pp}(\omega)} - i\omega. \tag{7.50}$$

Here both MFC and MAF are normalized to $C_{pp}(0)$. Computing $\tilde{v}(\omega)$ using Eq. (7.50) is straightforward since all the quantities involved are readily available from MD simulations.

References

1. C.A. Angell, Science **267**(5206), 1924 (1995)
2. C.A. Angell, Chem. Rev. **102**(8), 2627 (2002)
3. T. Hecksher, A.I. Nielsen, N.B. Olsen, J.C. Dyre, Nat. Phys. **4**(9), 737 (2008)
4. M. Ozawa, C. Scalliet, A. Ninarello, L. Berthier, J. Chem. Phys. **151**(8), 084504 (2019)
5. V. Lubchenko, P.G. Wolynes, Annu. Rev. Phys. Chem. **58**(1), 235 (2007)
6. L. Berthier, G. Biroli, J.P. Bouchaud, L. Cipelletti, D.E. Masri, D. L'Hôte, F. Ladieu, M. Pierno, Science **310**(5755), 1797 (2005)
7. J. Frenkel, *Kinetic Theory of Liquids* (Oxford University Press, Oxford, 1955)
8. L. Landau, E. Lifshitz, *Theory of Elasticity: Volume 6* (Pergamon Press, Oxford, 1986)
9. K. Trachenko, V.V. Brazhkin, Rep. Prog. Phys. **79**(1), 016502 (2015)
10. J.C. Dyre, J. Non-Cryst. Solids **235–237**, 142 (1998)
11. W.L. Johnson, M.D. Demetriou, J.S. Harmon, M.L. Lind, K. Samwer, MRS Bull. **32**(8), 644 (2007)
12. J. Krausser, K.H. Samwer, A. Zaccone, Proc. Natl. Acad. Sci. **112**(45), 13762 (2015)
13. A.E. Lagogianni, J. Krausser, Z. Evenson, K. Samwer, A. Zaccone, J. Stat. Mech. Theory Exp. **2016**(8), 084001 (2016)
14. P. Lunkenheimer, F. Humann, A. Loidl, K. Samwer, J. Chem. Phys. **153**(12), 124507 (2020)

15. G. Porpora, F. Rusciano, V. Guida, F. Greco, R. Pastore, J. Phys. Condens. Matter **33**(10), 104001 (2020)
16. M.I. Ojovan, K.P. Travis, R.J. Hand, J. Phys. Condens. Matter **19**(41), 415107 (2007)
17. J. Mattsson, H.M. Wyss, A. Fernandez-Nieves, K. Miyazaki, Z. Hu, D.R. Reichman, D.A. Weitz, Nature **462**(7269), 83 (2009)
18. N. Gnan, E. Zaccarelli, Nat. Phys. **15**(7), 683 (2019)
19. R. Higler, J. Krausser, J. van der Gucht, A. Zaccone, J. Sprakel, Soft Matter **14**, 780 (2018)
20. J. Hansen, I. McDonald, *Theory of Simple Liquids* (Elsevier Science, Amsterdam, 2006)
21. D.J. Evans, G. Morriss, *Statistical Mechanics of Nonequilibrium Liquids*, 2nd edn. (Cambridge University Press, Cambridge, 2008)
22. M.P. Allen, D.J. Tildesley, *Computer Simulation of Liquids*, 2nd edn. (Oxford University Press, Oxford, 2017)
23. A. Zaccone, E. Scossa-Romano, Phys. Rev. B **83**, 184205 (2011)
24. C. Stamper, D. Cortie, Z. Yue, X. Wang, D. Yu, J. Phys. Chem. Lett. **13**(13), 3105 (2022)
25. T. Keyes, J. Phys. Chem. A **101**(16), 2921 (1997)
26. R.M. Stratt, Accounts Chem. Res. **28**(5), 201 (1995)
27. A. Zaccone, M. Baggioli, Proc. Natl. Acad. Sci. **118**(5), e2022303118 (2021)
28. R. Zwanzig, J. Stat. Phys. **9**(3), 215 (1973)
29. G. Jung, M. Hanke, F. Schmid, J. Chem. Theory Comput. **13**(6), 2481 (2017)
30. F. Gottwald, S.D. Ivanov, O. Kühn, J. Chem. Phys. **144**(16), 164102 (2016)

Plastic Deformation

8

Abstract

The plastic deformation, or plasticity, of crystals has been rationalized in terms of dislocations since the 1930s, and current computational protocols based on accelerated dislocation dynamics provide an accurate description, and prediction, of plastic material failure for crystalline solids. For amorphous solids, theoretical approaches have been traditionally much more heuristic and based on geometrically not well-defined concepts such as shear transformation zones, with consequently poor predictive power. A more microscopic theory of plasticity in amorphous solids will be formulated here, thanks to the recent discovery of dislocation-type topological defects in the displacement field of glasses under deformation. These defects are closely related to the nonaffine displacements based on which the nonaffine elasticity theory has been developed in Chaps. 2 and 3. Widely observed phenomena such as yielding and shear banding can also be understood within this framework.

8.1   Plasticity of Crystals

The most important type of topological defects that influences the mechanical properties of crystals, in particular at large strain, is called a dislocation. There is a very large literature on dislocations in crystals, and the term was introduced by G. I. Taylor in 1934 in a fundamental study [1] where the self-organization of dislocations into slip planes was identified as the main mechanism leading to plastic flow of crystals under applied loads.

© The Author(s), under exclusive license to Springer Nature Switzerland AG 2023
A. Zaccone, *Theory of Disordered Solids*, Lecture Notes in Physics 1015,
https://doi.org/10.1007/978-3-031-24706-4_8

Fig. 8.1 An edge dislocation is represented by an extra half-plane being inserted in the crystal lattice. In the figure, this extra half-plane is the upper half of the yz plane. Since there are many such extra half-planes along the z direction, one speaks of a dislocation "line" along the z direction

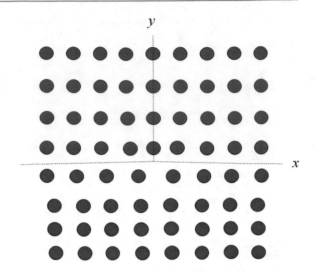

8.1.1 Dislocations

A most common type of dislocation is the edge dislocation, whereby an extra half-plane is inserted in the crystal lattice, starting from a certain atom, which represents the center of the dislocation. The situation is depicted in Fig. 8.1.

The other common type of dislocation is the so-called screw dislocation, whereby the crystal lattice is cut along an atomic half-plane, and the two parts originating from the cutting are then shifted with respect to each other parallel to the edge of the cut, in opposite directions, by a distance equal to a lattice spacing, approximately.

Geometrically, the most important consequence arising from the presence of a dislocation in the atomic lattice is as follows. Upon drawing a closed contour $\mathcal{L}$ (the so-called Burgers loop or Burgers circuit) that goes around the dislocation line D, the displacement vector $\mathbf{u}$ of elasticity theory (cfr. Appendix A and Chap. 2) receives an increment given by a vector $\mathbf{b}$. Clearly, this vector must be equal, or approximately equal, in both magnitude and orientation, to a lattice vector. The vector $\mathbf{b}$ is known as the Burgers vector and obeys the following equation:

$$b_i \equiv - \oint_{\mathcal{L}} du_i = - \oint_{\mathcal{L}} \frac{du_i}{dx^k}\, dx^k , \qquad (8.1)$$

where Cartesian components are denoted by Latin indices. The direction along which the closed contour $\mathcal{L}$ is traversed and the vector $\boldsymbol{\tau}$ tangent to the dislocation line D are related via the corkscrew rule. Mathematically, the above relation implies

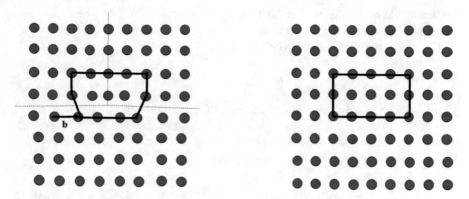

Fig. 8.2 The left panel shows the Burgers vector around an edge dislocation of the type previously depicted in Fig. 8.1. The Burgers loop around the dislocation is a deformed version of the original rectangular circuit shown in the right panel, when the dislocation is absent. The distortion of the rectangular circuit is measured by the Burgers vector **b**

that the displacement vector field is a multi-valued function: every time we go around the loop, it picks up an additional increment equal to **b**.[1]

This situation is schematically illustrated in Fig. 8.2. In a crystal with no dislocations, a rectangular circuit would start on one atom of the lattice and after a closed loop would end on the same atom (right panel in Fig. 8.2). Upon introducing a dislocation, the rectangular circuit gets deformed (it is no longer a rectangle), and the extent of deformation is measured, in terms of the displacement vector **u**, by the Burgers vector **b** (left panel in Fig. 8.2).

The stress field generated by an edge dislocation can be derived using standard continuum mechanics [2]. In polar components, it is found that the stress decays with distance from the dislocation core as $\sim 1/r$: $\sigma_{rr} = \sigma_{\phi\phi} = -(bC/r)\sin\phi$, and $\sigma_{r\phi} = (bC/r)\cos\phi$, where $C = G/2\pi(1-\nu)$, with G the shear modulus and ν the Poisson's ratio.

The stress field around a screw dislocation is more easily characterized since the displacements around it give rise to shear stresses only, and one finds $\sigma_{xy} = Gb/2\pi r$. This shear stress acts in the slip direction, and it contributes to the total stress that another dislocation in its vicinity will experience.

The Burgers vector **b** identifies the direction along which the dislocation can glide and is normal to the edge dislocation line. The dislocation line and the Burgers vector, in turn, identify the slip plane. So if τ is a vector tangent to D, then $\mathbf{b} \times \tau$ defines a vector normal to the slip plane.

The interaction of a stress field with a dislocation line is calculated by considering the elastic energy, i.e., the reversible work, associated with the stress field acting on

---

[1] Mathematically this is analogous to a multi-valued function in complex analysis, e.g., the complex logarithm $\ln z$: if we go in a closed loop around the origin $z = 0$, the imaginary part of $\ln z$ picks up an increment equal to 2π.

the strain field of the dislocation. By then replacing the strain field with that of the dislocation, one obtains the following form of the force acting on the dislocation, also known as the Peach-Köhler (PK) force:

$$f_i = \epsilon_{ikl} \tau_k \sigma_{lm} b_m \tag{8.2}$$

where ϵ_{ikl} is the Levi-Civita symbol and σ_{lm} are the components of a generic stress field. If this stress field is the one of a second dislocation in the vicinity of the first one, we get the interaction force between two dislocations. For example, for two screw dislocations, the PK force formula gives:

$$f = \pm \frac{Gb_1 b_2}{2\pi r} \tag{8.3}$$

where r is the distance between the cores of the two dislocations and the force is repulsive $(+)$ if the two Burgers vectors are parallel and attractive $(-)$ if the two Burgers vectors are anti-parallel.

If there is a non-zero component of the PK force in the slip plane, this force component can trigger the glide of the dislocations. A typical example is when the slip planes of the two dislocations contain the line connecting their cores. The dislocations will glide away from each other, if the PK force is repulsive; instead, they will come together and cancel each other out, if the PK force is attractive.

From the above, it is quite evident that dislocations can thus form dipoles and even multipoles, in particular, quadrupoles [3]. There are also mechanisms, such as the Frank-Read source, by which dislocations can multiply. Through their interactions outlined above, and also thanks to multiplication mechanisms such as the Frank-Read source, dislocations can self-organize into arrays and other assemblies leading to a local increase of the dislocations density in a certain region of the material. In particular, this may lead to slip systems, i.e., macroscopic bands along which the material can flow with little resistance, thus leading to plastic material failure via the onset of plastic flow.

Accelerated dislocation dynamics computational protocols have been developed for crystals with good predictive power. In practice, one generates crystal structures and identifies all the dislocations present. Then all the forces on dislocation segments are computed via the PK force, and one then solves the corresponding equations of motion to determine the rate of change of dislocation structure. Input data (shear modulus, Poisson ratio, dislocation mobility, interatomic distance) for solving the PK equation of motion are easily obtained from the MD simulations.

Then, one identifies all the Burgers vectors and associated line vectors in the displacement field, giving a set of dislocation lines and a set of nodes, which represent the points connected by dislocation segments. Using suitably defined mobility laws which express the velocity of each node as a function of the PK forces (by means of, e.g., Peierls-Nabarro type stress concepts[2]), this leads to a set of overdamped equations for the evolution of the nodes. Solving these equations can be done using massively parallelized computers, which make it possible to compute far-field, dipole, and multipole interactions in a fast way.

The final result of these calculations is the evolution of the dislocations density as a function of strain, which strongly mimics the stress versus strain curve and allows one to predict the material failure at the yield point. Also, these calculations allow one to visualize the complex network of self-organized dislocations leading to failure.

8.1.2 Schmid's Law

Under a unidirectional tensile load, the slip planes are oriented at approximately 45° with respect to the direction of tensile strain.

Several decades before the advent of computer simulations, this fact was predicted by the following argument, due to Erich Schmid.

Let $\sigma = F/A_0$ be the tensile stress acting on the sample, in the case of a uniaxial stress, with F the applied tensile force and A_0 the cross-section area of the sample. Denoting with ϕ the angle between the normal direction to the slip plane and the direction of the tensile force F and with λ the angle between the slip direction and the direction of F, the slip plane area is thus given by $A_s = A_0/\cos\phi$.

Based on this, the tensile force resolved in the slip direction, $F\cos\lambda$, produces a resolved shear stress given by the so-called Schmid's law [4,5]:

$$\sigma_{RSS} = \sigma\cos\phi\cos\lambda. \tag{8.4}$$

In general, the three directions are not coplanar, which implies that $\phi + \lambda \neq 90°$, while $\phi + \lambda = 90°$ is the minimum possible value [4, 5]. Dislocations will give rise to slip systems that, in general, are randomly oriented. For a given load σ, slip systems will therefore be initiated by facilitated motion of dislocations systems self-organizing in a slip plane, which experiences the largest resolved shear stress σ_{RSS}. This is reminiscent of what happens with avalanches that initiate in a spatial direction where the resolved stress is largest and thus can overwhelm frictional resistive forces. The largest resolved stress clearly corresponds to the maximum value of $\cos\phi\cos\lambda$. Under the constraint $\min(\phi+\lambda) = 90°$, this, therefore, happens for $\phi = \lambda = 45°$.

---

[2] The Peierls-Nabarro stress is the force required to move a dislocation within a plane of atoms in the unit cell and is given by the following expression: $\sigma_{PN} \propto Ge^{-2\pi W/b}$, where $W = d/(1-v)$ with d the interplanar spacing, while the other symbols have the usual meaning.

8.1.3   Dislocations as Topological Defects

Dislocations are topological line defects. A topological defect occurs in a continuum whenever a localized lack of order (in a point or along a line) generates a non-vanishing strain in the far field. In the case of dislocations, the local order which is destroyed along the dislocation line is the inversion symmetry around the atom representing the tip of the extra half-plane. Other well-known examples of topological defects are vortices in superfluid helium and disclinations in nematic liquid crystals.

Similarly to a point charge in electrodynamics,[3] the presence of a topological defect can be detected by measuring an appropriate field around a closed contour or surface enclosing the core of the topological defect. These kinds of defects are associated with topological invariants, i.e., quantities that characterize the topological space and are invariant under homeomorphisms i.e., under continuous deformations of the space.[4] Topological invariants are usually normalized to be integers, such as the winding numbers, which characterize the topological defect.

For example, a vortex is formally defined as the mapping from a closed loop $\mathcal{L}$ onto the order parameter space given by the real line starting from the origin and terminating at $2m\pi$, where $m = 0 \pm 1, \pm 2, \ldots$ is the winding number:

$$\int_{\mathcal{L}} d\theta = \int_{\mathcal{L}} \frac{d\theta}{ds} ds = 2m\pi \tag{8.5}$$

where s is the parameter of the closed curve $\mathcal{L}$.[5] In other words, m is the number of times that the closed loop $\mathcal{L}$ has to be traversed so that the mapping becomes single-valued. The winding number m is a topological invariant in the following sense. All closed curves with the same winding number m can be deformed continuously into each other. Conversely, it is impossible to smoothly deform a mapping belonging to a class with winding number m into another mapping belonging to a class with $m' \neq m$.

Another well-known example of topological invariant is the genus g of a closed surface in $\mathbb{R}^3$, which is related to the surface integral of the Gaussian curvature via the Gauss-Bonnet theorem:

$$\int_S K dA = \int_S (\kappa_1 \kappa_2) dA = 2\pi (2 - 2g) \tag{8.6}$$

where g is the genus of the surface S and $K = \kappa_1 \kappa_2$ is the Gauss curvature, given by the product of the two principal curvatures κ_1 and κ_2. It is straightforward, by

---

[3] Note that they also have in common the $1/r$ behavior of the field.

[4] A typical homeomorphism between two surfaces in $\mathbb{R}^3$ is the one by which a doughnut can be continuously transformed into a coffee mug.

[5] In the complex plane, the winding number of a closed path $\mathcal{L}$ around the origin is given by $\frac{1}{2\pi i} \oint_{\mathcal{L}} \frac{dz}{z}$.

applying Eq. (8.6) to a sphere, to show that the result of the integral will be 4π, and hence $g = 0$. For a torus or a doughnut, one instead gets $g = 1$. Hence it is clear that a sphere with $g = 0$ cannot be smoothly deformed into a torus or a coffee mug, which has $g = 1$, while the latter two can be smoothly deformed into each other. States of the same system, which are characterized by different values of the topological invariant, are separated via a topological phase transition.

Going back to dislocations, Eq. (8.1) can be written in vector notation as:

$$\oint_{\mathcal{L}} d\mathbf{u} = \oint_{\mathcal{L}} \frac{d\mathbf{u}}{ds} ds = \mathbf{b}, \tag{8.7}$$

where, clearly, $\mathbf{u}$ plays an analogous role as that of θ in Eq. (8.5). For ordinary crystals, the winding number is equal one in units of lattice spacing, and $|\mathbf{b}| = d$, where d is the lattice spacing. The situation is a bit different for smectic liquid crystals, where dislocations can have also winding numbers different than one, and

$$\oint_{\mathcal{L}} d\mathbf{u} = \mathbf{b} = ma\mathbf{e}_z, \tag{8.8}$$

where $m = 0, \pm 1, \pm 2, \ldots$ and $\mathbf{e}_z$ is the unit vector along the z-axis.

8.1.4 Volterra Construction

Vito Volterra, in 1907, at a time when the existence of atoms in solids was still a controversial topic, devised a process to analyze the state of deformation in materials produced by internal defects, such as dislocations. In a typical Volterra construction, one takes a macroscopic cylinder of material oriented along the z-axis and applies a radial cut to the cylinder along the yz-plane, indicated as Σ surface in Fig. 8.3. The two sides (Σ' and Σ'' in Fig. 8.3) of the cut plane are then displaced by $\mathbf{b}$ with respect to each other and glued together. This cut-and-weld process introduces internal stresses, which Volterra analyzed with the tools of continuum mechanics, and which turn out to correspond to states of stress and strain around dislocations in solids.

Examples of Volterra constructions for screw and edge dislocations are shown in Fig. 8.3.

8.2 Theory of Plastic Deformation in Amorphous Solids Mediated by Dislocation-Type Defects

Despite much theoretical and numerical research devoted to identifying dislocations in amorphous solids [7,8], developing a framework for the plasticity of glasses based on topological defects has remained elusive [9]. Evidence has been collected of a correlation between plastic activity and vibrational modes [10], and recently this,

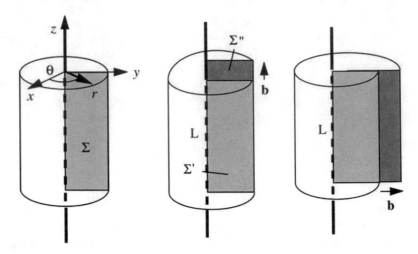

Fig. 8.3 A cylinder of material (left) with the corresponding Volterra constructions for screw dislocations (center) and edge dislocations (right). The material is cut along the surface Σ. The two sides Σ' and Σ'' are then displaced with respect to each other and then glued together (with additional material inserted or removed if needed). In the figure L represents the line or core of the defects, e.g., the dislocation line. The process induces a state of stress and deformation described by the same continuum equations, which describe the stress and strain fields around dislocations [2]. Adapted from Ref. [6] with permission of Springer, NY

together with the evolution of INMs under a shear strain, has been used within the nonaffine response framework to quantitatively predict the plastic yielding transition of model glasses [11]. However, in spite of much effort [12], the geometric nature of the underlying "defects" has remained unclear.

In recent years, in Ref. [13], evidence of the possible presence of edge dislocation-type defects in amorphous solids has been suggested based on experiments and differential geometry.

Sheets of temperature-responsive N-isopropylacrylamide (NIPA) polymer gel were studied experimentally. The gel shrinks at a temperature about 34 °C, with a shrinkage factor that depends on cross-linking density. A non-Euclidean reference metric with Gaussian curvature was prescribed via UV-controlled spatial variation of the cross-linking ratio across the sheet.[6] This resulted in a state of deformation of the material corresponding to the presence of edge dislocations, which was confirmed via application of the Volterra construction (right panel in Fig. 8.3).

As we shall see in the following, a microscopic justification for the presence of edge dislocation-type defects in amorphous solids is provided by the analysis of nonaffine displacements introduced in Chap. 2. This led to the discovery of well-

---

[6] The elastic deformation energy can be written in terms of a metric tensor, whereby defect-free states of deformation correspond to Euclidean flat-space metric, whereas curved non-Euclidean metrics correspond to internal stresses associated with topological defects such as dislocations.

defined topological defects as the mediators of plasticity in glasses in Ref. [14] and to a corresponding field-theoretic framework [15].

8.2.1   Dislocation-Type Defects in Amorphous Solids from Nonaffine Displacements

As discussed in Chap. 2, nonaffine displacements originate directly from the lack of centrosymmetry of the lattice. This simple fact makes them ubiquitous in glasses, liquids, and all disordered states of matter. Nonaffine displacements are clearly important also near topological defects, since, e.g., atoms that lie on an edge dislocation line are not centers of symmetry [16]. Furthermore, it has also been noted that nonaffine atomic motions in crystalline materials lead to so-called incompatible deformations described by multi-valued displacement fields, which are described, at a continuum level, by a formalism that has a lot in common with that of dislocations [17, 18].

These observations suggest a deeper link between nonaffine displacements and topological defects, which we shall now consider more carefully.

Just like the effect of a dislocation shows up in the circulation of the displacement vector field $\mathbf{u}$, cfr. Eq. (8.1), one can perform a similar type of the analysis for the displacement field around a particle that is nonaffinely displaced from its rest position in an amorphous solid. This analysis reveals that each particle whose neighbors perform nonaffine motions acts like a dislocation point in a 2D slice of the material.

The situation, for dislocation-type defects in amorphous solids, is schematically depicted in Fig. 8.4.

It is clear from the figure that the mere fact that one of the nearest neighbors was displaced with a nonaffine component leads to the appearance of the equivalent of a Burgers vector, measured with respect to an ideal affine deformation. Of course, the same remains true if also other nearest neighbors are displaced in a nonaffine manner.

From a geometric point of view, the circuit distortion induced by nonaffinity in Fig. 8.4 is analogous to that induced by an edge dislocation in Fig. 8.2. Indeed, in the case of the crystal dislocation, the original rectangle is transformed into a different geometric figure, i.e., a trapeze. In the case of nonaffinity in the amorphous solids, the original trapezium circuit in the undeformed solid (left panel) would be mapped onto another trapezium (with two parallel sides) by an affine deformation. In the presence of nonaffinity, it is instead mapped onto a quadrilateral that is no longer a trapezium since there are no two sides which are parallel.

This Burgers vector is thus a measure of the local nonaffinity of deformation and induces a distortion of the strain field, which is analogous to that of a dislocation, as already noted in the continuum mechanics literature [17, 18].

Importantly, unlike previous attempts, this dislocation-type defect is identified not in the static snapshot of the lattice at rest but in the deformed lattice.

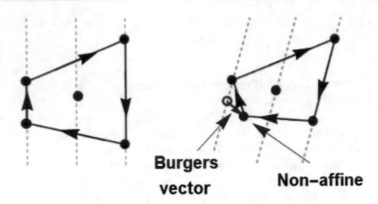

Fig. 8.4 The Burgers circuit or Burgers loop, cfr. Fig. 8.2, of the topological singularity associated with the nonaffine displacement field in an amorphous solid, leading to the Burgers vector $-\mathbf{b}$, around a tagged atom at the center of the circuit. For simplicity, and without loss of generality, we consider the case of the left bottom atom undergoing a nonaffine displacement, while the other atoms move affinely. The red circle indicates the affine position where the left-bottom atom would have moved to, had it been displaced affinely. The red line, which measures the nonaffine displacement u_i^{NA}, as defined in Eq. (8.10), represents the Burgers vector for the nonaffine displacement field in a disordered solid. Adapted from Ref. [15] with permission from the American Physical Society

To understand also from a mathematical point of view the analogy between nonaffine displacement fields and the displacement fields around dislocations, let us go back to the definition of nonaffine displacements, and let us derive the full consequences of nonaffinity on the field-theoretic description of the deformation field. We will follow the treatment of Ref. [15].

The mechanical deformation in a material can be characterized by the displacement vector field, in Cartesian components ($i = x, y, z$), given by u_i [2, 19], which describes the deviations of the material points from the original positions x_i that they have in the undeformed frame:

$$x_i' = x_i + u_i . \tag{8.9}$$

The total displacement vector u_i can be splitted into its affine and nonaffine contributions [20]:

$$u_i = u_i^A + u_i^{NA} = \Lambda_{ki} x_k + u_i^{NA} \tag{8.10}$$

where Λ_{ki} is a matrix of constants naturally related to the macroscopically applied strain. As discussed in Chap. 2, nonaffine displacements u_i^{NA} originate in disordered systems from the need to preserve mechanical equilibrium in the affine positions and throughout the deformation process. In ordered crystals, the symmetrized strain tensor $e_{ij} \equiv \partial_{(i}u_{j)} = \frac{1}{2}(\partial_i u_j + \partial_j u_i)$ obeys the so-called compatibility condition [21, 22]:

$$\nabla \times \nabla \times \mathbf{e} = 0, \qquad (8.11)$$

which is equivalent to saying that du_i is a closed differential form, as we shall demonstrate in the following.

In mathematical physics and in classical field theory, the language of differential forms was introduced by E. Cartan as a useful tool for the manipulation of objects that are invariant under coordinate transformations [23]. A generic p-form is defined as $K = \frac{1}{p!}K_{i_1,\dots i_p}dx^{i_1} \wedge \dots dx^{i_p}$, where the anti-symmetric wedge product between two coordinate differentials is defined to be the directed (parallelogram) area spanned by the two coordinate differentials, $dx \wedge dy = -dy \wedge dx$ and $dx \wedge dx = dy \wedge dy = 0$ [23]. The factorial in the denominator is needed to avoid double counting of terms in the summation over repeated indices due to the anti-symmetry of the wedge product. Hence, given a vector A_i, its 1-form is given by:

$$A = A_j dx^j \qquad (8.12)$$

where to make contact with the standard notation of covariant electrodynamics, we use covariant (subscript) and contravariant (superscript) indices.

Next, we introduce the exterior derivative d, which differentiates and brings a differential factor:

$$df = \frac{\partial f}{\partial x^j}dx^j \qquad (8.13)$$

where f is a 0-form (i.e., simply a scalar function). Hence, the exterior derivative transforms a p-form into a $p + 1$ form, for example, a 1-form into a 2-form, as follows:

$$dA = d(A_j dx^j) = (\partial_i A_j)dx^i \wedge dx^j. \qquad (8.14)$$

Furthermore, a p-form H is *closed* if $dH = 0$. For example, in covariant electrodynamics, if A_j are the components of the 4-dimensional vector potential in Minkowski space, its relation to the Faraday tensor F (a 2-form) is given by the familiar Bianchi identity:

$$dF = ddA = 0 \qquad (8.15)$$

with $A = A_k dx^k$ and $F = \frac{1}{2}F_{ij}dx^i dx^j$. In components, the Bianchi identity can be written as $\epsilon_{lijk}\partial_i F_{jk} = 0$, where ϵ_{lijk} is the four-dimensional Levi-Civita symbol.

The compatibility condition expresses the single-valuedness of the displacement field [17, 18]. For small strain steps, these conditions are equivalent to the fact that the displacements can be obtained by integrating the strains, and they can be expressed as:

$$\oint_{\mathcal{L}} du_i = 0. \tag{8.16}$$

i.e., the integral of the displacement fields around a close loop $\mathcal{L}$ must vanish. In other words, according to Eq. (8.16), the displacement field u_i is an *exact* differential form,[7] and we know from differential geometry that "every exact form is also closed."[8]

Without loss of generality, the displacement field can be written as:

$$du_i = \left(e_{ij} + \omega_{ij}\right) dx_j \tag{8.17}$$

where the first term in bracket is the linearized strain tensor of linear elasticity theory and is the symmetric part of du_i, while the second term is the anti-symmetric part. After some algebraic manipulations, one can prove that:

$$\oint \omega_{ij} \, dx_j = - \int x_l \, \omega_{ij,l} \, dx_j \tag{8.18}$$

and, using standard tensor identities, that:

$$\omega_{ij,l} = \epsilon_{mil} \, \epsilon_{mpq} \, e_{pj,q} \tag{8.19}$$

where $e_{pj,q}$ denotes the derivative along Cartesian component q of the strain tensor components e_{pj}.

We can now write:

$$\oint du_i = \oint \left(e_{ij} - x_l \, \epsilon_{mil} \, \epsilon_{mpq} \, e_{pj,q}\right) dx_j. \tag{8.20}$$

By invoking Stokes' theorem $\oint \mathbf{F} \cdot d\mathbf{x} = \int_S \nabla \times F \cdot \mathbf{n} \, dS$, the above becomes:

$$\oint du_i = - \int \int_S n_r \, \epsilon_{mil} \left(\epsilon_{rsj} \, \epsilon_{mpq} \, e_{pj,qs}\right) x_l \, dS. \tag{8.21}$$

The term inside the bracket is then the curl of the curl of the strain tensor:

$$\epsilon_{rsj} \, \epsilon_{mpq} \, e_{pj,qs} \equiv \nabla \times \nabla \times \mathbf{e}. \tag{8.22}$$

---

[7] A p-form H is exact if there exists a $p + 1$-form K whose differential is given by $H = dK$.

[8] A lemma by Poincaré shows that every closed form is also locally exact [24].

Therefore, we have just demonstrated that the condition of compatible deformation is equivalent to the requirement:

$$\nabla \times \nabla \times \mathbf{e} = 0,$$

which was Eq. (8.11) and that:

$$\nabla \times \nabla \times \mathbf{e} = 0 \;\Leftrightarrow\; \oint_{\mathcal{L}} du_i = 0 \;\Leftrightarrow\; ddu_i = 0. \tag{8.23}$$

Let us go back to the generic splitting of the total displacement into an affine part and a nonaffine part (Eq. (8.10)):

$$u_i(\mathbf{x}) = \underbrace{\Lambda_{ij}\, x^j}_{\text{affine}} + \underbrace{u_i^{\text{NA}}(\mathbf{x})}_{\text{nonaffine}}, \tag{8.24}$$

where Λ_{ij} is a a constant tensor, while the second nonaffine term u_i^{NA} does not obey any specific requirements and displays a random behavior (cfr. Fig. 2.4 in Chap. 2).

The first quantity we want to compute is the circulation of the displacement field, which defines the Burgers vector:

$$\oint_{\mathcal{L}} du_i = -b_i. \tag{8.25}$$

The left-hand side can be rewritten using the deformation gradient tensor $F_{ij} \equiv \frac{\partial u_i}{\partial x^j}$ as:

$$\oint_{\mathcal{L}} \mathbf{F} \cdot d\mathbf{s} = \oint_{\mathcal{L}} \frac{\partial u_i}{\partial x_j}\, dx^j, \tag{8.26}$$

where the deformation gradient tensor $\mathbf{F}$ has two contributions:

$$F_{ij} = \Lambda_{ij} + \frac{\partial u_i^{\text{NA}}}{\partial x^j}. \tag{8.27}$$

The first term on the right-hand side represents the affine contribution; hence, it is represented by a matrix of constants and defines a conservative field $\oint_{\mathcal{L}} \mathbf{\Lambda} \cdot d\mathbf{s} = 0$. This also means that the associated tensor Λ_{ij} is irrotational and the corresponding deformation is a compatible deformation. This, in turn, demonstrates that the Burgers vector $\mathbf{b}$ receives no contribution from the affine part of the displacement field and is, instead, exclusively determined by the nonaffine part of the displacement field u_i^{NA}, and therefore:

$$\oint_{\mathcal{L}} du_i^{\text{NA}} = -b_i. \tag{8.28}$$

This analysis thus fully supports the geometric picture of Fig. 8.4 and the related qualitative discussion.

We can recast the above findings in terms of field-theoretic language as follows.

Introducing a continuous field ϕ_m representing the position inside the material as a continuous field,[9] the curl of the strain tensor can be written as:

$$\nabla \times \mathbf{e} = \epsilon_{ijk}\partial_i\, e_{mj} = \epsilon_{ijk}\partial_i\, \partial_m\, \phi_j\,. \tag{8.29}$$

The curl of the strain tensor being equal to zero is then equivalent to the single-valuedness condition for the continuous matter fields:

$$\left[\partial_i, \partial_j\right]\phi_k = 0, \tag{8.30}$$

where the square bracket indicates an anti-symmetric combination of the indices and where we introduced the two-form current:

$$J_I^{\mu\nu} = \epsilon_{\mu\nu\rho}\partial_\rho\phi_I\,, \tag{8.31}$$

and therefore, it is evident that Eq. (8.30) and the vanishing of Eq. (8.29) both express the conservation of the two-form current J:

$$\partial_\mu J_I^{\mu\nu} = 0. \tag{8.32}$$

Hence, we have just demonstrated that an affine deformation is always a *compatible* deformation (in the language of continuum mechanics) and is associated with both a vanishing Burgers vector, expressed by $\oint_{\mathcal{L}} du_i = -b_i = 0$, and with the conservation of the two-form current Eq. (8.32).

Conversely, deformations in amorphous solids, being inherently nonaffine, are *incompatible*. Furthermore, they are associated with a non-vanishing Burgers vector $\mathbf{b} \neq 0$ and with the non-conservation of the two-form current: $\partial_\mu J_I^{\mu\nu} \neq 0$. This is a broken symmetry caused by the presence of the topological defect (i.e., the dislocation-type defect given by the non-zero Burgers vector). In particular, using field theory in the hydrodynamic limit, it can be shown (see Ref. [15] for the full derivation) that:

$$\partial_\mu J_I^{\mu\nu} = -\Omega J_I^{t\nu}, \tag{8.33}$$

where Ω is called "phase relaxation" and Eq. (8.33) is sometimes referred to as modified Josephson relation. This relation can also be understood in the spirit of the

---

[9] We use the Greek letter ϕ to be consistent with the notation typically used in field theories. In particular, $\langle\phi_I\rangle = x_I$ holds at equilibrium, with x_I the coordinate position inside the material and index I running on spatial directions only.

standard relaxation-time approximation widely used in field theory and statistical mechanics.

Furthermore, for completeness, also the following relations using the language of differential forms must hold:

$$J_i = *du_i \tag{8.34}$$

where $*$ is the standard three-dimensional Hodge dual [23].[10] This implies that:

$$b_i = \oint_{\mathcal{L}} du_i = \oint_{\mathcal{L}} *J_i = \int_{\Sigma} d*J_i \tag{8.35}$$

where the loop $\mathcal{L}$ is the boundary of the surface Σ, and we used the Stokes theorem. Finally, one can notice that:

$$d*J_i = 0 \longrightarrow \partial_\mu J_i^{\mu\nu} = 0. \tag{8.36}$$

This last equality implies that a non-zero Burger vector $b_i \neq 0$ is in 1-to-1 relation with the non-conservation of the two-form current $J_i^{\mu\nu}$. It can also be shown (see Ref. [15]), by means of hydrodynamics, that the nonaffine part of the shear modulus is proportional to the phase relaxation for transverse modes, i.e $G_{NA} \propto \Omega_\perp$.

To summarize what we have obtained so far, the nonaffinity of the displacement field u_i in amorphous solids implies that the displacement field is a multi-valued function and therefore du_i neither an exact nor a closed form, i.e., $du_i \neq 0$, where d denotes the exterior derivative. This in turn means that the associated Burgers vector is not zero and there is a topological symmetry breaking of the two-form current conservation.

Now, let us go back to Eq. (8.28), and let us define the nonaffine part of the tensor F_{ij} as $\mathcal{N}_{ij} \equiv \frac{\partial u_i^{NA}}{\partial x^j}$. Using Stokes theorem, we can rewrite the left-hand side of Eq. (8.28) as:

$$\int\int_S \nabla \times \mathcal{N} \cdot \hat{n} \, dS \tag{8.37}$$

where S is the surface enclosed by the loop $\mathcal{L}$ and $\hat{n}$ is the unit vector orthogonal to the surface. Notice that this term is not zero since $\mathcal{N}$ is not irrotational, and in components, it reads as:

$$\int\int_S \epsilon_{abj} \, \partial_b \, \mathcal{N}_{ij} \, n_a \, dS. \tag{8.38}$$

---

[10] For example, the Hodge dual of a 1-form A is given by $*A = A_x \, dy \wedge dz + A_y \, dz \wedge dx + A_z \, dx \wedge dy$.

We can now assume that this integral is non-zero for any surface S enclosed by any loop $\mathcal{L}$. This implies that:

$$\epsilon_{abj} \, \partial_b \, \partial_j \, u_i^{\mathrm{NA}} \neq 0 \tag{8.39}$$

which corresponds to the statement that the Nye tensor [25], measuring the density of elastic defects (nonaffinity, in our case), has some non-zero component:

$$\epsilon_{abj} \, \partial_b \, \partial_j \, u_i^{\mathrm{NA}} \equiv -\alpha_{ai} \neq 0 . \tag{8.40}$$

Upon recalling the definition of the two-form $J^{\mu\nu}$, we can rewrite the last expression as:

$$\alpha_i^a = \partial_\mu J_i^{\mu a} = -\Omega \, J_i^{ta} \neq 0 \tag{8.41}$$

which indicates the presence of phase relaxation in the system caused by dislocation-type defects.

Having defined the rank-2 dislocation density tensor, α_{ai}, we write, for the displacement field in an amorphous solid, the Burgers-type relation in the form:

$$\oint_{\mathcal{L}} du_i = \oint_L \frac{\partial u_i}{\partial x_k} dx_k = -b_i \tag{8.42}$$

which can be recast in differential form using Stokes' theorem as:

$$\epsilon_{ilm} \, \partial_l \, \partial_m \, u_k = -\alpha_{ik} , \tag{8.43}$$

where α_{ik} is the so-called dislocation density tensor, related to the Burgers vector via $db_i = \alpha_{mi} dA_m$, with A_m being the axial vector orthogonal to the area element enclosed by the path $\mathcal{L}$ [2, 26] and dA_m the corresponding area element.

The fact that α_{ik} is not zero, which is yet another manifestation of the incompatibility of deformation induced by nonaffinity via dislocation-type defects, has interesting connections with Einstein-Cartan theories of gravitation. In particular, the mapping between the reference frame or coordinate system prior to deformation and that after deformation can be written by introducing a metric tensor, which, as mentioned already in the previous section, is non-Euclidean in the presence of topological defects such as dislocations. As argued in Ref. [26], the non-Euclidean metric tensor in the presence of dislocation-type defects is similar to that of gravitational theories with torsion, also known as Einstein-Cartan theories. This becomes evident upon parallel-transporting vectors from a local flat frame to a nearby one, and the affine[11] connection Γ_{il}^k is thus not a Levi-Civita (torsion-free)

---

[11] Here, the word "affine connection" is used with the usual meaning it has in differential geometry, and affine does not refer to "affine deformations" introduced in Chap. 2.

connection, but rather has antisymmetric lower indices and a non-zero torsion tensor $T_{bc}^a = \Gamma_{bc}^a - \Gamma_{cb}^a \neq 0$.

In the next section, we shall set the above somewhat abstract results in practice, and we will see how the Burgers vector defined above can actually be measured in simulations of glasses under deformation. Furthermore, the above concepts can be used to formulate a mechanism of plasticity in amorphous solids mediated by dislocation-type defects arising from nonaffinity.

8.3   Plasticity Mediated by Dislocation-Type Defects in Amorphous Solids: Polymer Glasses

In Ref. [14] the vector displacement field u_i obtained from MD simulations of model Kremer-Grest polymer glasses (a similar system to the one introduced in Chap. 3, Sect. 3.3) was analyzed in order to verify the concepts and mechanism presented in the previous section for the plasticity of amorphous solids.

Also in this case, polymer chains are fully flexible and are made of spherical beads connected via the FENE potential introduced in Sect. 3.3 in Chap. 3 and the Lennard-Jones potential active also between beads belonging to different chains. The simulated glasses were then submitted to quasi-static athermal shear (AQS) deformations using a standard simulation protocol described, e.g., in [27]. A glass sample initially quenched down to zero temperature is deformed by AQS procedure consisting in the relaxation of the system after each strain step ($\delta\gamma_{xz} = 0.001$).

The displacement field u_i was measured, step by step in the stepwise increasing shear strain γ, from 2D slices of the three-dimensional MD simulation of AQS of the polymer glass at temperatures well below the glass transition temperature T_g. The discrete data points of u_i from the simulations were then subjected to an interpolation procedure in order to obtain a smooth field for further formal calculations, and it was checked that the different choices of interpolating functions were not changing the final result. In order to compute the Burgers vector, two different loop geometries for $\mathcal{L}$ were used, i.e., circles and squares. The insensitivity of the computed Burgers vector $\mathbf{b}$ to whether circles or squares is used and to their sizes demonstrates the topological invariant nature of $\mathbf{b}$.

Results of this analysis are shown in Fig. 8.5.

The numerical results fully confirm the existence of dislocation-type defects in the displacement field u_i of polymer glasses under deformation and further confirm that their origin is due to nonaffine particle displacements. Further evidence is presented in Fig. 8.6.

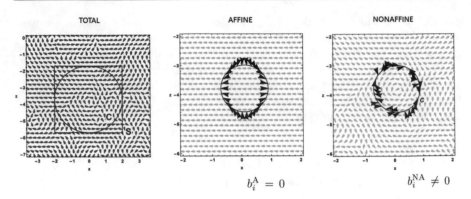

$$b_i^A = 0 \qquad\qquad\qquad b_i^{NA} \neq 0$$

Fig. 8.5 Rendering of the displacement vector field u_i for a shear deformation of a model Kremer-Grest polymer glass. The left panel shows the total displacement field u_i with a clear dislocation-type topological defect, giving rise to a non-zero Burgers vector b_i, which is fairly invariant with respect to the chosen shape of the loop $\mathcal{L}$ for the Burgers integral, cfr. Eq. (8.42). The central panel shows only the affine part of the displacement field (cfr. Eq. (8.10)), u_i^A, which gives an identically zero contribution to the Burgers vector, i.e., $b_i^A \equiv \oint_{\mathcal{L}} du_i^A = 0$. Consistent with the theory presented in the previous section, it is only the nonaffine part of the displacement field, u_i^{NA}, which produces a non-zero Burgers vector, as shown in the right panel, and $b_i^{NA} \equiv \oint_{\mathcal{L}} du_i^{NA} \neq 0$. Adapted, with modifications, from Ref. [14], with permission from the American Physical Society

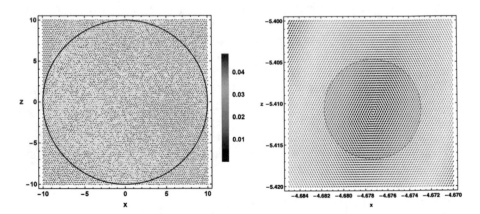

Fig. 8.6 Left panel: a 2D snapshot or slice of the interpolated displacement field u_i for a single simulated replica of the deformed polymer glass at a macroscopic shear strain $\gamma = 0.8$. The colors indicate the amplitude of the displacement field $|\mathbf{u}|$. The red curve is the closed Burgers loop $\mathcal{L}$ with radius $R = 10$ on which the Burgers vector is computed using Eq. (8.42). Right panel: zoom-in of a strongly nonaffine region with vortex-like shape that illustrates the singular nature of the displacement field around the dislocation-type topological defect. Adapted from Ref. [14], with permission from the American Physical Society

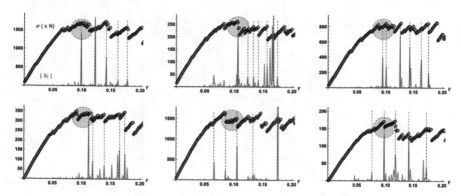

Fig. 8.7 Stress-strain curves for different replicas of the same simulated polymer glass system under athermal quasi-static shear (AQS). Blue circles represent stress vs strain, while the orange spikes represent the average modulus of the Burgers vector in the system, $|b_i|$. The spikes correspond to plastic instabilities, which may occur already well before the yielding point. The latter typically corresponds to a very large spike occurring at strain around $\gamma = 0.1$. Adapted from Ref. [14], with permission from the American Physical Society

The average of the absolute value of the Burgers vectors present in the system at given strain value γ during a quasi-static ($\dot{\gamma} \rightarrow 0$) strain ramp provides, pretty much like the dislocations density in the case of crystals, an excellent predictor of plastic instabilities. This includes also prediction of the yielding transition point $d\sigma/d\gamma = 0$, at which the material failure occurs and the plastic flow sets in. In typical stress-strain curves of amorphous materials measured experimentally at a finite deformation rate $\dot{\gamma}$, the yielding transition at which $d\sigma/d\gamma = 0$ leads to the onset of pseudo-Newtonian flow with $\sigma = \eta\dot{\gamma} = const$ as a function of γ. In the case of quasi-static athermal shear deformation, since $\dot{\gamma} = 0$, the plateau in stress σ vs strain γ cannot be interpreted as a Newtonian plateau; nonetheless, the yielding point is clearly signalled by the onset of flattening in the stress-strain curve.

In Fig. 8.7, the stress versus strain curve (blue circles) of the model polymer glass is shown for different replicas of the system together with the average absolute value of the Burgers vector (orange spikes).

Finally, Fig. 8.8 shows the averaged stress-strain curve and averaged modulus of Burgers vector over ten different realizations (replicas) of the polymer glass.

The existence of such well-defined topological defects as mediators of plastic deformation in glasses, first discovered in 2021 in Ref. [14], has been later independently confirmed in simulations of model glasses in Ref. [28].

8.4   Shear Banding and Eshelby-Like Quadrupoles

As mentioned in Sects. 8.1.1 and 8.1.2, dislocations in solids display the tendency to self-organize into slip systems or slip planes. For tensile deformation, these slip planes are oriented at $45°$ with respect to the direction of tensile strain, as prescribed by Schmid's law, since that is the direction of maximum resolved shear stress.

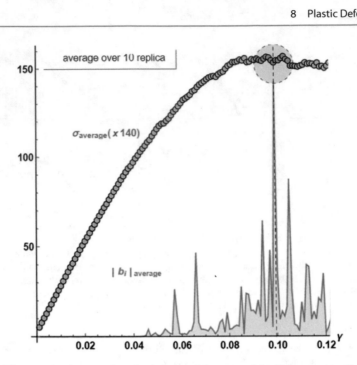

Fig. 8.8 The stress-strain curve (purple circles) and the norm of the Burgers vector $|b_i|$ (orange) averaged over ten independent replicas. The vertical dashed line indicates the location of the main peak corresponding to the yield point (material failure). The gray shaded area emphasizes the position of the yielding point. Adapted from Ref. [14], with permission from the American Physical Society

In amorphous solids, slip systems are ubiquitously observed at large deformations and near yielding. They have a similar appearance to slip systems in crystals and are referred to as shear bands. These are flowing bands within the material, which may extend across the whole sample at the point of yielding but remain "localized" in the direction perpendicular to the slip direction.

The microscopic origin of shear bands in amorphous solids has been the topic of much research and debate over the past few decades, with various competing phenomenological models that have been proposed to describe them, including shear transformation zones [29] and elegant field-theoretic models [30].

A popular concept that emerged in recent years is that of Eshelby quadrupoles. J. D. Eshelby was the first to solve analytically for the stress and strain fields around a spherical or ellipsoidal inclusion in solids [31]. The Eshelby tensor relates the strain tensor inside the inclusion to the strain tensor in the unstressed material. From that one can also compute the stress and strain field in the elastic medium outside the inclusion, which turns out to have a quadrupolar form.

A breakthrough came in 2012 when Procaccia and co-workers were able to compute the strain field of a line array of Eshelby inclusions (Eshelby quadrupoles) and demonstrated that the elastic deformation energy is minimized when the

Fig. 8.9 An array of seven aligned Eshelby quadrupoles in a simulated model athermal amorphous solid in 2D where spherical particles interact via a modified Lennard-Jones potential. Reproduced from Ref. [32] with permission from the American Physical Society

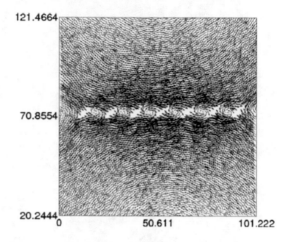

quadrupoles align in an array at 45° to the principal stress axis. The alignment of quadrupolar fields was also detected in numerical simulations, an example of which is shown in Fig. 8.9.

Although the calculations were done in 2D, the analytical expressions for the strain field are too cumbersome to be quoted.

The interpretation of shear bands in amorphous solids as the result of the alignment of Eshelby quadrupoles has found confirmation in experiments on metallic glasses [33] and in numerical simulations [34].

The alignment of local force quadrupoles can be schematized as shown in Fig. 8.10.

Using the formal analogy between electrostatics and elasticity, the displacement field along the shear banding direction can be, much more simplistically, evaluated based on the above sketch, as done in Ref. [33]. The result is a sinusoidal spatial variation of the displacement field as a function of the spatial coordinate along the band. This is connected with a sinusoidal variation of density within the band, which has been observed experimentally by means of transmission electron microscopy.

While there is strong evidence for the alignment of Eshelby-type quadrupoles as the mechanism for shear banding in amorphous solids, there are of course no well-defined "inclusions," in amorphous solids, which may lead to or justify elastic fields such as those predicted by Eshelby.

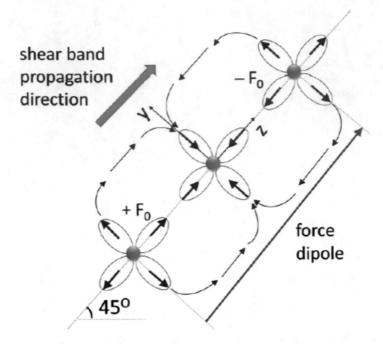

Fig. 8.10 Schematic illustration of force profiles in and around an alignment of Eshelby-like quadrupoles along the 45-degree direction in tensile strain. As stressed in the figure, in the 2D slice of material, net dipoles arise and alternate along the 45-degree direction. Adapted from Ref. [33] with permission from the American Physical Society

However, the quadrupoles observed in simulations and speculated to arise from continuum Eshelby arguments can be interpreted from the point of view of the dislocation-type defects introduced in the previous section.

In particular, in Ref. [14], the spatial variation of the absolute value of the Burgers vectors, $|\mathbf{b}|$, was mapped spatially at different values of shear strain, before and after the yielding transition. The map is shown in Fig. 8.11.

This shows that dislocation-type defects tend to concentrate or "coagulate" into bands where the local value of $|\mathbf{b}|$ is larger than in the surrounding matrix.

A possible mechanism for the concentration and localization of dislocation-type defects in the shear bands shown in Fig. 8.11 is as follows. Each dislocation-type defect produces a strain field that behaves like a monopole, $\sim b/4\pi r$, and we have seen already (cfr. Eq. (8.3)) that anti-parallel dislocation-type defects attract each other and thus may form higher-order multi-poles. In particular, they can also form arrays of quadrupoles, a phenomenon already described in standard crystalline solids [3], which can then align in the direction of the largest resolved stress, which is at 45° to the tensile direction, as prescribed by Schmid's law (cfr. Sect. 8.1.2) and by the energy-minimization argument of [35].

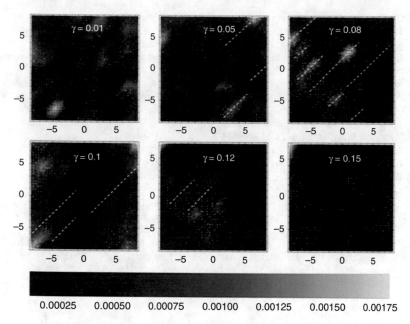

Fig. 8.11 Spatial map showing the evolution of the displacements vector **u** upon increasing the external strain γ. The background color correlates with the Burgers vector norm $|\mathbf{b}|$, and the arrow correlates with its direction. The dashed white lines guide the eye toward shear band forming at $45°$. Adapted from Ref. [14], with permission from the American Physical Society

References

1. G.I. Taylor, Proc. Roy. Soc. London. Ser. A Containing Papers Math. Phys. Char. **145**(855), 362 (1934)
2. L. Landau, E. Lifshitz, *Theory of Elasticity: Volume 6* (Pergamon Press, Oxford, 1986)
3. J. Moore, D. Kuhlmann-Wilsdorf, J. Appl. Phys. **41**(11), 4411 (1970)
4. E. Schmid, W. Boas, *Theorien der Kristallplastizität und -festigkeit* (Springer, Berlin, 1935), pp. 279–301
5. T.H. Courtney, *Mechanical Behavior of Materials* (Waveland Press, Long Grove, 2005)
6. M. Kleman, O. Lavrentovich, *Soft Matter Physics* (Springer, New York, 2003)
7. P. Chaudhari, A. Levi, P. Steinhardt, Phys. Rev. Lett. **43**, 1517 (1979)
8. P.J. Steinhardt, P. Chaudhari, Philos. Mag. A **44**(6), 1375 (1981)
9. A. Acharya, M. Widom, J. Mech. Phys. Solids **104**, 1 (2017)
10. A. Tanguy, B. Mantisi, M. Tsamados, Europhys. Lett. **90**(1), 16004 (2010)
11. I. Kriuchevskyi, T.W. Sirk, A. Zaccone, Phys. Rev. E **105**, 055004 (2022)
12. S. Wijtmans, M.L. Manning, Soft Matt. **13**, 5649 (2017)
13. M. Moshe, I. Levin, H. Aharoni, R. Kupferman, E. Sharon, Proc. Natl. Acad. Sci. **112**(35), 10873 (2015)
14. M. Baggioli, I. Kriuchevskyi, T.W. Sirk, A. Zaccone, Phys. Rev. Lett. **127**, 015501 (2021)
15. M. Baggioli, M. Landry, A. Zaccone, Phys. Rev. E **105**, 024602 (2022)
16. P. Dederichs, C. Lehmann, H. Schober, A. Scholz, R. Zeller, J. Nuclear Mat. **69–70**, 176 (1978)
17. J.A. Zimmerman, D.J. Bammann, H. Gao, Int. J. Solids Struct. **46**(2), 238 (2009)
18. A. Acharya, J. Bassani, J. Mech. Phys. Solids **48**(8), 1565 (2000)

19. P. Chaikin, T. Lubensky, *Principles of Condensed Matter Physics* (Cambridge University Press, Cambridge, 2000)
20. B.A. DiDonna, T.C. Lubensky, Phys. Rev. E **72**, 066619 (2005)
21. A.E.H. Love, *A Treatise on the Mathematical Theory of Elasticity* (Cambridge University Press, Cambridge, 1892)
22. E. Beltrami, Il Nuovo Cimento (1877–1894) **20**(1), 5 (1886)
23. K. Cahill, *Physical Mathematics*, 2nd edn. (Cambridge University Press, Cambridge, 2019)
24. F.W. Warner, *Foundations of Differentiable Manifolds and Lie Groups* (Springer, New York, 1983)
25. J. Nye, Acta Metallurgica **1**(2), 153 (1953)
26. M.L. Ruggiero, A. Tartaglia, Amer. J. Phys. **71**(12), 1303 (2003)
27. C.E. Maloney, A. Lemaître, Phys. Rev. E **74**, 016118 (2006)
28. Z.W. Wu, Y. Chen, W.-H. Wang, W. Kob, L. Xu, Nat. Commun. **4**, 2955 (2023)
29. Y. Cheng, E. Ma, Progr. Mater. Sci. **56**(4), 379 (2011)
30. R. Benzi, M. Sbragaglia, M. Bernaschi, S. Succi, F. Toschi, Soft Matt. **12**, 514 (2016)
31. J.D. Eshelby, Proc. Roy. Soc. London. Ser. A, Math. Phys. Sci. **241**(1226), 376 (1957)
32. R. Dasgupta, H.G.E. Hentschel, I. Procaccia, Phys. Rev. E **87**, 022810 (2013)
33. V. Hieronymus-Schmidt, H. Rösner, G. Wilde, A. Zaccone, Phys. Rev. B **95**, 134111 (2017)
34. D. Şopu, A. Stukowski, M. Stoica, S. Scudino, Phys. Rev. Lett. **119**, 195503 (2017)
35. R. Dasgupta, H.G.E. Hentschel, I. Procaccia, Phys. Rev. Lett. **109**, 255502 (2012)

Confinement Effects

9

Abstract

In many situations of practical and technological interest, amorphous systems are strongly confined along, e.g., the vertical direction. This is the case of thin films, nanometer-thick slabs, and quasi-2D systems. These settings are especially important for nanotechnology. It is therefore desirable to develop a mechanistic understanding of how wave propagation, including phonons and vibration modes, and elasticity depend on the thickness of the film. We shall study a mathematical confinement model, which, combined with the nonaffine elasticity and viscoelasticity framework of Chaps. 2 and 3, is able to rationalize finite-size effects in random packing's elasticity and the confinement-dependent shear modulus of various experimental soft matter systems, including the surprisingly large, solid-like, shear modulus of confined simple liquids under good wetting conditions. The same confinement model leads to a universal law for the vibrational density of states of confined solids with a ω^3 frequency dependence instead of Debye's ω^2 law.

9.1 Elasticity and Waves Under Confinement

In Chap. 3, Sect. 3.2.2, we derived an equation of motion for the displacement of a particle in an amorphous medium under an external strain; recall Eq. (3.30), which we rewrite here in the Markovian limit as:

$$\frac{d^2\mathbf{s}_i}{dt^2} + \nu\frac{d\mathbf{s}_i}{dt} + \mathbf{H}_{ij}\mathbf{x}_j = \mathbf{\Xi}_{i,\kappa\chi}\boldsymbol{\eta}_{\kappa\chi} \tag{9.1}$$

where $\boldsymbol{\eta}_{\kappa\chi}$ is the (Green-Saint Venant) strain tensor and ν is a microscopic friction coefficient, which arises from dynamical couplings mediated by the anharmonicity of the pair potential. The last term on the right-hand side represents the net force

© The Author(s), under exclusive license to Springer Nature Switzerland AG 2023
A. Zaccone, *Theory of Disordered Solids*, Lecture Notes in Physics 1015,
https://doi.org/10.1007/978-3-031-24706-4_9

acting on the particle i in its affine position and which therefore triggers the ensuing nonaffine displacement. Using this equation of motion as the starting point, we then arrived at expressions for the viscoelastic moduli, Eq. (3.35), which contain the effect of nonaffinity as an integral (with a minus sign in front of it) over eigenfrequency ω_p, which we rewrite here in discretized form (in units of mass $m = 1$) as:

$$C_{\mu\nu\kappa\chi}(\omega) = C_{\mu\nu\kappa\chi}^A - \frac{1}{V} \sum_p \frac{\hat{\Xi}_{p,\mu\nu}\hat{\Xi}_{p,\kappa\chi}}{\omega_p^2 - \omega^2 + i\omega\nu} \tag{9.2}$$

and we specialize on the shear deformation, $\mu\nu\kappa\chi = xyxy$ and $G^*(\omega) \equiv C_{xyxy}(\omega)$.

Next, we take advantage of the fact that long-wavelength vibrational modes in isotropic media can be split into longitudinal (L) and transverse (T) modes (see, e.g., [1]) and arrive at:

$$G^*(\omega) = G_\infty - A \sum_{\mathbf{k}\lambda} \frac{\omega_{p,\mathbf{k}\lambda}^2}{\omega_{p,\mathbf{k}\lambda}^2 - \omega^2 + i\omega\nu} \tag{9.3}$$

where $\lambda = L, T$ and A is a dimensionful prefactor. Then, we move to continuous variables for the eigenfrequencies $\omega_p(k)$, by assuming appropriate dispersion relations $\omega_{p,L}(k)$ and $\omega_{p,T}(k)$ for L and T modes, respectively.

At this point, we proceed with the standard replacement of the discrete sum over eigenfrequencies with a continuous integral in momentum space or k-space, $\sum_{\mathbf{k}} \cdots \rightarrow \frac{V}{(2\pi)^3} \int \ldots d^3k$:

$$G^*(\omega) = G_\infty - B \int_0^{k_D} \frac{\omega_{p,L}^2(k)}{\omega_{p,L}^2(k) - \omega^2 + i\omega\nu} k^2 dk \tag{9.4}$$

$$- B \int_0^{k_D} \frac{\omega_{p,T}^2(k)}{\omega_{p,T}^2(k) - \omega^2 + i\omega\nu} k^2 dk,$$

the upper limit of the integral being given by the Debye cutoff wavevector k_D (cfr. Chap. 5, Sect. 5.1.2).[1]

For confined systems, the lower integration limit is not zero, because of the finite system size. On dimensional grounds, one thus expects the lower integration limit

---

[1] One should note that while k is, in general, not a good quantum number for amorphous materials (as the connection between energy and wavevector is no longer single-valued as it is in crystals where Bloch's theorem holds), it still can be used to provide successful descriptions of the properties of amorphous materials and liquids [2].

to scale with $\sim 1/L$, leading to [3]:

$$G^*(\omega) = G^A - B \int_{\frac{1}{L}}^{k_D} \frac{\omega_{p,\lambda}^2(k)}{\omega_{p,\lambda}^2(k) - \omega^2 + i\omega\nu} k^2 dk \tag{9.5}$$

where B is a multiplicative prefactor.

For simplicity, let us first focus on liquids, where, in most cases, only the longitudinal branch, $\lambda = L$, survives, for reasons inherent to liquid dynamics.[2] Also, we know from Chap. 2 that, for bulk liquids, $G' = 0$ in the limit $\omega \to 0$.

We take the real part of G^*, which gives the storage modulus G', and focus on comparatively low external oscillation frequencies $\omega \ll \omega_p$ used experimentally in dynamical mechanical analysis (DMA) (cfr. Chap. 3). In both integrals, numerator and denominator cancel out, leaving the same expression in both integrals. Therefore, as anticipated above, the final low-frequency result does not depend on the form of $\omega_{p,L}(k)$.

Therefore, we arrive at the very simple form for the shear modulus of confined systems:

$$G' = G_A - \alpha \int_{1/L}^{k_D} k^2 dk = G_A - \frac{\alpha}{3}k_D^3 + \frac{\beta}{3}L^{-3}. \tag{9.6}$$

As derived in Sect. 2.8.3 in Chap. 2, using the stress-fluctuation version of the nonaffine response formalism and equilibrium statistical mechanics, the affine term G_A and the negative nonaffine term $(-\frac{\alpha}{3}k_D^3)$ cancel each other out exactly, such that $G'(\omega \to 0) = 0$ for $L \to \infty$, i.e., for bulk liquids with no confinement. For liquids under sub-millimeter confinement, only the third term in the above equation survives, and we obtain the fundamental scaling:

$$G' \approx \beta'L^{-3} \tag{9.7}$$

where $\beta' = \beta/3$ is a numerical prefactor. This is a remarkable result, which shows that even simple liquids are not always liquid-like and can behave in a solid-like manner under confinement (a fact which is confirmed by many experiments).[3] In simple words, the underlying mechanism resides in the effective "cutting off" of low-energy modes due to the confinement, which are no longer available for the nonaffine relaxations. Since the net effect of the nonaffine motions is chiefly to reduce the shear modulus, the confinement-induced reduction of nonaffinity of the

---

[2] And, in particular, to the k-gap phenomenon, i.e., the absence of transverse modes in a range (gap) of wavevector k from 0 to a certain value k_g [4].

[3] Solid-like behavior is much more common for complex fluids, which, differently from simple liquids, contain mesoscopic particles such as colloidal particles or long polymer chains. These mesoscopic objects can then self-organize into stress-bearing structures under the external load under given conditions, thus giving rise to apparent solid-like mechanical behavior.

elastic response directly translates into the liquid, picking up shear rigidity, even at low frequencies.

It should be noted that G_A does not depend on L, as one can realize, e.g., by recalling the Born-Huang formulae introduced in Chap. 2 where none of the factors depend on the system size L.

For confined amorphous solids, instead, $G_A - \frac{\alpha}{3}k_D^3 > 0$, and one has the final scaling on L given by:

$$G' = G'_{\text{bulk}} + \beta' L^{-3}, \tag{9.8}$$

where G'_{bulk} is the value of shear modulus for unconfined, bulk samples.

The above results have been derived in Ref. [3], without providing a rigorous proof for the k-integral over the confined system. Such proof involves the computation of the volume of occupied states in k-space, which goes as follows. Let us consider the geometry sketched in Fig. 9.1.

Fig. 9.1 (a) Sketch, in real space, of the confined sample. (b) Geometry of the different regions over which the k-space integral can be taken. This is not to scale; in fact $k_D \gg 2\pi/L$. Both parts of this diagram have full rotational symmetry about the z axis. Adapted from Ref. [5] with permission from the American Physical Society

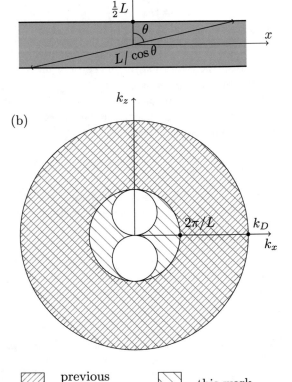

We consider a cylindrical system confined to length L in the z direction. We let its extent in the orthogonal directions (i.e., the cylinder's diameter) be infinite, although analogous results can be obtained in the finite case [5]. We employ spherical polar coordinates, measuring the polar angle θ from the vertical z axis (Fig. 9.1a). Since the system has cylindrical symmetry, no quantities depend on the azimuthal angle ϕ, the origin of which remains arbitrary.

The volume element in k-space is $dV_k = k^2 dk \sin\theta d\theta d\phi$. If an integrand does not depend on θ or ϕ, then $dV_k = 4\pi k^2 dk$, which is another statement of the fact that the integral in Eq. (9.5) represents a volume in k-space, to within a constant factor.

If the system were not confined along the z direction, clearly the lower integration limit on k would be zero. In that case, the occupied states belong to a spherical volume, which is the standard Debye sphere [6, 7].

In the confined system, instead, the maximum allowed wavelength along the z-axis is $\lambda_{max} \approx L$, giving a minimum wavevector $k_{min} \approx 2\pi/L$. In the above more simplified treatment, we used the approximation that the lower limit of the k-space integral in Eq. (9.5) is k_{min} regardless of the direction of propagation of the wave. In this approximation, the lower limit is a spherical surface in k-space with radius $2\pi/L$, and the integral is taken over the pink narrow hatched volume in Fig. 9.1b.

In reality, the lower limit of the integral must vary with the angle θ. Depending on the value of θ, the maximum length over which a wave can propagate within the confined medium is $L/\cos\theta$ (see Fig. 9.1a).

Using this value to set the maximum allowed wavelength in the generic direction θ, we now have:

$$k_{min} = 2\pi \cos\theta/L. \tag{9.9}$$

In the range $0 \leq \theta \leq \pi$, this equation describes two spheres with radius π/L, centred at $(0, 0, \pm\pi/L)$ in k-space. The integral in Eq. (9.5) must now be taken over the wide blue hatched volume in Fig. 9.1b. A 3D rendering is shown in Fig. 9.2.

It is evident that the volume of the two small spheres is given by:

$$V_{k,min} = 2 \tfrac{4}{3}\pi \left(\frac{\pi}{L}\right)^3 = \frac{8\pi^4}{3 L^3}. \tag{9.10}$$

The allowed volume of occupied states in k-space is therefore:

$$V_k = \tfrac{4}{3}\pi k_D^3 - \tfrac{8}{3}\pi^4 L^{-3}, \tag{9.11}$$

which exhibits the same $\sim L^{-3}$ scaling as derived previously.

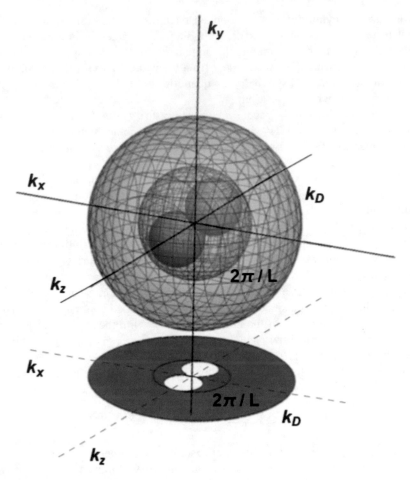

Fig. 9.2 3D rendering of the geometry of integration in k-space for the confined system of Fig. 9.1. Adapted from Ref. [5] with permission from the American Physical Society

In the next section, we shall see how the above predicted trends compare with data from experiments and simulations.

9.2   Comparison with Experimental and Simulations Data

9.2.1   Confined Liquids

The scaling $G' \sim L^{-3}$ for confined liquids is found to be obeyed by a wide range of materials, provided that the liquid and the confining solid surface are in good

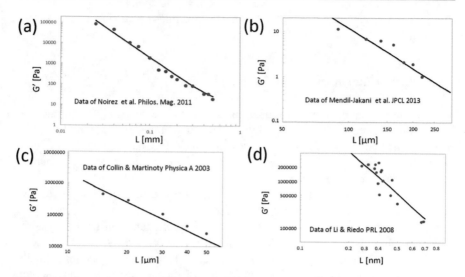

Fig. 9.3 Experimental data of low-frequency shear modulus G' plotted as a function of confinement length L for different systems: (**a**) short-chain (non-entangled) polybutylacrylate [10], (**b**) an ionic liquid [11], (**c**) short-chain (non-entangled) polystyrene melts [12], and (**d**) nanoconfined water [13]. Circles represent experimental data, while the solid line is the law $G' \sim L^{-3}$, with an adjustable prefactor. Adapted from Ref. [5] with permission from the American Physical Society

wetting conditions.[4] A standard setup is a rheometer with a submillimeter gap between two parallel solid plates, inside which the liquid is placed. The movement of the plates controlled by a force transducer then leads to the shear deformation. The good wetting reflects the near immobilization of the liquid molecules at the solid surface owing to strong attractive interactions. This, in turn, results in good stress transmission from the moving solid surface of the plates to the liquid. In the opposite case, if the liquid molecules are not well anchored to the solid surface, they are dragged away, and the solid-like response is not measurable [8].[5]

Fittings of experimental data with the $G' \sim L^{-3}$ law derived above are shown in Fig. 9.3.

Although experimental data at the nanometer scale (Ref. [13]) appear to be compatible with the L^{-3} law, it is still an open question whether a deviation

---

[4] Wetting, as measured by the contact angle, is the ability of a liquid to remain in contact with a solid surface, resulting from the balance or competition between intermolecular cohesive forces of the liquid molecules with each other and the adhesive forces between the liquid molecules and the solid surface. In particular, good wetting occurs when there is a strong attractive interaction between the liquid molecules and the solid surface, resulting in the near immobilization of the liquid interfacial layer, i.e., the so-called no-slip boundary condition as it is often referred to in fluid mechanics.

[5] Instead, the standard Newtonian viscous liquid behavior (cfr. Chap. 4) is measured [9].

from this power law, e.g., saturation to a plateau, may occur under very strong confinement.

9.2.2 Confined Amorphous Solids

The effect of confinement in amorphous solids is most cleanly studied in numerical simulations of random jammed packings. This was done, indeed, in Ref. [14], where jammed packings of frictionless soft spheres (cfr. Chap. 2, Sect. 2.4.2) were systematically studied upon varying the size of the simulation box and then sheared to measure the linear elastic response.

Upon decreasing the system's size N, it was observed that the (low-frequency) shear modulus G increases by a positive correction that scales with $1/N$, where N is the number of particles in the simulation box. For Euclidean (non-fractal) systems in $d = 3$, obviously $N \sim L^3$. Therefore, the correction $\sim 1/N \sim L^{-3}$ exactly coincides with the correction for amorphous solids that was obtained analytically from the confined version of nonaffine elasticity theory earlier in this chapter, which reads as $G' \sim G'_{\text{bulk}} + \beta L^{-3}$. Hence, nonaffine elasticity theory, combined with the confinement model developed in this chapter, provides a rationalization of the fundamental finite-size correction $\sim 1/N$ of the shear modulus ubiquitously observed in simulations of amorphous elasticity.

9.3 Vibrational Density of States of Confined Solids

The same confinement model introduced above in Sect. 9.1 can be used to derive the vibrational density of states of any quasiparticle excitations that live in the confined sample [15]. We shall see how this works on the example of phonons (vibrational excitations), although the same concept is applicable also to electrons [16], etc. We will see how the confinement leads to a low-energy scaling of the vibrational density of states substantially different from the Debye $\sim \omega^2$ law (cfr. Chap. 5, Sect. 5.1.1), for both amorphous and crystalline solids.

Consistent with the derivation above, we shall continue treating k as a continuous variable. While it is true that one may expect a quantization of k, there are, however, various reasons why this is not the case. First of all, the system is confined along the vertical z direction, but it still remains large in the other two directions, such that it is overall macroscopic. This alone is enough to guarantee that k can be treated as a continuous variable (consistent with, e.g., Kittel's derivation of the Debye k-sphere for phonons [7]). Furthermore, the system is never exactly 2D, and smooth hard-wall boundary conditions (BCs) do not apply due to the importance of atomic-scale roughness in real systems. Finally, this has been carefully justified in numerical simulations of both amorphous and crystalline ice confined in graphene oxide layers in Ref. [15], where it was found that the measured lowest value of k, i.e., $\approx \frac{1}{4}\frac{\pi}{L}$, is much smaller than the lowest non-zero value predicted by hard-wall BCs, which is $k = \pi/L$.

The number of states dN with wavenumber in a spherical shell in $k-$space between k and $k + dk$ is given by:

$$dN = V_{k,1}\, 4\pi\, k^2 dk\,.\tag{9.12}$$

where $V_{k,1} = (2\pi)^3/L^3$ is the k-state volume occupied by a single wavevector. By assuming a linear dispersion relation for sound, the same treatment of Chap. 5, Sect. 5.1.1, leads to the standard Debye $\sim \omega^2$ law for the VDOS.

Let us now consider a confined solid, same setup and geometry depicted in Fig. 9.1a. As discussed in Sect. 9.1, at a fixed generic angle θ, the maximum allowed wavelength is $\lambda_{max} = L/\cos\theta$, due to the vertical confinement. In momentum space, this implies a minimum allowed wavevector equal to $2\pi \cos\theta/L$. In other words, below a certain crossover momentum $k_\times \equiv 2\pi/L$, the k-space is not completely available because of confinement. As computed in Sect. 9.1, the two little spheres in Fig. 9.2 represent the portion of k-space in the region $k < 2\pi/L$, which cannot be occupied due to the confinement.

It should be noted that in the non-confined direction, $\theta = \pi/2$, there is no cutoff on k, which can virtually go down to $k = 0$.

Under these conditions, it is clear that angular integration in k-space is no longer isotropic for $k < k_\times$, and it cannot yield the prefactor 4π as in Eq. (9.12), which then leads to Debye's law. Upon inverting Eq. (9.9), we obtain the following bound on the angle θ:

$$\theta_{min} = \cos^{-1}(L\,k/2\pi)\,.\tag{9.13}$$

This can be used to perform the angular integration under the constraints on the allowed k-space topology imposed by the confinement. Instead of the trivial 4π factor, this time we get:

$$2\int_{\cos^{-1}(L\,k/2\pi)}^{\pi/2} \sin\theta\, d\theta \int_0^\pi d\phi = 2\,L\,k\tag{9.14}$$

which has an explicit dependence on both k and the extent of confinement L. The fact that the integral in the azimuthal ϕ angle remains unaltered under confinement reflects the $SO(2)$ rotational invariance in the horizontal (x, y) plane.[6]

Hence, we obtain that, below a certain threshold $k = k_\times$, the number of states with wavenumber in the shell $\in [k, k + dk]$ is given by:

$$dN \sim L\,k^3 dk\,.\tag{9.15}$$

which clearly differs from the Debye result.

---

[6] It should be noted that Eq. (9.14) recovers the standard Debye result when the wavevector approaches the crossover value $k_\times$. In particular, when $k = k_\times$, the lower limit of integration tends to zero.

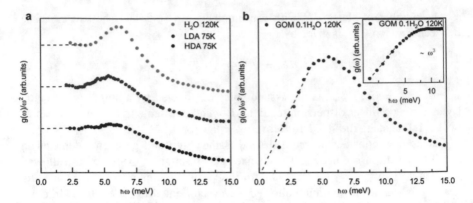

Fig. 9.4 Panel (**a**): Debye-normalized experimental VDOS of unconfined high-density amorphous (HDA) ice and low-density amorphous (LDA) ice, both following the standard Debye law at low ω; the original data are from Ref. [17]. Panel (**b**): Debye-normalized experimental VDOS of nano-confined amorphous ice sandwiched between two graphene oxide layers. The linear slope with ω clearly unveils the law $\sim\omega^3$ for the vibrational spectrum of confined solids derived in this section. Adapted from Ref. [15]

Following the same steps as in Sect. 5.1.1 (cfr. Eqs. (5.2)–(5.3)), one can deduce the VDOS at low frequency as:

$$g(\omega) \sim \omega^3 \tag{9.16}$$

which is the main result of this section. Compared to Debye's $\sim \omega^2$ law for the unconfined solids, this is a fundamentally different law, which bears many implications for the thermal and viscoelastic properties of the confined systems.[7]

The scaling law for the low-energy vibrational spectrum of confined solids, Eq. (9.16), has been validated with experiments (inelastic neutron scattering) and MD simulations in Ref. [15], on the example of nano-confined amorphous ice sandwiched between two graphene oxide sheets. The results are reproduced in Fig. 9.4.

This result has been further verified by means of atomistic MD simulations, as shown in Fig. 9.5, for the same system (ice confined in graphene oxide).

These data show, additionally, that the $\sim\omega^3$ law, just like the Debye law for the unconfined solids, applies to amorphous as well as to crystalline samples.[8]

---

[7] Recall (Chap. 3) that the viscoelastic moduli of amorphous solids depend on the VDOS, and the same applies to the specific heat and to the thermal conductivity (Chap. 6).

[8] The making of thin confined layers in the lab is normally much easier for amorphous slabs than for crystalline ones, due to the fact that crystallization under confinement is more difficult for experimental systems. This difficulty is not present in simulations.

Fig. 9.5 Debye-normalized simulated VDOS of confined/unconfined ordered/amorphous ice. Adapted from Ref. [15]

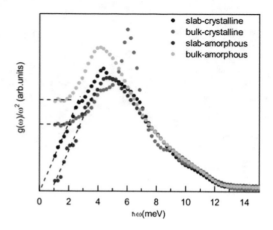

References

1. L. Landau, E. Lifshitz, *Theory of Elasticity: Volume 6* (Pergamon Press, Oxford, 1986)
2. J. Hansen, I. McDonald, *Theory of Simple Liquids* (Elsevier Science, Amsterdam, 2006)
3. A. Zaccone, K. Trachenko, Proc. Natl. Acad. Sci. **117**(33), 19653 (2020)
4. K. Trachenko, V.V. Brazhkin, Rep. Progr. Phys. **79**(1), 016502 (2015)
5. A.E. Phillips, M. Baggioli, T.W. Sirk, K. Trachenko, A. Zaccone, Phys. Rev. Mater. **5**, 035602 (2021)
6. M. Born, K. Huang, *Dynamical Theory of Crystal Lattices* (Clarendon Press, Oxford, 1954)
7. C. Kittel, *Introduction to Solid State Physics. Eighth edition* (Wiley, Hoboken, 2005)
8. L. Noirez, P. Baroni, J. Phys. Condensed Matt. **24**(37), 372101 (2012)
9. L. Noirez, P. Baroni, J. Mol. Struct. **972**(1), 16 (2010). Horizons in Hydrogen Bond Research 2009
10. L. Noirez, H. Mendil-Jakani, P. Baroni, Philosoph. Mag. **91**(13–15), 1977 (2011)
11. H. Mendil-Jakani, P. Baroni, L. Noirez, L. Chancelier, G. Gebel, J. Phys. Chem. Lett. **4**(21), 3775 (2013)
12. D. Collin, P. Martinoty, Phys. A Statist. Mech. Appl. **320**, 235 (2003)
13. T.D. Li, E. Riedo, Phys. Rev. Lett. **100**, 106102 (2008)
14. C.P. Goodrich, A.J. Liu, S.R. Nagel, Nat. Phys. **10**(8), 578 (2014)
15. Y. Yu, C. Yang, M. Baggioli, A.E. Phillips, A. Zaccone, L. Zhang, R. Kajimoto, M. Nakamura, D. Yu, L. Hong, Nat. Commun. **13**(1), 3649 (2022)
16. R. Travaglino, A. Zaccone, J. Appl. Phys. **133**, 033901 (2023)
17. M.M. Koza, B. Geil, K. Winkel, C. Köhler, F. Czeschka, M. Scheuermann, H. Schober, T. Hansen, Phys. Rev. Lett. **94**, 125506 (2005)

A Brief Reminder of Elasticity Theory

In this appendix, we briefly recap key concepts and definitions of linearized elasticity theory, with a focus on the fundamental relations between elastic stiffnesses, elastic constants, and the shear and bulk moduli for isotropic solids. Basic treatments of linear elasticity theory can be found in [1] and [2].

In elasticity theory, deformations are described by strain tensors and the force per unit area that results after a deformation is described by stress tensors. Stress and strain are related by the constitutive relation of the material, which are obtained from the free energy of deformation, since stresses are first derivatives of the energy with respect to the strain. As previously noted, there are different definitions of strains, which lead to different definitions of the stress tensors.

Typically one starts with the definition of the deformation gradient tensor $\mathbf{F}$, which relates the position vector of a material point i after the deformation ($\mathbf{r}_i$) and in the rest frame prior to the deformation ($\mathbf{r}_{i,0}$) via:

$$r_i^\alpha = F^{\alpha\beta} r_{i,0}^\beta \qquad (A.1)$$

where $\alpha, \beta = x, y, z$ as usual denote Cartesian components. It is important to note that here i does not denote a particle (atom, molecule, grain) but a generic material point inside a continuum.[1] All the most used definitions of strain tensor follow from the above definition of deformation gradient tensor.

The Cauchy-Green strain tensor η is thus defined as $\eta = \frac{1}{2}\left(\mathbf{F}^T \mathbf{F} - \mathbf{1}\right)$, where $\mathbf{1} = \delta_{\alpha\beta}$ is the identity (Kronecker) tensor. By differentiating the energy with respect to η, we obtain the second Piola-kirchhoff stress $\mathbf{t}$. For zero stress in the reference frame, the linearized constitutive relations are described by the 4-th rank elastic stiffness tensor $C_{\iota\xi\kappa\chi}$ with $t_{\iota\xi} = C_{\iota\xi\kappa\chi}\eta_{\kappa\chi}$. Since the second Piola-kirchhoff stress has no mechanical interpretation, it is common to use the Cauchy stress tensor σ, which

---

[1] On a formal level, a deformation described by $\mathbf{F}$ is a diffeomorphism of $\mathbb{R}^3$.

© The Author(s), under exclusive license to Springer Nature Switzerland AG 2023
A. Zaccone, *Theory of Disordered Solids*, Lecture Notes in Physics 1015,
https://doi.org/10.1007/978-3-031-24706-4

gives the force acting on a surface when contracted on the normal vector of the surface. For small deformations (linearized elastic theory), it is common to use the linearized strain tensor of elasticity theory (cfr. page 2–3 in [1]) $\mathbf{e} = \frac{1}{2}\left(\mathbf{F} + \mathbf{F}^T\right) - \mathbf{1}$ (first-order approximation of $\boldsymbol{\eta}$ for small $\mathbf{F}$) and the linear constitutive equation $\boldsymbol{\sigma}(\mathbf{e})$, called generalized Hooke's law, is

$$\sigma_{\iota\xi} = C_{\iota\xi\kappa\chi} e_{\kappa\chi} \tag{A.2}$$

where $C_{\iota\xi\kappa\chi}$ denotes the elastic stiffnesses that are second derivatives of the energy with respect to the strain tensor $\mathbf{e}$, $C_{\iota\xi\kappa\chi} = \frac{1}{V}\frac{\partial^2 U}{\partial e_{\iota\xi}\partial e_{\kappa\chi}}$. Also note that in the limit of small deformations, there is no difference in taking the derivative with respect to $\mathbf{e}$ or with respect to $\boldsymbol{\eta}$, since $\mathbf{e}$ is de facto the linearization of $\boldsymbol{\eta}$. Also note that $\mathbf{u} = \mathbf{F} - \mathbf{1}$, from which the standard definition of linearized strain tensor (cfr. again [1]) follows as:

$$\mathbf{e} = \frac{1}{2}\left(\mathbf{u}^T + \mathbf{u}\right). \tag{A.3}$$

As we previously noted (for details see [3]), we have for the case of no stresses ($\boldsymbol{\sigma}(0) = 0$) in the reference undeformed frame that the elastic stiffnesses are equal to the elastic constants.

For an isotropic homogeneous medium, because of rotation and translation invariance, we have that the linear constitutive equation between stress and strain is fully described by two constants, and thus the generalized Hooke's law can be written as:

$$\boldsymbol{\sigma} = 2\mu\mathbf{e} + \lambda\, tr(\mathbf{e})\mathbf{1} \tag{A.4}$$

$$= 2G\mathbf{e} + \left(k - \frac{2}{3}G\right) tr(\mathbf{e})\mathbf{1}$$

where we introduced the two Lamè coefficients (μ, λ), the shear modulus $G \equiv \mu$ and the bulk modulus $K = \lambda + \frac{2}{3}\mu$. Note that in two dimensions, we have to define $K = \lambda + \mu$.

The shear modulus μ is concerned with the deformation of a solid when it experiences a force parallel to one of its surfaces, while its opposite face experiences an opposing force. To visualize it, we consider a simple shear deformation where the undeformed cell, whose shape is described by tree vectors $(l, 0, 0)$, $(0, l, 0)$, $(0, 0, l)$, is deformed into the parallelepiped cell described by $(l, 0, 0)$, $(l\gamma, l, 0)$, $(0, 0, l)$ as illustrated in Fig. A.1.

The deformation gradient tensor for a simple shear deformation is given by:

$$\mathbf{F}(\gamma) = \begin{pmatrix} 1 & \gamma & 0 \\ 0 & 1 & 0 \\ 0 & 0 & 1 \end{pmatrix}$$

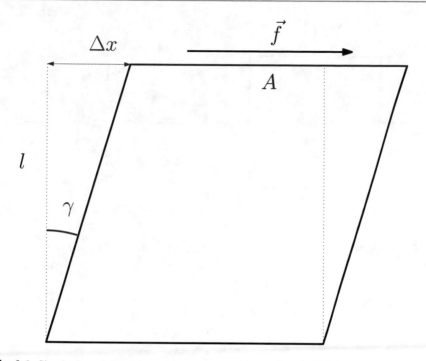

Fig. A.1 Simple shear deformation with parameters, the meaning of which is clarified in the text. Compare with the case of pure shear deformation presented in Fig. 2.1 in Chap. 2

and it follows that the linearized strain tensor is:

$$\mathbf{e} = \begin{pmatrix} 0 & \gamma/2 & 0 \\ \gamma/2 & 0 & 0 \\ 0 & 0 & 0 \end{pmatrix}$$

where we have that $2\,e_{xy} = \gamma$.

Then according to Formula (A.4), the only non-zero elements of the Cauchy-stress tensor are:

$$\sigma_{xy} = \sigma_{yx} = G\gamma = 2G\,e_{xy}$$

and thus we have that the force acting on the surface A is equal to $(f, 0, 0)$ with f given by:

$$f = A\sigma_{xy}(\mathbf{e}) = AG\gamma.$$

Thus to measure the shear modulus, we apply a force in the x direction on the surface A, and then we measure the displacements Δx. The shear modulus is then

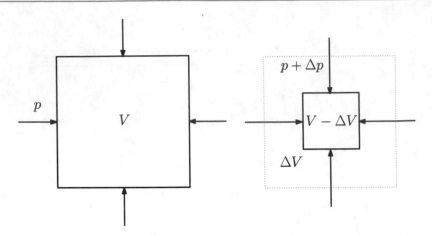

Fig. A.2 Isotropic or hydrostatic compression accompanied by change of volume from V to $V - \Delta V$

given by the following equation:

$$G = \frac{\sigma}{\gamma} = \frac{f\, l}{A\, \Delta x}\,.$$

The bulk modulus K of a substance measures the substance's resistance to hydrostatic compression, and it is defined as the pressure increase needed to cause a given relative decrease in the volume. To visualize it, we can consider an isotropic compression where a homogeneous hydrostatic pressure is applied to the surface of the body (see Fig. A.2). This situation is described by the diagonal stress tensor:

$$\sigma_{\alpha\beta} = -\Delta p\, \delta_{\alpha\beta}.$$

The increase of pressure causes a deformation (which can be obtained from the inverse of Eq. (A.4)) and thus a change of the body's volume. The volume difference ΔV between the undeformed and deformed body is given by:

$$\Delta V = V tr(\mathbf{e})\,.$$

With the last two equations and $tr(\sigma) = K\, tr(\mathbf{e})$ (trace of Eq. A.4), we can relate the change in volume to the change in pressure:

$$\frac{\Delta V}{V} = -K\, \Delta p$$

and thus we obtain the definition of the bulk modulus K.

Comparing the two equations (A.4) and (A.2), we have that the elastic stiffnesses (or elastic constants) $C_{\iota\xi\kappa\chi}$ are related to the Lamè constants by:

$$C_{\iota\xi\kappa\chi} = \frac{\partial\sigma_{\iota\xi}}{\partial\epsilon_{\kappa\chi}} = \begin{cases} \mu & \iota = \kappa, \ \xi = \chi \ \iota \neq \xi \\ \lambda + 2\mu & \iota = \xi = \kappa = \chi \\ \lambda & \iota = \xi \ \kappa = \chi, \ \iota \neq \kappa \\ 0 & \text{else} . \end{cases}$$

With these relations, we have that the shear modulus for an isotropic solid is given by:

$$G = C_{xyxy} \tag{A.5}$$

and the bulk modulus is given, in d-dimensional space, by:

$$K = \frac{C_{xxxx} + (d-1)\,C_{xxyy}}{d} . \tag{A.6}$$

Lattice Dynamics of Metallic Glasses with the EAM Potential

B

In this appendix, analytical formulae are provided, which allow one to evaluate the elements of the Hessian matrix $\mathbf{H}_{ij}$ and the affine-force vectors Ξ_i for atomic dynamics in metals using the embedded atom method (EAM) potential.

We recall from Chap. 1 (Sect. 1.1.3) the form of the EAM potential: the total potential energy acting on a tagged atom i is given by:

$$U_i = F_A \left(\sum_{j \neq i} \rho_{AB}(r_{ij}) \right) + \frac{1}{2} \sum_{j \neq i} \psi_{AB}(r_{ij}). \tag{B.1}$$

The many-body nature of the EAM potential is a result of the embedding energy term (i.e., the first term on the r.h.s.). Both summations in the above formula are over all neighbors j of particle i within the cutoff distance [4]. Then we can get the net force acting on a tagged atom with the aid of the following set of relations:

$$\mathbf{n}_{ij} = \frac{\mathbf{r}_{ij}}{r_{ij}}; \quad \bar{\rho}_i = \sum_{j \neq i} \rho_{AB}(r_{ij})$$

$$Z_{ij} = \frac{\partial U_i}{\partial r_{ij}} = \frac{1}{2} \frac{\partial \psi_{AB}(r_{ij})}{\partial r_{ij}} + \frac{\partial F_A}{\partial \bar{\rho}_i} \frac{\partial \rho_{AB}(r_{ij})}{\partial r_{ij}}$$

$$\mathbf{f}_i = -\frac{\partial U}{\partial \mathbf{r}_i} = -\frac{\partial U_i}{\partial \mathbf{r}_i} - \frac{\partial \sum_{k \neq i} U_k}{\partial \mathbf{r}_i}$$

$$= -\frac{\partial U_i}{\partial \mathbf{r}_i} - \frac{\partial \sum_{k \neq i} U_k}{\partial r_{ik}} \frac{\partial r_{ik}}{\partial \mathbf{r}_i}$$

© The Author(s), under exclusive license to Springer Nature Switzerland AG 2023
A. Zaccone, *Theory of Disordered Solids*, Lecture Notes in Physics 1015,
https://doi.org/10.1007/978-3-031-24706-4

$$= -\frac{\partial U_i}{\partial \mathbf{r}_i} + \frac{\partial \sum_{k \neq i} U_k}{\partial r_{ik}} \frac{\mathbf{r}_{ik}}{r_{ik}}$$

$$= -\frac{\partial U_i}{\partial \mathbf{r}_i} + \sum_{k \neq i} Z_{ki} \frac{\mathbf{r}_{ik}}{r_{ik}}.$$

The Hessian is then written for $i \neq j$ as:

$$\mathbf{H}_{ij}|_{i \neq j} = \frac{\partial^2 U}{\partial \mathbf{r}_i \partial \mathbf{r}_j} = \frac{\partial \frac{\partial U_i}{\partial \mathbf{r}_i}}{\partial \mathbf{r}_j} - \frac{\partial \sum_{k \neq i} Z_{ki} \frac{\mathbf{r}_{ik}}{r_{ik}}}{\partial \mathbf{r}_j}$$

$$= \frac{\partial^2 U_i}{\partial \mathbf{r}_i \partial \mathbf{r}_j} - \frac{\partial Z_{ji}}{\partial \mathbf{r}_j} \frac{\mathbf{r}_{ji}}{r_{ji}} - Z_{ji} \frac{\partial \frac{\mathbf{r}_{ij}}{r_{ij}}}{\partial \mathbf{r}_j} - \frac{\partial \sum_{k \neq i, k \neq j} Z_{ki} \frac{\mathbf{r}_{ik}}{r_{ik}}}{\partial \mathbf{r}_j}$$

$$= \frac{\partial^2 U_i}{\partial \mathbf{r}_i \partial \mathbf{r}_j} - \frac{\partial Z_{ji}}{\partial r_{ij}} \frac{\partial r_{ij}}{\partial \mathbf{r}_j} \otimes \frac{\mathbf{r}_{ij}}{r_{ij}} - Z_{ji} \frac{\partial \frac{\mathbf{r}_{ij}}{r_{ij}}}{\partial \mathbf{r}_j} - \sum_{k \neq i, k \neq j} \frac{\partial Z_{ki}}{\partial \mathbf{r}_j} \otimes \frac{\mathbf{r}_{ik}}{r_{ik}} \qquad (\text{B.2})$$

with $d = 3$:

$$\frac{\partial \frac{\mathbf{r}_{ij}}{r_{ij}}}{\partial \mathbf{r}_j} = \frac{I_{3 \times 3}}{r_{ij}} - \frac{\mathbf{r}_{ij} \otimes \mathbf{r}_{ij}}{r_{ij}^3}, \qquad (\text{B.3})$$

and:

$$\mathbf{H}_{ii} = \frac{\partial^2 U}{\partial \mathbf{r}_i \partial \mathbf{r}_i} = \frac{\partial^2 U_i}{\partial \mathbf{r}_i^2} - \frac{\partial \sum_{k \neq i} Z_{ki}}{\partial \mathbf{r}_j} \frac{\mathbf{r}_{ik}}{r_{ik}} - \sum_{k \neq i} Z_{ki} \frac{\partial \frac{\mathbf{r}_{ik}}{r_{ik}}}{\partial \mathbf{r}_i}$$

$$= \frac{\partial^2 U_i}{\partial \mathbf{r}_i^2} + \frac{\partial \sum_{k \neq i} Z_{ji}}{\partial \mathbf{r}_j} \frac{\mathbf{r}_{ik}}{r_{ik}} \otimes \frac{\mathbf{r}_{ik}}{r_{ik}} - \sum_{k \neq i} Z_{ki} \frac{\partial \frac{\mathbf{r}_{ik}}{r_{ik}}}{\partial \mathbf{r}_i}$$

$$= \frac{\partial^2 U_i}{\partial \mathbf{r}_i^2} + \frac{\partial \sum_{k \neq i} Z_{ji}}{\partial \mathbf{r}_j} \frac{\mathbf{r}_{ik}}{r_{ik}} \otimes \frac{\mathbf{r}_{ik}}{r_{ik}} + \sum_{k \neq i} Z_{ki} \left(\frac{I_{3 \times 3}}{r_{ik}} - \frac{\mathbf{r}_{ik} \otimes \mathbf{r}_{ik}}{r_{ik}^3} \right) \qquad (\text{B.4})$$

for the diagonal $i = j$ elements. To find $\Xi_{i, \kappa \chi} = \sum_j \Xi_{ij, \kappa \chi}$, we write (in components):

$$\Xi_{ij, \kappa \chi}^{\mu} = -S_{ij, \mu \nu} \frac{\partial r_{ij}^{\nu}}{\partial \eta_{\kappa \chi}} = -\frac{1}{2} S_{ij, \mu \nu} (\delta_{\nu \kappa} r_{ij}^{\chi} + \delta_{\nu \chi} r_{ij}^{\kappa}) \qquad (\text{B.5})$$

with:

$$\mathbf{S}_{ij} = \frac{\partial^2 U_i}{\partial \mathbf{r}_{ij} \partial \mathbf{r}_{ij}} = \frac{\partial}{\partial \mathbf{r}_{ij}} \left(\frac{\partial U}{\partial \mathbf{r}_{ij}} \right) = \frac{\partial}{\partial \mathbf{r}_{ij}} \left(\sum_k \frac{\partial U_k}{\partial \mathbf{r}_{ij}} \right)$$

$$= \frac{\partial}{\partial \mathbf{r}_{ij}} \left(\sum_{k,l \neq k} \frac{\partial U_k}{\partial r_{lk}} \frac{\partial r_{lk}}{\partial \mathbf{r}_{ij}} \right)$$

$$= \frac{\partial}{\partial \mathbf{r}_{ij}} \left(\frac{\partial U_i}{\partial r_{ji}} \frac{\partial r_{ji}}{\partial \mathbf{r}_{ji}} + \frac{\partial U_j}{\partial r_{ji}} \frac{\partial r_{ji}}{\partial \mathbf{r}_{ji}} \right) \frac{\partial U_i}{\partial r_{ji}} \frac{\partial r_{ji}}{\partial \mathbf{r}_{ji}}$$

$$= \frac{\partial}{\partial \mathbf{r}_{ij}} \left(Z_{ij} \frac{\mathbf{r}_{ij}}{R_{ij}} + Z_{ji} \frac{\mathbf{r}_{ij}}{R_{ij}} \right)$$

$$= \frac{\partial}{\partial \mathbf{r}_{ij}} \left(Z_{ij} \mathbf{n}_{ij} + Z_{ji} \mathbf{n}_{ij} \right)$$

$$= \frac{\partial Z_{ij}}{\partial \mathbf{r}_{ij}} \mathbf{n}_{ij} + Z_{ij} \frac{\partial \mathbf{n}_{ij}}{\partial \mathbf{r}_{ij}} + \frac{\partial Z_{ji}}{\partial \mathbf{r}_{ij}} \mathbf{n}_{ij} + Z_{ji} \frac{\partial \mathbf{n}_{ij}}{\partial \mathbf{r}_{ji}}$$

$$= \frac{\partial}{\partial \mathbf{r}_{ij}} \left(\frac{\partial U_i}{\partial \mathbf{r}_{ij}} \right) \mathbf{n}_{ij} + Z_{ij} \frac{\partial}{\partial \mathbf{r}_{ij}} \left(\frac{\mathbf{r}_{ij}}{r_{ij}} \right)$$

$$+ \frac{\partial}{\partial \mathbf{r}_{ij}} \left(\frac{\partial U_j}{\partial \mathbf{r}_{ij}} \right) \mathbf{n}_{ij} + Z_{ji} \frac{\partial}{\partial \mathbf{r}_{ij}} \left(\frac{\mathbf{r}_{ij}}{r_{ij}} \right)$$

$$= \sum_k \frac{\partial \left(\frac{\partial U_i}{\partial \mathbf{r}_{ij}} \right)}{\partial r_{ik}} \frac{\partial r_{ik}}{\partial \mathbf{r}_{ij}} \mathbf{n}_{ij} + Z_{ij} \frac{r_{ij} - \mathbf{r}_{ij} \frac{\partial r_{ij}}{\partial \mathbf{r}_{ij}}}{R_{ij}^2}$$

$$+ \sum_k \frac{\partial \left(\frac{\partial U_j}{\partial \mathbf{r}_{ij}} \right)}{\partial r_{jk}} \frac{\partial r_{jk}}{\partial \mathbf{r}_{ij}} \mathbf{n}_{ij} + Z_{ji} \frac{r_{ij} - \mathbf{r}_{ij} \frac{\partial r_{ij}}{\partial \mathbf{r}_{ij}}}{r_{ij}^2}$$

$$= \frac{\partial^2 U_i}{\partial^2 r_{ij}} \mathbf{n}_{ij} \mathbf{n}_{ij} + Z_{ij} \frac{(1 - \mathbf{n}_{ij} \mathbf{n}_{ij})}{r_{ij}} + \frac{\partial^2 U_J}{\partial^2 r_{ij}} \mathbf{n}_{ij} \mathbf{n}_{ij}$$

$$+ Z_{ji} \frac{(1 - \mathbf{n}_{ij} \mathbf{n}_{ij})}{r_{ij}}. \tag{B.6}$$

To distinguish **S** from **H**, one can rewrite $\mathbf{H}(i \neq j)$ as:

$$\mathbf{H}_{ij} = \frac{\partial^2 U}{\partial \mathbf{r}_i \partial \mathbf{r}_j} = \frac{\partial}{\partial \mathbf{r}_i} \left(\sum_k \frac{\partial U_k}{\partial \mathbf{r}_j} \right) = \frac{\partial}{\partial \mathbf{r}_i} \left(\sum_{k,l \neq k} \frac{\partial U_k}{\partial r_{kl}} \frac{\partial r_{kl}}{\partial \mathbf{r}_j} \right)$$

$$= \frac{\partial}{\partial \mathbf{r}_i} \left(\sum_{l \neq j} \frac{\partial U_j}{\partial r_{jl}} \frac{\partial r_{jl}}{\partial \mathbf{r}_j} + \sum_{k \neq j, l \neq k} \frac{\partial U_k}{\partial r_{kl}} \frac{\partial r_{kl}}{\partial \mathbf{r}_j} \right)$$

$$= \frac{\partial}{\partial \mathbf{r}_i} \left(\sum_{l \neq j} \frac{\partial U_j}{\partial r_{jl}} \frac{\partial r_{jl}}{\partial \mathbf{r}_j} + \sum_{k \neq j} \frac{\partial U_k}{\partial r_{kj}} \frac{\partial r_{kj}}{\partial \mathbf{r}_j} \right)$$

$$= \frac{\partial}{\partial \mathbf{r}_i} \left(\sum_{l \neq j} \frac{\partial U_j}{\partial r_{jl}} \frac{\mathbf{r}_{jl}}{\partial r_{jl}} + \sum_{k \neq j} \frac{\partial U_k}{\partial r_{kj}} \frac{\mathbf{r}_{jk}}{\partial r_{jk}} \right)$$

$$= \sum_{k \neq j} \frac{\partial}{\partial \mathbf{r}_i} \left(\frac{\partial U_j}{\partial r_{jk}} \frac{\mathbf{r}_{jk}}{r_{jk}} + \frac{\partial U_k}{\partial r_{kj}} \frac{\mathbf{r}_{jk}}{r_{jk}} \right)$$

$$= \sum_{k \neq j} \left(\sum_{l \neq j} \frac{\partial}{\partial r_{jl}} \left(\frac{\partial U_j}{\partial r_{jk}} \right) \frac{\partial r_{jl}}{\partial \mathbf{r}_i} \frac{\mathbf{r}_{jk}}{r_{jk}} + \frac{\partial U_j}{\partial r_{ji}} \frac{\partial}{\partial \mathbf{r}_i} \left(\frac{\mathbf{r}_{ji}}{r_{ji}} \right) \right)$$

$$+ \sum_{k \neq j} \left(\sum_{l \neq j} \frac{\partial}{\partial r_{kl}} \left(\frac{\partial U_k}{\partial r_{jk}} \right) \frac{\partial r_{kl}}{\partial \mathbf{r}_i} \frac{\mathbf{r}_{jk}}{r_{jk}} + \frac{\partial U_i}{\partial r_{ji}} \frac{\partial}{\partial \mathbf{r}_i} \left(\frac{\mathbf{r}_{ji}}{r_{ji}} \right) \right)$$

$$= \sum_{k \neq j} \left(\frac{\partial}{\partial r_{ji}} \left(\frac{\partial U_j}{\partial r_{jk}} \right) \frac{\partial r_{ji}}{\partial \mathbf{r}_i} \frac{\mathbf{r}_{jk}}{r_{jk}} \right) + Z_{ji} \frac{(-1 + \mathbf{n}_{ij} \mathbf{n}_{ij})}{r_{ij}}$$

$$+ \sum_{k \neq j} \left(\sum_{l \neq k} \frac{\partial}{\partial r_{kl}} \left(\frac{\partial U_k}{\partial r_{kj}} \right) \frac{\partial r_{kl}}{\partial \mathbf{r}_i} \frac{\mathbf{r}_{jk}}{r_{jk}} \right) + Z_{ij} \frac{(-1 + \mathbf{n}_{ij} \mathbf{n}_{ij})}{r_{ij}}$$

$$= \sum_{k \neq j} \left(\frac{\partial}{\partial r_{ji}} \left(\frac{\partial U_j}{\partial r_{jk}} \right) \mathbf{n}_{ij} \mathbf{n}_{jk} \right) + Z_{ji} \frac{(-1 + \mathbf{n}_{ij} \mathbf{n}_{ij})}{r_{ij}}$$

$$+ \sum_{k \neq j, i} \frac{\partial}{\partial r_{ki}} \left(\frac{\partial U_k}{\partial r_{kj}} \right) \mathbf{n}_{ik} \mathbf{n}_{jk} + \sum_{k \neq i} \frac{\partial^2 U_i}{\partial r_{ik} \partial r_{ij}} \mathbf{n}_{ik} \mathbf{n}_{ji}$$

$$+ Z_{ij} \frac{(-1 + \mathbf{n}_{ij} \mathbf{n}_{ij})}{r_{ij}}. \tag{B.7}$$

These formulae have been used in Ref. [5] and in Chap. 3 to compute the viscoelastic moduli of metallic glasses based on configurations from MD simulations where the EAM potential was used (in the particular case of $Zr_{50}Cu_{50}$). These formulae are completely generic and can be used to compute the lattice dynamics and elastic or viscoelastic moduli in a fully microscopic way for any metal alloy.

Generalized Langevin Equation Derived from Caldeira-Leggett Hamiltonians

C.1 Caldeira-Leggett Hamiltonian

In condensed matter physics, the Zwanzig-Caldeira-Leggett (ZCL) particle-bath model is widely applied to low-temperature quantum physics problems, especially in quantum tunnelling in superconductors and in chemical reaction rate theory. We study the classical version of the Caldeira-Leggett coupling between the tagged particle (system, S) and a bath (B) of harmonic oscillators, which was originally proposed in [6]:

$$\mathcal{H} = \mathcal{H}_S + \mathcal{H}_B \tag{C.1}$$

where $\mathcal{H}_S = P^2/2M + \mathcal{U}(Q)$ is the Hamiltonian of the tagged particle (P is momentum and Q is position). The second term on the r.h.s. is the standard Hamiltonian of the bath of harmonic oscillators that are coupled to the tagged particle:

$$\mathcal{H}_B = \sum_{m=1}^{N} \frac{1}{2} \left[\frac{P_m^2}{M_m} + M_m \omega_m^2 \left(X_m - \frac{F_m(Q)}{\omega_m^2} \right)^2 \right], \tag{C.2}$$

consisting of the standard harmonic oscillator expression for each bath oscillator m and of the coupling term between the tagged particle and the m-th bath oscillator, which contains the coupling function $F_m(Q)$.

In the Caldeira-Leggett model, the classical friction force arises from a linear coupling to a population of harmonic oscillators, which represent the other "particles" in the system. The coupling function is hence taken to be $F_m(Q) = \gamma_m Q$, where γ_m is known as the strength of coupling between the tagged particle and the m-th bath oscillator. In a physical system, liquid or solid, the coupling is expected to be large for nearby particles and small for particles far away in the material, and

© The Author(s), under exclusive license to Springer Nature Switzerland AG 2023
A. Zaccone, *Theory of Disordered Solids*, Lecture Notes in Physics 1015,
https://doi.org/10.1007/978-3-031-24706-4

hence γ_m is different for all the different oscillators the tagged particle is interacting with. This supports the argument that microscopic friction originates from coupling (via harmonic spring) between tagged particle and bath oscillators, which only depends on their relative distances.

The standard Euler-Lagrange equations of motion for the above Hamiltonian $\mathcal{H}$ can be readily obtained:

$$\frac{dQ}{dt} = \frac{P}{M}; \quad \frac{dP}{dt} = -\mathcal{U}'(Q) + \sum_m M_m \gamma_m \left(X_m - \frac{\gamma_m Q}{\omega_m^2} \right)$$

$$\frac{dX_m}{dt} = \frac{P_m}{M_m}; \quad \frac{dP_m}{dt} = -M_m \omega_m^2 X_m + M_m \gamma_m Q.$$

(C.3)

C.2   Derivation of the Generalized Langevin Equation

From the second line of Eq. (C.3), upon solving the second-order inhomogeneous ODE with the Green's function method, we get:

$$X_m(t) = X_m(0) \cos(\omega_m t) + \frac{P_m(0) \sin(\omega_m t)}{M_m \omega_m} + \int_0^t \gamma_m Q(t') dt'$$

(C.4)

The integral $\int_0^t \gamma_m Q(t') \sin(\omega_m(t - t'))/\omega_m dt'$ can be evaluated using integration by parts:

$$X_m(t) - \frac{\gamma_m Q(t)}{\omega_m^2} = \left(X_m(0) - \frac{\gamma_m Q(0)}{\omega_m^2} \right) \cos(\omega_m t)$$

$$+ P_m(0) \frac{\sin(\omega_m t)}{M_m \omega_m} - \int_0^t \frac{\gamma_m P(t') \cos(\omega_m(t - t'))}{M \omega_m^2} dt'$$

(C.5)

Substituting Eq. (C.5) into the equation for $P(t)$ in Eq. (C.3), we derive the following equation of motion for the tagged particle:

$$\frac{dP}{dt} = -\mathcal{U}'(Q(t)) - \sum_m \int_0^t \frac{M_m \cos(\omega_m(t - t'))}{M \omega_m^2} \gamma_m^2 P(t') dt'$$

$$+ \sum_m \left\{ M_m \gamma_m \left[X_m(0) - \frac{\gamma_m Q(0)}{\omega_m^2} \right] \cos(\omega_m t) + \gamma_m P_m(0) \frac{\sin(\omega_m t)}{\omega_m} \right\}$$

$$= -\mathcal{U}'(Q(t)) - \int_0^t v(t') \frac{P(t - t')}{M} dt' + F_P(t).$$

(C.6)

The second line provides the generalized Langevin equation (GLE) that has been used in Chap. 3, Sect. 3.2, as a starting point to derive the microscopic nonaffine theory of viscoelasticity.

Following Zwanzig [7], we then define the thermal noise, also referred to as the stochastic force, $F_P(t)$, which is equal to the second line in Eq. (C.6):

$$F_P(t) = \sum_m \left\{ M_m \gamma_m \left[X_m(0) - \frac{\gamma_m Q(0)}{\omega_m^2} \right] \cos(\omega_m t) + \gamma_m P_m(0) \frac{\sin(\omega_m t)}{\omega_m} \right\}.$$

(C.7)

Furthermore, in the second line of Eq. (C.6) we identified the memory function for the friction [6, 7]:

$$\nu(t) = \sum_m M_m \frac{\gamma_m^2}{\omega_m^2} \cos(\omega_m t)$$

(C.8)

which has been used in Chap. 3.

C.3   The Fluctuation-Dissipation Theorem

The noise $F_P(t)$ is defined in terms of initial positions and momenta of bath oscillators. Following [7], we assume the initial conditions for the bath oscillators can be taken to be Boltzmann-distributed $\sim \exp(-\mathcal{H}_B / k_B T)$, where the bath is in thermal equilibrium with respect to a frozen or constrained system coordinate $X(0)$.

Then for the averaged X and P, we obtain:

$$\left\langle X_m(0) - \frac{\gamma_m Q(0)}{\omega_m^2} \right\rangle = 0, \quad \langle P_m(0) \rangle = 0.$$

(C.9)

The second moments are:

$$\left\langle \left(X_m(0) - \frac{\gamma_m Q(0)}{\omega_m^2} \right)^2 \right\rangle = \frac{k_B T}{M_m \omega_m^2}, \quad \langle P_m(0)^2 \rangle = M_m k_B T.$$

(C.10)

We now take the Boltzmann average of the stochastic force, Eq. (C.7), and find:

$$\langle F_P(t) \rangle = 0.$$

(C.11)

Now, by direct calculation, using Eqs. (C.9) and (C.10) and standard trigonometric identities, we can get the fluctuation-dissipation theorem (FDT) for the particle-bath Hamiltonian:

$$
\begin{aligned}
\langle F_P(t)F_P(t')\rangle &= \frac{1}{Z_N}\int F_P(t)F_P(t')\exp\left(-\frac{\mathcal{H}_B}{k_BT}\right)d\mathbf{X}(0)d\mathbf{P}(0)\\
&= \sum_m\left(M_m\gamma_m^2\frac{k_BT}{\omega_m^2}\cos(\omega_mt)\cos(\omega_mt')\right.\\
&\qquad\left.+M_m\gamma_m^2\frac{k_BT}{\omega_m^2}\sin(\omega_mt)\sin(\omega_mt')\right)\\
&= k_BT\sum_m\frac{M_m\gamma_m^2}{\omega_m^2}\cos(\omega_m(t-t'))\\
&= k_BT\nu(t-t')
\end{aligned}
\tag{C.12}
$$

where Z_N is the canonical partition function:

$$
Z_N = \int\exp\left(-\frac{\mathcal{H}_B}{k_BT}\right)d\mathbf{X}(0)d\mathbf{P}(0)
\tag{C.13}
$$

and $\mathbf{X}(0)=\{X_1(0),X_2(0),\ldots\}$, $\mathbf{P}(0)=\{P_1(0),P_2(0),\ldots\}$.

Equation (C.12) is the FDT associated with the GLE given by Eq. (C.6).

C.4   The Memory Kernel

The friction memory kernel $\nu(t)$ can be rewritten in continuum limit:

$$
\nu(t) = \int_0^\infty\frac{M_p\gamma^2(\omega_p)}{M\omega_p^2}\cos(\omega_pt)g(\omega_p)d\omega_p.
\tag{C.14}
$$

For any given (well-behaved) VDOS function $g(\omega_p)$, the existence of a well-behaved function $\gamma(\omega_p)$ that satisfies Eq. (C.14) is guaranteed by the fact that we can always decompose $\nu(t)$ into a basis of $\{\cos(\omega_pt)\}$ functions, by taking a cosine transform. The inverse cosine transform in turn gives the spectrum of coupling constants $\gamma(\omega_p)$ as a function of the memory kernel:

$$
\gamma^2(\omega_p) = \frac{2\omega_p^2}{\pi g(\omega_p)}\int_0^\infty\nu(t)\cos(\omega_pt)dt.
\tag{C.15}
$$

This coupling function contains information on how strongly the particle motion is coupled to the motion of the bath oscillators (i.e., the other particles) in a certain eigenmode with vibrational frequency ω_p. This is an important information and provides a measure of the degree of long-range anharmonic couplings in the motion of the molecules.

In general, the determination of the memory kernel is an open problem for which several approaches have been proposed very recently, most of which have been tested only on model systems so far; some examples can be found in [8–12].

References

1. L. Landau, E. Lifshitz, *Theory of Elasticity: Volume 6* (Pergamon Press, Oxford, 1986)
2. P. Chaikin, T. Lubensky, *Principles of Condensed Matter Physics* (Cambridge University Press, Cambridge, 2000)
3. A. Lemaître, C. Maloney, J. Statist. Phys. **123**(2), 415 (2006)
4. A.P. Sutton, J. Chen, Philos. Mag. Lett. **61**(3), 139 (1990)
5. B. Cui, J. Yang, J. Qiao, M. Jiang, L. Dai, Y.J. Wang, A. Zaccone, Phys. Rev. B **96**, 094203 (2017)
6. R. Zwanzig, J. Statist. Phys. **9**(3), 215 (1973)
7. R. Zwanzig, R.D. Mountain, J. Chem. Phys. **43**(12), 4464 (1965)
8. Z. Li, X. Bian, X. Yang, G.E. Karniadakis, J. Chem. Phys. **145**(4), 044102 (2016)
9. Z. Li, H.S. Lee, E. Darve, G.E. Karniadakis, J. Chem. Phys. **146**(1), 014104 (2017)
10. G. Jung, M. Hanke, F. Schmid, J. Chem. Theory Comput. **13**(6), 2481 (2017)
11. H. Meyer, T. Voigtmann, T. Schilling, J. Chem. Phys. **147**(21), 214110 (2017)
12. S. Izvekov, J. Chem. Phys. **146**(12), 124109 (2017)

Glossary

α-relaxation It is the main, cooperative, and collective process, in terms of cooperative particle rearrangements, through which a glassy system escapes from the metabasin in the energy landscape that it was originally occupying to a new one.

Affine deformation It is a deformation of a sample in which a material point or, at the microscopic level, a particle, identified by a vector, is displaced to a new position (the affine position) given by another vector. The latter, called affine position vector, is given by the vector of the original position left-multiplied by the macroscopic deformation tensor. The vector difference between the affine position vector and the vector of the original position is called affine displacement.

β-relaxation Faster relaxation process (compared to the α-relaxation) through which a molecule or a bunch of molecules rearrange themselves.

Boson peak Excess of vibrational modes, typically in the Terahertz regime, which manifests itself as a peak in the Debye-normalized vibrational density of states.

Carnahan-Starling equation of state It is an approximate equation of state for the fluid phase of the hard-sphere system. It also provides the contact value of the radial distribution function as a function of the packing fraction of hard spheres.

Clathrate It is a solid consisting of a crystalline lattice with "cages" in which other atoms or molecules are trapped. The trapped molecules typically perform "rattling" motions that are highly anharmonic.

Mode-coupling theory It is a theory of density fluctuations in liquids, which uses projection operator methods to arrive at a generalized Langevin equation for the intermediate scattering function, which can be solved approximately with certain assumptions on the memory kernel. It predicts a critical temperature for dynamical slow-down in liquids, which is somewhat higher than the glass transition temperature.

© The Author(s), under exclusive license to Springer Nature Switzerland AG 2023
A. Zaccone, *Theory of Disordered Solids*, Lecture Notes in Physics 1015,
https://doi.org/10.1007/978-3-031-24706-4

Nonaffine displacements They are extra displacements on top of the affine displacements (for the latter cfr. "affine transformation"). They are triggered by force imbalance in the affine position in the absence of inversion symmetry, e.g., in disordered solids, in non-centrosymmetric crystals or even in centrosymmetric crystals at finite temperature due to thermal fluctuations instantaneously breaking the inversion symmetry.

Thermoelectric materials Broad class of materials where either a temperature difference generates an electric potential or an electric current generates a temperature difference. To optimize the thermoelectric figure of merit one seeks to maximize the electric conductivity and to minimize the thermal conductivity.

Percus-Yevick theory It is a theory for the radial distribution function of hard-sphere liquids, which solves the Ornstein-Zernike equations by implementing the so-called Percus-Yevick closure.

Index

A
Acoustic attenuation, 154
Adam-Gibbs model, 225
Affine deformation, 58
Affine force, 62, 65, 66
Affine-force correlator, 137
Affine position, 60
Affine shear modulus, 59, 130, 231
Affine transformation, 58
Akhiezer damping, 156, 170, 204
Andrade creep, 148
Angell plot, 225
Anharmonic crystal, 8
Anharmonicity, 12, 13, 233
Atomic dynamics in metals, 285

B
Bond-bending, 4, 5, 59
Bond-depleted lattice, 40
Bond-orientational order parameter, 38
Bond stiffness, 57
Born-Huang formula, 58
Boson peak, 136, 146, 154, 183, 214
    damped phonon model, 201, 215
Brillouin zone boundaries, 180
Bulk modulus, 96
Burgers loop, 244, 252
Burgers vector, 244

C
Centrosymmetry, 41
Colloidal gelation, 31
Compatibility condition, 253
Compressibility equation, 16
Constraint-counting, 44
Coordination number, 17

Covalent bond, 3
Creep, 147

D
Damped harmonic oscillator, 35
Debye density of states, 179
Debye frequency, 173, 181
Debye law, 180
Debye model of solids, 179
Debye momentum, 181
Debye temperature, 173, 182
Defective fcc crystal, 40
Deformation gradient tensor, 58, 60, 280
de Gennes narrowing, 17
Degrees of freedom, 44
Diffusons, 216
Dislocations, 243
Dislocation-type defects in amorphous solids,
    251
Dynamical matrix, 57
Dyson equation, 14

E
Einstein-Cartan theories, 258
Elasticity theory, 279
Elastic networks, 73
Embedded atom method (EAM), 6, 285
Ergodicity, 36
Eshelby quadrupoles, 262, 263
Excluded-volume, 97, 103

F
Floppy modes, 44
Fluctuating elasticity theory, 204
Fluctuation-dissipation theorem, 291

© The Author(s), under exclusive license to Springer Nature Switzerland AG 2023
A. Zaccone, *Theory of Disordered Solids*, Lecture Notes in Physics 1015,
https://doi.org/10.1007/978-3-031-24706-4

Fractal dimension, 31
Fractal model, 30
Fragility, 224
      index, 224
Fragility index, 234

G
Galilean transformations, 128
Generalized Langevin equation, 127, 290
Glass melting, 85
Glass transition, 36, 85
Goldstone's theorem, 35, 195
Grüneisen parameter, 10, 11, 203
Green's function, 13

H
Hard-sphere liquids, 24
Hessian matrix, 57, 69
Hyperuniformity, 16

I
Instantaneous normal modes, 135
Intermediate scattering function, 34
Internal stresses, 74
Inversion symmetry, 41
Ioffe-Regel crossover, 174
Irving-Kirkwood formula, 108
Isostaticity, 22, 44
Isostatic point, 44, 190
      central-force lattice, 44
      covalent bonds, 45
      frictional random packing, 46
      frictionless ellipsoids, 45

J
Jamming, 66, 67, 96

K
Kauzmann temperature, 225
k-gap theory, 125
KSZ model, 233

L
Langevin thermostat, 136
Lindemann melting, 84
Linewidth, 13, 17, 35, 215
Lorentzian, 13
Loss modulus, 124

M
Marchenko-Pastur distribution, 151, 194, 195
Marginal stability, 198
Maxwell model, 122
Mechanical constraints, 44
Memory kernel, 144, 292
Metallic glasses, 5, 235

N
Nearest neighbors, 17
Newton law, 222
Nonaffine correction to the shear modulus, 65,
      68
Nonaffine deformation, 61
Nonaffine displacements, 61, 63, 68
Nonaffine positions, 60

P
Pair correlation function, 15
Particle-bath model, 289
Peach-Köhler (PK) force, 246
Peierls-Nabarro stress, 247
Plastic instabilities, 261
Plasticity, 243
      amorphous solids, 259
      crystals, 243
Plemelj identity, 205
Polydisperse random packings, 26
Potential of mean force, 47, 232
Propagator, 13
Pseudopotential, 5–7, 235

Q
Quasi-particle approximation, 13
Quasi-static deformation, 60

R
Radial distribution function, 15, 17, 22, 31, 231
      fractal aggregate, 32
      random sphere packings, 23
Random close packing, 22
Random matrix theory, 151, 193
Random network model, 19, 224
Random packings of soft spheres, 66
Rayleigh damping, 169
Rayleigh dissipation function, 222
Relaxation modulus, 147
Renormalized frequency, 13
Replica theory, 36

Replica trick, 115
Rigidity transition, 67

S
Schmid's law, 247, 261
Self-averaging, 115
Self-energy, 14
Self-stress, 44
Shear bands, 262
Shear modulus, 57, 280
    colloidal gels, 88
    confined amorphous solids, 270
    confined systems, 269
    dense gels, 91
    fractal gels, 89
    glasses, 80, 111
    Lennard-Jones glasses, 87
    liquids, 109
    liquids under sub-millimeter confinement,
        269
    perfect crystals, 109
    polymer glasses, 80
    random jammed sphere packings, 66
Sound attenuation in amorphous solids, 160
Specific heat, 213, 218
Spontaneous symmetry breaking, 36
Spring constant, 57
Static structure factor, 15
Storage modulus, 124, 131
Strain-rate tensor, 128
Strain tensor, 58
Stress-fluctuation formalism, 105
Stress *vs.* strain curve, 261

T
Thermal conductivity, 215
Thermal expansion, 85
Thermal expansion coefficient, 11, 37, 231

Topological defects, 248
Topological phase transition, 249
Two-level states (TLS), 188, 218

V
Van Hove correlation function, 33
Van Hove singularities, 182
Vibrational density of states, 130, 186, 195,
        205
    confined solid, 275
    simple lattices with randomness, 190
Viscoelasticity, 119
Viscoelastic moduli, 128
    metallic glasses, 140
    polymer glasses, 132
Viscosity, 221
    Dyre's shoving model, 229
    Frenkel's theory, 226
    Green-Kubo relation, 238
    KSZ model, 231
    nonaffine response theory, 238
Viscous shear flow, 222
Vogel-Fulcher-Tammann (VFT) equation, 224
Volterra construction, 249

W
Winding number, 248
Wishart ensembles, 194

Y
Yielding point, 261
Yielding transition, 261

Z
Zener model, 126

Printed in the United States
by Baker & Taylor Publisher Services